A
Brief
History
Of Time #2

Creation or Evolution?

Darwin was right!

Eureka!

Now Praise Ye All Darwin!

Are you confused by all the claims about Evolution and Creation?
Was Darwin right about monkey to man Evolution ?
Or did he scientifically prove
Earth is young and Creation is true?
For too long Darwin has been denounced by Fundamental
Christians as the agent of Satan for his
1859 Origins of Species book.
He has been held up as the destroyer of beliefs and
human souls by his claims that all
life began by chance over
3 billion years ago on a 4.8 billion year old Earth and
GOD, Jesus and the Bible is myth and fairy tales!
Did Darwin's work really earn him fame and burial with
the greats in Westminster Abbey – or did
he fail to see The Truth of Creation!
This book will prove that Darwin was correct and should
be praised for his work by all Evoquacks – even
though many Christian Creationists sneer!

Darwin was more accurate than Hawking!

Creationists and Evolutionists may be astounded but
prepare your mind to be shocked as new research
confirms Darwin and his meticulous studies
prove The True History of Time, Earth's age,
Evolution and Creation – and Hawking is a fool!

EUREKA!

EUREKA MOMENTS!

Eureka! - Creation Day Four and First Law of Thermodynamics!

Eureka! - Textonic Plates is impossible – Newton Physics!

Eureka! - Trillions of jellyfish are baked in The White Cliffs!

Eureka! - Why Trilobites didn't evolve!

Eureka! - Alice cannot do simple sums!

Eureka! - Earth can easily support 50 Billion Humans!

Eureka! - Hitler's concrete bunkers prove Creation!

Eureka! - There's gold in them rocks!

Eureka! - Earth's new and old magnetic poles!

Eureka! - Gas in the Arctic muck!

Eureka! - Oil and Noah's waterproofing!

Eureka! - Volcanoes and Black Smokers Fuel mystery!

Eureka! - Mid-Ocean Ridges calculations!

Eureka! - Comet disappearances!

Eureka! - Coal from wood in a few days!

Eureka! - Fossil fish shoals electrocuted!

Eureka! - Worms prove it: Earth is Young and Flood was real!

Eureka! - GOD designed our electrical sparking heart!

Eureka! - Easter Island statues prove young Earth!

Eureka! - Mammoth massacre in mucky loess explained!

Eureka! - Shellfish on Everest!

Eureka! - Radiocarbon lies and fairytales!

Eureka! - Gabon nuclear reactor is Faraday's Law!

Eureka! - Why do Christians still proclaim GOD and Creation?

Eureka! - Evolutionists love Woozle!

Introduction

In the Dark Ages back in 1988 a guy called
Stephen Hawking decided to write a novel
filled with a mix of sci-fi and fool's filosophy
for the great unwashed masses heavily
brainwashed by evolutionist films like
StarWars, Planet of Apes, Return of the Dedi.
Hawking named it 'A Brief History of Time'.
The book distorted everything about time and
its ludicrous content is certainly perfect
proof Hawking is possessed by demons.

Now in November 2017 Hawking's demons have once again
used him as a reverse SETI to tell humans we must
evacuate Earth in 600 years – this we know is a
satanic lie as GOD says Earth will endure forever!

In fact Alfred Russell Wallace was given
'The Theory of Evolution' by Satan – the
father of the lie – when he snared Wallace
in 1844 at a demon raising mesmerist's meeting
after which Wallace became a spirit medium!
GOD says seances attract Satan's evil spirits!

Evolution is thus Satan's lying contradiction of
GOD's Creation as recorded in The Bible!

Science reveals that practically everything
written about humans evolving from
monkeys/fish/rocks is Satanic nonsense!

With the benefit of my decades of dedicated
study, The Bible and the vast unlimited store of
knowledge freely given to all Christians here is
my real history of the time and the human race:
'A Brief History of Time #2'

ABHOT#2 raison d'etre is to educate

Arthur Schopenhauer stated succinctly:

All truth passes through three stages.
First, it is ridiculed.
Second, it is violently opposed.
Third, it is unlikely to be accepted as being self-evident.

My book is full of truth but you will not like it!

Earthworms fool all Evoquacks!

Worms + 6KY + Flood =Young Earth

Evolution or EVONONSENSE?

Index

Chapter One

Big Bang EVONONSENSE?

BigBang EVONONSENSE is based on the silly fairy tale of once upon a time there was nothing but a SINGULARITY at the centre of a NOTHING and it blew up blasting heat and dust across what we now call The Universe!

EVONONSENSE – I capitalise EVONONSENSE throughout the book to emphasize how silly evolutions are - means the whole idea, writings, beliefs, and evidences put forward as 'evolution' as a totally nonsensical fantasy cooked up in the centuries before Hollywood began making movies featuring dinosaurs, giants and fairy princesses. Evolution plus Nonsense = EVONONSENSE.

The 'Singularity' was originally claimed to be millions of miles across, then thousands, then just a few feet across until finally, on Wiki as I type this, just a trillionth of a trillionth of a millimetre – which is highly significant as most people don't know or cannot visualise a millimetre let alone a tenth, hundredth, thousandth, millionth of one!

No-one can explain how the entire Universe became crushed down into that invisible 'singularity.' This simple fact is brushed aside by the high priests of EVONONSENSE just as is the fact that there has never been a single verifiable example of evolution ever found – or likely to be found!

Then as so often happens I followed a Wiki lead and came across this gem of Evoquackery: Question to BigBnger: *"Did the BigBang start from a point? "*

Answer: *"The simple answer is that, No, the Big Bang did not happen at a point. Instead it happened everywhere in the universe at the same time."*

Another 'science' website admits: *'Our Universe started as a singularity 13.7 Billion Years Ago. What is a "singularity" and where does it come from? Well, to be honest, we don't know for sure. Singularities are zones which defy our current understanding of physics. They are thought to exist at the core of "black holes." Our universe is thought to have begun as an infinitesimally small, infinitely hot, infinitely dense, something - a singularity. Where did it come from? We don't know. Why did it appear? We don't know.'*

Right there we can be sure all BigBang Evoquacks believe Woozled fairy tales! **ONE EVOQUACK CLAIMS THE BIGBANG STARTED MICROSCOPIC AND ANOTHER SAYS IT STARTED SQUIDZILLION MILES ACROSS!** That sums up EVONONSENSE perfectly! The BigBang was not a BigBang as the Universe was already existing – or something! Bundle all their BigBang Woozle up and label it 'A Singularity of EVONONSENSE.' If you follow the events, fossils names and ages promoted by the world's Evoquacks you wander off through a looking glass worse than anything Alice ever stepped into.

EVONONSENSE'S scientists have great faith in their religion!

They adore their high priests and flock to worship them with praises that reach the sky!

They lie, cheat and swindle their governments out of vast sums of money to pursue their false gods and build ever greater temples to them and pour ever greater awards on each other for inventing the silliest EVONONSENSE.

And they sneer at Christians who claim GOD created everything! They are crazy! Or possessed by demons?

Anyway BigBangEVONONSENSE goes something like this:

Once upon a time there was nothing but that minutely minute speck of super concentrated matter - the festering Singularity. (Or maybe the entire visible Universe was festering to burst?)

13.8 BYA Big Bang occurred. BYA = billion years ago. 13,800,000,000 years is a long time!
For 9.3 billion years red hot dust blew around organising itself into stars, galaxies and such.
4.5 bya Dust came our way and clung together to make Earth.
3.8 bya Moisture appeared and life sparked on Earth with wet rock using Hawking's Gravities.
3.46 bya Life starts eating methane. Sterile rock plus water = methane? Methane is organic!
3 bya Viruses appeared! Viruses before cells! We really don't stand a chance do we?
2.4 bya Oxygen built up and made all the iron we see.
2.16 bya FIRST UNDISPUTED FOSSIL! Christened STROMATOLITES.
2 bya Eukaryotic cells with bellies full of bacteria appeared.
1.5 bya Eukas divide into three groups: fungi, plants and animals. Eukas opposite nature?
900 MYA = millions years ago. Eukas became multi celled. See above -don't laugh!
800 MYA Eukas turn into sponges.
780 MYA Eukas turn into PLACOZOAs which evolved into common ancestor of mammals.
770 MYA Earth goes into deep freeze. No-one knows why but it's a good Woozle?
730 MYA Earth comes out of deepfreeze and comb-jellies stick to the seabed and enjoy life.
680 MYA Ordinary jellyfish separate from those Placozoas.
630 MYA The creatures now learn they have top and bottom and, presumably, front and rear.
590 MYA Creatures split into two groups, one has babies that grow anuses first, other mouths.
580 MYA Jellyfish and corals appear. See 730MYA don't laugh!
575 MYA EDIACARANS appear! Seen as STRANGE! Some Evoquacks think they is ferns.
570 MYA Starfish and worms that can walk about appear.
565 MYA Some walkers leave fossil trails.
550 MYA Ferns aka bracken is dying leaving fossils.
540 MYA The walkers have to grow backbones or they cannot walk far.   Don't laugh!
535 MYA Cambrian Explosion makes vast numbers of new creatures in Wales!
530 MYA First proper bony fish and hard shelled trilobites appear and breed like flies!
520 MYA Eels appear slithering about on the seabed and bogs.
515 MYA Cambrian explosion runs out of steam and nothing new appears for quite a while.
500 MYA Some creatures climb onto dry land – they were half insects and half crabs.
489 MYA Wales gets overrun with Ordovician Biodiversings. Welsh still has OB complex.
465 MYA Plants start growing on land – ferns? Plus what did the insect/crabs eat?
460 MYA Fish separate into those that are grow bony like kippers and the rubbery like sharks.
440 MYA Fish separate again into today's fish and today's land creatures.
425 MYA Lovely fish CEOLOCANTH appears :amazingly swims upside down, backward!
417 MYA LUNGFISH appears after separating from other fish that intend to become animals.
400 MYA Insects appears after getting good report from colonisers 100 million years earlier.
397 MYA Four legs animals appear in freshwater? They walk out, dry off, conquer the land.
385 MYA First tree appears as fossils seems to indicate. No monkeys yet as they need trees?
375 MYA Fish with legs seems to wandering about. No arms so cannot climb the new trees.
340 MYA Turtles separate from the animals that don't want to get wet or swim in deep water.
320 MYA DIMTRODON appeared looking and thinking itself a dinosaurs but its not due yet!
310 MYA Creatures spilt into today's snakes, birds, dinosaurs and some of today's mammals.
250 MYA Mass extinction wipes out lots of things including all those trillions of trilobites!
250 – 210 MYA Things improve,ammonites, dinosaurs appear and go to Cambria aka Wales.
210 MYA A bird some people say is a dinosaur wanders around leaving footprints in wet mud.
210 MYA A Human the Evoquacks say isn't a human walks with that dinosaur in same mud.
200 MYA ANOTHER MASS EXTINCTION! Dinosaurs survive this! Not sure if humans did!
199 MYA Mammals realise need to become warm blooded so they invent thermostat controls.
199 MYA Human Drynose with dinosaurs gets warm blood so its muddy feet will dry faster.
180 MYA The opossum with selfish junk DNA split from placental lineage say Scientists.

180 MYA Duck Bill Platypus appears. Lives in Australia because Aussies like weird things.
168 MYA Flightless dinosaur Epidexipteryx appears in China. Strangely not mass produced!
150 MYA Archaeopteryx in Europe on holiday from China testing market for Epi knock-offs?
140 MYA Kangaroos split off, goes Waltzing Matilda walkabout, finally settles in Australia.
131 MYA Chinese improve Archaeopteryx, name it Eoconfusciusorni to honour old man.
130 MYA Pretty flowers appear after rapid evolution and petition from starving bumblebees.
105 MYA Mammals split, whales and dogs one group, monkeys and rats another.
100 MYA Biggest dinosaur Argentosaurus claimed as living now. He just grew and grewed?
93 MYA 27% of marine softies wiped out as big volcano uses all the oxygen. Volcano lost.
75 MYA Rodents separate from rabbits and become very successful due to their dirty habits.
70 my Grasses EVOLVE in isolated clumps. TRex, all dinosaurs eat huge quantities of grass.
68 MYA Cretinous-Terty Extinction wipes out dinosaurs, ammonites. Mammals, rodents rule!
63 MYA Primates evolved from nowhere and split into dry noses and wet noses again.
58 MYA Tarsier, a Wet, splits off from rest of Wets and EVOLVES big eyes to see at night.
55 MYA Paleocene/Eocene EXTINCTION! Another! Boils all sea life dead but not land life!
50 MYA ARTIODACTYLS, with wolf/tapir DNA EVOLVE INTO WHALE! I kid you not!
48 MYA INDOHYUS small deer decided to EVOLVE INTO WHALE as it liked swimming.
47 MYA IDA, a monkey lived in sea but came ashore to have babies. It became famous for it!
40 MYA South America received first shipment of higher monkeys for The Matto Grosso zoo.
25 MYA Apes separated from European monkeys and went to live in Belgian Congo places.
18 MYA Gibbons evolved off from other apes and became gibbons.
14 MYA Orang Utans evolved off ape and went to South Asia where we find them today.
7 MYA Gorillas stopped being apes.
6 MYA Humans EVOLVED off from chimps and bozos. It was drynosed as previous ones.
2 MYA Josephartigasia, monster polar bear rodent EVOLVES in South America.
1.7MYA Humans evolved off Apes in Olduvai Gorge. The things could even think and plan!
800TYA- Thousand years ago. Humanoid learned to cook. Previously ate only McDonalds .
200TYA Humans still hulking brutes given to wife beating, living in caves.
 From 200,000 years ago to perhaps 20,000 years ago is a mass of EVONONSENSE about
how humans evolved from monkeys, apes or similar through all sorts of imaginatively named
subhumans such as Cro-Magnons, Australopithicus, Neanderthals and similar which I can't be
bothered to list and mention as the fossils show they were either extinct monkeys and apes or
poor specimens of diseased, deformed, crippled or malnourished humans caught out, starved
and frozen by the post-Flood ice age. GOD made sure no pre-flood human fossils remain.
 Another well publicised chart of the fantasy timeline of EVONONSENSE as favoured by
Evoquacks is basically this but note: they lost a billion years!
Archeozoic Era 1.8 – 1 BYA - Those billion years again! No fossils but plenty daydreams.
Proterozoic Era 1BYA - 500 MYA. Foraminiferas and water plants aka Creation Day Four.
Paleozoic Era 500 MYA - 200 MYA. This is more Day Four.
Mesozoic Era 200MYA - 70MYA. This is actually Day Five Creation. Check the Bible.
Cenozoic Era 70MYA – 12MYA. This is more Day Five blended with Day Six.
Commercial break. Here evolution had a long siesta for about 11 million years.
Plastocine Era 975,000 – 25,000 years ago. More Day Five and some Day Six.
The Plastocine Era aka Pliestocene Era is actually the 6,000 years from Creation to today.
To Present Day – humans seem settled into tribes based on colour, language and religion or
none.
 Most Evoquacks are united in thinking that humans settled into our present form something
like 40,000 years ago after a bit of breeding with Neanderthals. Several cliques of Evoquacks
subscribe to the idea Martians came in UFOs and impregnated willing and unwilling women.
 I have to apologise for my whimsical comments to some of the above dates and events but
the list - from New Scientist and Smithsonian! - shows the EVONONSENSE Woozles the

Evoquacks have to swallow and preach to make a living in their religion. (Note: The Smithsonian is a sinister museum institution run by Satanists who deliberately suppress all evidence for Creation while praising all EVONONSENSE.)

Look on Wiki for the Smithsonian's idiotic and Satanic timeline for EVONONSENSE from Hawking's wet sparking rocks right down to humans. The Leakey clan claim humans evolved off apes in Olduvai Gorge, East Africa a long time ago. The clan claim they found the bones by old campfires! Now it seems the bits of bones may have been caused recently by hungry humans to get at the stuff inside monkey skulls. Eating brain causes KURU as Muslims prove.

All Evoquacks stretch, mould, bend, twist and reshape the Plastocine like kids in playschool to try fit it to whatever they see in Planet of Apes and similar movies. Most Evoquacks are men who were fooled into playing with Plastocine by seeing Raquel Welch in a skimpy rabbit skin outfit pursued by TRexes in the old One Million BC film! Mary Schweitzer was brainwashed into her particular padded cell at five years old by being given a book about a dinosaur that wanted to become a chicken. Alice Roberts believes millions of feet of Darwin's topsoil disappeared as if by magic because she sat enraptured at the feet of Dawkins as he spouted his vitriolic Satanisms.

Recorded history shows human tribes hate each other for local nuances of colour, foods, language and religion. Humans has struggled up to today with lots of wars, pestilences, crime, pollutions and famines keeping the population very low and needing considerable wealth diverted into armed forces, police states and health services to ensure life that is pretty grim with most humans seeing all the world's troubles and their own as due to inability of the nation's leaders to get things sorted out and make the streets safe and everyone prosperous.

Countless colleges, professors, authors and television programs shove EVONONSENSE at students and the public and reinforce it with very intricate and careful calculations, fossils, illustrations and animations that strive to make EVONONSENSE real - but when analysed by a clear thinker all their claims are just hot air as not one of them can provide a single missing link – or the smallest shred of evidence of anything having evolved or is evolving today!

Turn the television on to constantly hear claims that life began billions of years ago or one creature evolved into another. Reports of a new fossil find are good for news headlines and pushed as more proof of ancient EVONONSENSE but invariably the bones are from a creature that was complete in itself with no need to evolve and no sign of having evolved – and of course the age of the fossil is that of the rocks around it as fossils date the rocks the fossils is found in...I've just had one idiot parrot that EVONSENSE is proven by giraffes get laryngitis just like humans – or some such idiocy!

Lots of scientists think the last 4.5 billion years have been constant upward evolution and tomorrow science will dispel all our troubles.

They haven't a clue of how bad things really are!

Like monkeys they cannot think above the level of 4 years old children but can be taught simple tricks.

Here is a typical childish sneering response to me from an Evoquack thoroughly bamboozled by Satan:

"The Theory of Evolution in no way involves believing that anyone evolved from monkeys - the theory instead suggests that apes and humans, based on similarities (real or apparent) in genetics and biology, likely evolved from a common ancestor which in turn would be expected to share some similarities to both humans and apes, but not actually be either human or ape. Unlike physics, it's a lot harder to use replicable mathematics in studying and experimenting with the Theory of Evolution, and it's difficult to find models for Evolution that can be practically tested in the laboratory. Comparing Physics and Evolution is a bit like comparing apples and oranges. In any event, watching Flat-Earthers attempt to ape an understanding of physics without actually understanding physics might be enough to make one wonder if there are more similarities between apes and Flat-Earthers than meets the eye..."

I replied to the Woozler asking if he was aware that if all the DNA in a single cell was stretched out it would be six feet long. I doubt if he will understand the implications of that as he puts his eye to the telescope and imagine the stars is billions of years old.

Why did he like practically all Evoquacks like to think they are evolved from chimps? Is it because they saw chimps tea parties on telly when they were children and the memory stuck and they have some deeply repressed hope that reincarnation is true and they can come back as chimps and hold their own tea parties?

Alice Roberts – no professor she! - agrees that Darwin's METICULOUS SCIENTIFIC RESEARCH proves worms build topsoil at the rate of one inch per five years but when I asked her why she claims the Chixlub crater is 66 million years old which means that there should be about 13million inches of topsoil everywhere her brain refused to accept the logic of my question so she went into full Schopenahuer Stage One mode like a typical spoiled brat.

Poor Alice's head is so stuffed with woozled EVONSENSE that she is quite happy to show herself a fool. Oh, by the way, she says GOD and Satan do not exist...

Childishly truculent Hugh Ross attempted to debate Kent Hovind the Dinosaur Man – and came across as the typical Evoquack with a head stuffed full of Planet of Apes and StarWars crap as he prattled on about the stars being billions of light years away...but then he really made a fool of himself by saying that if the Bible was true then space would be filled with hydrogen! It is filled with hydrogen – as science confirmed – and that hydrogen is what split off from all the water on Day Four of Creation! Ross needs to get his head out of those old Planet of Apes videos...

The high priest of EVONONSENSE, Hawking, stated in 2017 that he fears an invasion of aliens from other worlds 'as they could be two billion years more evolved than me.' This surely is him regurgitating Pierre Tilehard de Chardin's 'noosphere' fantasy?

It was about the most stupid sentence ever uttered in the history of humankind and clearly shows Hawking thinks that two billion years ago there was nothing on Earth except microbes slopping round in the primordial soup that had somehow found some moisture despite that billion degree BigBang heating... But what does he think evolving for an extra two billion years would produce – diapered chimps able to sit round a mini table and eat fairy cakes?

Hawking is a fool.

Chapter Two
Einstein's Fudge and other dark matters.

While doodling his sums on his blackboard Einstein kept coming to an unacceptable result – the answers showed that the Universe should have been contracting back to that minutely minute singularity despite astronomy with the existing telescopes quite clearly revealing it was expanding at a fast rate. He didn't want any painful contractions as having himself and all the creatures squashed to pulp would make a bloody mess of the singularity and it might happen before he could be nominated for a Nobel Prize! So he fudged his sums by introducing his 'Cosmological Constant'! This is like as what means there is an invisible nothing that is preventing the universe contracting by blocking all the gaps between heavenly bodies the way cement holds bricks apart. The more he thought about it the better it sounded. The first time I thought about I thought it sheer stupidity. But even better: the charlatan claimed and got all his worshippers to believe that his nothing can be imagined as something big or bigger or really big so when added into their formulas the sums can make whatever answer is wanted!

Einstein's invisible nothingness is 'hypothesized' .i.e imagined, as 'dark matter' or 'dark energy' or 'inflations' like as what makes the universe spread out and appear the same in all directions – just as GOD intended. Einstein was a good salesman and his acolytes ensured his silly ideas were sold to the unwashed masses just as successfully as earlier salesmen sold 'The Emperors New Clothes!' Million of Einsteinites now parade about like naked fools...with millions of acolytes eagerly parading through universities to exchange their own clothes for Einstein's Fudge!

Neither Einstein or anyone else has ever seen any dark matter but the high priests of the worldwide EVONONSENSE Religion accepts the bigbang-dark matter hypothesis as its core belief. In 2012 Alan Guth and Andrei Linde were awarded the lucrative Breakthrough Prize for some fairy tale to do with the first few seconds of the conjectured, fantasised, speculated, posited, BigBang's inflation. What did I say about D-K Inferiors being susceptible to Woozle? I checked Youtube but cannot find any video of Guth and Linde walking naked to the podium to thank the Nobel committee. I feel sure many devoted worshippers have bought into Guth and Linde's New Clothes Bubble just as eagerly as in 1720 investors pushed the South Sea Bubble's fantasy to new heights before it as exposed as a sham for the gullible.

Evoquacks denounce some evangelists for getting rich but the money sloshing around the temples of EVONONSENSE are almost as exorbitant as in megachurches. Guth and Linde may have received $3 million for their fantasy of cosmological inflation which was basically the same as written in the Bible at Job 9:8 'He alone stretches out the heavens.' 'He' being GOD. Job is guaranteed eternal life while Guth and Linde are dead men walking to permanent extinction for calling GOD a liar and leading millions of acolytes to destruction.

Once Old Charlatan Einstein showed that scientists will accept his 'Emperor's New Clothes' it opened the floodgates to all manner of shysters to pedal a cornucopia of ever more fantastic 'Emperor's New Clothes' to the gullible worshippers of EVONONSENSE. Monkey to man idiocy pales beside the wolf/tapir to whale fairy tale, or the latest I heard today: 'Giraffes get laryngitis because they evolved from humans!'

Naturally being a keen professor of the natural world I quickly researched this giraffe laryngitis business and as expected it included a lot of ridiculous mumbo jumbo to appeal to the evolved monkeys. It's originator probably fudged some research into DNA to arrive at his comedy of errors: viz: 'Over the millions of years as the giraffes needed to stretch its neck to survive famines and competition for food it realised that each incremental extension would demand longer vertebra, muscles, tendons, more volume of blood, more area of skin and its pigments and follicles etc but the most pressing need would be for many extra feet of laryngal nerve.' At a childish level this being perfectly logical as after all if an adult female giraffe was to find herself starving as TRexes or Brontosauruses were eating all the lower leaves of trees it behooved her to ensure her babies when weaned would be able to reach the topmost leaves of the tastiest trees - so she duly ordered eggs from her ovaries that included the script aka DNA for longer necks - and in a flash of amazing foresight ordered the super clever DNA to include surplus laryngal nerve to cater for many billions of years future neck stretch!

Evoquacks believe that in another million years giraffe necks will be a hundred feet long and all zoos will have to heighten their giraffe houses. The big flaw in this ointment, apart from the jumping ability of lions and tigers, is the neck length of Brontosaurs as of course these would realise that to survive they too would have to browse the very same leaves the giraffe favoured. The other flaw is the giraffe's front hips as each time the giraffe's neck stretched it would also have to splay its front legs more to drink from pools or streams inevitably below its front feet...maybe lots of giraffe's died of thirst while waiting for their hip splay to catch up to their head? EUREKA! The originator of the giraffe larynge fAllahcy skipped over the TRex competition matter by fondly imagining TRex was already a dim and distant memory when the first giraffe baby evolved from a fourlegged chicken or whatever? His trick worked as now the Giraffe Laryngal Nerve Syndrome is accepted Woozle in EVONONSENSE circles. Maybe the Nobel Prize Committee is even now preparing to cascade millions of dollars on the giraffe's larynx woozler?

Sensible people and especially Creationists know that giraffes were designed as long necks to delight humans as there is nothing more endearing than a baby giraffe trying to gain control of its legs in the first hour after birth – well, apart from a baby elephant or kitten - and surely Eve would have thought tiny TRexes ever so cute as they emerged from their shells into the soft pearly light?

Perhaps the root of the giraffe laryngal nerve woozle is inability of its originator to accept that at Creation GOD had ensured the Earth would have a superabundance of grass and green food and He intended all creatures to live, munch grass and play happily together with none being vicious killing carnivores? Obviously He decided we'd like to watch giraffes browsing the tops of trees to encourage the trees to produce more fruit lower down? Why do wine growers trim their vines back each year?

Creationists have no need to resort to fudging anything as the Bible is a perfect record of the universe and all the creatures in it. Although we have a very small selection of living creatures at this sad stage we do have countless fossils from The Flood to show the amazing variety of land, water and air creatures GOD originally created. Just get a book of famous people and see what amazing variety of looks, sizes, colourings and attributes humans have and think about Adam and Eve's DNA having such countless permutations coded into it! The land, seas and air that sustains all creatures and plants show how good Creation was and will be again. Our bodies and the growth of a new layer of skin every day prove we are intended to live forever with no evolution needed or possible!

Why do Evoquacks fatally gorge on their junk filled fairytales and scornfully reject the plain and simple Creation and GOD's Eternity promise?

Giraffe's have always had long necks! EUREKA!

Chapter Three

Hitler's bunker gave me the keys to how worms prove Earth is young!
Worms make fools of Hawking, Dawkins, Attenborough!

Crippled to immobility by a demon Hawking fantasizes flying off to distant planets while the equally possessed Dawkins sneers at GOD and Jesus and claims himself evolved from a monkey. Hawking backs the SETI club that tries to hear signals from other worlds despite that being impossible. Angels passing through the Universe on GOD's business might receive SETI's signals but they are banned from responding to them though how long it will be before we have proof of some of Satan's UFO's tickling SETI's ears is an interesting thought. Demons can beep beeps to SETI but are they are hopping about the SETI receivers or wandering about the building enjoying themselves. Hawking actually acts as a SETI receiver as his demon make him utter sci-fi nonsense.

Just today November 2017 Satan's slave, Hawking has claimed that humans will have to abandon Earth in 600 years as it will be too polluted and exhausted to support life. This is yet another lie from his demons as GOD said Earth will remain forever as spoken at Ecclesiastes 1:4 'A generation goes and a generation comes, But the earth remains forever'. GOD designed worms to ensure Earth will remain ever fertile thanks to the industrious worms and GOD designing plants to grow profusely with nothing but sunlight, water and carbon dioxide – and the amazing ability He designed into their roots to dissolve rock to bring up essential minerals for all humans and creatures!

Likewise, David Attenborough is Satan's best promoter as in every nature video or book he makes he says this or that creature 'evolved from rudimentary life many millions of years ago!' But he claims to be a Christian! Christians by definition believe the Bible and descent from Adam and Eve – that makes him an hypocrite and tool of Satan's. He and Dawkins really signed up for Hawking's gravity power!

Yet if only these three could open their eyes and think they would realise that the humble earthworm slithering about the soil is a perfect example of GOD designing for purpose.

That purpose is to maintain Earth's cleanliness and fertility by clearing away and recycling all dead vegetation while simultaneously aerating and irrigating the soil! Hawk, Dawk and Attie together could not design either the worm or its functions! Nor can any of them understand how worms prove Earth is young.

Satan is delighted with thee three promoting him and his agenda.

Naturally all Evoquacks violently ridicule me saying worms prove Earth is young and The Flood really was a worldwide event just about 4,350 years ago.

The industrious worm has totally undermined the whole edifice of EVONSENSE and but the armies of fanatical Evoquacks are frantically shoving themselves and bundles of other people's money under it to prevent final collapse a little while longer!

I just discovered a video of abandoned World War 2 forts of Hitler's Atlantic Wall and growing directly on the solid concrete are good amounts of vegetation building up mats of solid soil.

Read on to learn how Hitler's bunker revealed exactly how worms prove Earth is young and how GOD intended the Earth to be a wonderful fertile home for we humans!

Chapter Four
Cenes. Genre: Comedy: Act One, Scene One, Take One.

Set props: Pre-Cambrian fossils. Fantasy Fossil record. TRex.

Lights! Action! Act 1, Scene 1, Take ... take all those 'cenes' and throw them on the cutting room floor!

All the great films - The Thomas Crown Affair with Steve and Faye, The Ten Commandments with Charlton Heston, Wuthering Heights with Lawrence and Merle, Gone With The Wind with Clark and Vivien - had directors who read a filmscript and broke the story down into scenes to show the progress of the story.

Mention the chess scene in Thomas Crown and we instantly visualize Steve and Faye across a chessboard in a firelit study; The Parting of the Red Sea in Ten Commandments brings to mind Charlton at his most shaggy majesty with face upturned to tumultuous clouds; Wuthering Heights will ever be Lawrence and Merle laughing on the high moors; Vivien Leigh in a buggy racing down a street of blazing houses is Gone With The Wind's scene for many. We've all seen too many great scenes in too many great films to call them all to mind without some prompting trick of memory. If given a hint to stir our braincells we can all instantly recall our favourite film's scene and replay it in full colour and sound. Actually I've just read that most people cannot do this! Can this really be true?

We know these memorable film scenes are often not in the original story nor are they always filmed in the order we finally sit and enjoy them. Directors can get precious and add scenes to make a personal point or clarify a plot. Economics dictate that if a scene calls for a multitude of people in shot and a similar crowded scene is needed later in the film it is better to shoot the early and later scenes together. If a highlight or climactic moment is an expensive set, vehicles, planes or property to be destroyed it is best to shoot it last even if the story is concerned with past events or flashbacks. So long as the final film is an enjoyable experience we the public are happy and if the film makes money the studio is happy too. (Hollywood is run by Jews and they are notorious for cheating the US Treasury by accounting tricks – in defiance of GOD's Commandment Seven.) There may not have been any crime and romance events exactly like that enjoyed by the fictitious Thomas Crown but plenty of bored men have plotted robberies and become entangled in romances; maybe no Heathcliff ever loved a Cathy as Lawrence and Merle did but we all know what it is like to be free in the sun with a lover; we've all known the fear of a near accident and burglary, and so we see the film director's scenes as fictitious and grant him or her our good wishes if the film gives us a happy experience. Quite often the author of the original story is not happy with the way the story has deviated from what was slaved over and speak their mind at awards ceremonies but by then it is too late and we the public don't get terribly upset at some wealthy writer's masterpiece being altered.

So it is with the story of Creation. The original author, GOD YHWH, produced a simple straightforward story easy enough for anyone to follow and understand. It needed no fancy camera angles as that horrible creepy Hitchcock was wont to use, it has not needed sequels and prequel, it did not need extra storylines adding as the awful soap operas must constantly feed their enslaved adherents. If that original script had been followed to the letter the world would be a far better place and we would all spend every day in the delightful company of something around 50 billion other happy people and lots of amazing animals, birds and fish!

But Satan rebelled and ruined GOD's script by perverting all humans with the lure of sex, drugs and idolatry, and admixing the genes of humans and creatures to create hybrids that hated and fought each other until eventually GOD saw the Earth was ruined and corrupt apart from the faithful untainted Noah and his family.

Angered, GOD acted. After ensuring Noah and a pedigree selection of airbreathing creatures were safely on the Ark He smashed Earth mightily and made boiling waters erupt out of it and vastly more icy waters fall from the sky. The hybrid bodies of Satan's Nephilim children were drowned along with every creature that had been altered away from GOD's simple designs as well as every human that had enjoyed sinning with Satan and his gang. Satan's army of fallen angels was imprisoned until Judgement Day when they will be released for s short while before being totally destroyed. Unfortunately for us humans, killing the Nephilim bodies did not kill or capture the half-angel spirit and it is these invisible entities that have since the Flood aided Satan in creating the illusions of UFOs, ghosts, demons, poltergeists and all other occult and demonic manifestation.

If only these simple facts could be accepted how much simpler life would be!

But film 'scenes' are not the same as the 'cenes' the geologists and anthropologists speak of - though these people are just as inventive or as devious as the best and worst directors and producers Hollywood has ever employed.

No, the 'cenes' the GOD denying Evoquacks - geologists - Satanists - bandy about so much are just like a good novel: figments of imagination. We cannot blandly excuse these people for the 'cenes' they force on the public! Their ranks are crammed with more quacks than ever was the snake oil business but while the snake oil would only have its buyers running for the lavatory the 'cenes' sold by the gists and quacks are pure fiction intended to deceive and harm us and ensure buyers follow Satan to destruction!

And how these quacks do love to come up with new names for the same old sedimentation period! Look at all these silly names for the 371 days of the Flood!

Jurassic, Triassic, Cenozoic, Paleogenic, Cretaceous, Carboniferous, Mesozoic, Quaternary– and let's not forget Silurian, Eocene, Devonian and Pliocene – and while I am at it I might as well toss in Miocene and Permian and Ordovican – and to 'cap' them all I'll add Cambrian and Pre-Cambrian. Oh dear, I missed out Neocene, Holocene, Upper Paleolithic and Mesozoic Marine Revolutions, Oligocene and Pliestocene and Hettangian, Sinemurian and Pliensbachian!

Now I've just seen Neoproterozoic while surfing La Brea Tar Pits.

And my own area supposedly had a Permo-Triassic era!

These are all contrived nonsense names that desperate Evoquacks put on the various strata in which they discover various fossils they find in deep holes or exposed along natural rock faces of coasts or canyons. Type any of those names into your web browser and you will spark a flood of woozle that will bury you as densely and effectively as Noah's Flood buried the Nephilim bodies and the fossils that are claimed to be from the fairy tale cene researched .

A cene is nothing but a layer of sediment from a particular flow of water. The initial Flood water first washed the light loose stuff- the humans, animals, small buildings, vegetation - off the land and sent it swirling down the nearest slope or valley into the great depressions and cracks in the Earth's crust, then the waters scoured the subsoils and made that all rush away to expose the footings of the bigger stuff - the great trees, the rocks and mountains which in turn succumbed to the pounding waters and went crashing and tumbling as gravity urged it on.

During those turbulent days and weeks all the creatures, vegetation and artefacts were torn and pummelled into unrecognisable disconnected parts thoroughly mixed and tangled in their resting places. Recovered bodies of sailors unfortunate to be cast onto rocks in today's storms and those of victims of recent flashfloods and tsunamis mutely testify that human bodies especially are unable to withstand the pounding and mincing action of powerful water, debris and hard rock. All the billions of tons of initial Flood debris settled into layers with many other layers added as the waters kept surging over the planet and stripping deeper layers. First area topsoil debris was overlain with first subsoil then second area topsoil fell and was followed by second area subsoil or perhaps first area sub-subsoil. As the rains continued the topsoil, subsoil and sub-subsoils from all the land hurried along down the nearest slope pummelling everything like the most efficient mincing machine. Rock was ground into sand and fine dust and all mixed with calcium carbonate and other binders and eventually settled into layers that we now see as rocks, slates, clays, shales, conglomerate and puddingstones all layered neatly or higgledy-piggledy on the shattered and jagged bedrock. Today Evoquacks label the layers cenes, epochs and eras.

This explanation for the process of fossilization leaves a number of questions. The delicate and intricate structures revealed by many of these fossilized marine organisms indicates that the fossilization process must have been associated with short periods of time. In fact, billions of fossil fish worldwide reveal that death occurred in a split second of time.

For an example we could examine the numerous fossils in a location known as The Red Sandstone, an area covering half of Scotland. Observers have described this region as an aquatic graveyard with thousands of different localities disclosing the same scene of destruction. The red sandstone deposit, that covers an area of approximately 10,000 square miles and over 150 feet in thickness, screams with the evidence of a wide-scale catastrophe. Fossilized fish are found there contorted into abnormal shapes. Others have died suddenly in swimming positions.

Google 'Images of Erosion' and you will get hundreds of photos of eroded canyons with all their amazing colours, sediments and twisting courses complete with rills, gullies and scree slopes along the sides and spires, peaks and ridges along the canyon bottom – but some of these amazing canyons are actually just a few feet wide even though they look as majestic as the Grand Canyon or Bryce Canyon! Zoom in and out on each picture without reading the name or location and many times you will think you are looking at Grand or Bryce when the picture is in a farmer's field where heavy rain cut a mini canyon in just a few hours or a couple of days! Add to this the impressive canyon topography formed during the few hours of the Condit Dam draining and you will have to come to the unavoidable conclusion : the Grand Canyon was made as the Flood waters drained off Earth about 4,350 years ago!

One aspect of The Flood that is obviously beyond the understanding of Evoquacks is that subsoil from afar would arrive at a deep abyss after the nearest topsoil had made one layer and the nearest subsoil had laid another and so on until the furthest lowest soils finally arrived at the lowest depth leaving exposed bedrock behind it. Just try to think through the erosion on one side of hills verses the other. As the watery mess reached the lowest point of the ruined Earth it systematically dropped its debris in the strata we see today. Great rocks at the bottom, smaller boulders and stones graded by size, gravels, sands, heavy organic matter such as bones, lighter organic matter like tree bark, leaves and grasses, buoyant wood, fruit and light organic matter floated over the suspended fine dusts that eventually settled and locked everything in place.

There were vast quantities of green leaved plants in the water and they had been pounded to shreds to release vast quantities of calcium carbonate. Calcium carbonate makes sand set into sandstone and ensured some sediments set into hard rock while others became crumbly shales or coals or puddingstone or just rested as loose marls. That nearer the top of the flood waters didn't harden as much as lower stratas. When the Flood water finally started draining away to expose dry land it easily cut deep channels through these layers and the loosened debris surged down to the sea to add more layers. Watch the video of the draining of Condit Dam to see how a miniature Grand Canyon forms in sediments in just an hour or so.

A film director would break the Creation and Flood story down into many shooting scenes. Scene One is a water covered ball appearing in a unlimited nothingness; Scene Two is the ball drying to leave some water full of all manner of aquatic creature, Scene Three is the land blooming as plants, creatures and humans sprouted all over it; Scene Four as the initial happy easy life giving way to lust, envy, strife, disease and disgusting idolatries; Scene Five is a man chopping great trees to build a huge box and an assortment of creatures entering the box before its door was shut to the catcalls and derision of the masses of evil onlookers; Scene Six is a seven day pause as bored with the lack of action the people drifted away from the box; Scene Seven is sudden roars and shudderings shake the Earth, torrents burst from the ground and slam down from the sky - and so one for many scenes.

The Evoquacks base their religion on the fact that any eroded cliff face will show various strata. Some layers will be hard rock, some softer rock, some fragile shale stuff, some hard granites, some soft and loose gravels and pebbles, some sands, some limestones, some chalk, some coal and so on.

These eroded layers can be seen everywhere exposed either by nature or man. When I was little I sent many happy hours in the local brickworks quarry picking through the various layers and watching the mechanical digger excavating material for making bricks – grey shale was preferred as just like the shales being explored for potential fracking it contains a lot of carboniferous material i.e finely ground up plant and animal matter and burns hot without need for much fuel – but above and below that layer was hard rock, soft loose gravels, yellow clay and coal. Half a mile from my present apartment a road cutting has exposed strata almost like the old quarry but lacking the yellow clay. The babbling brook six metres from my front door tumbles over millstone grit which it cannot erode though after storms it can toss large rocks about to eradicate the trout lies. Ten miles away is another stream pouring between two grit boulders onto some yellow clay and I am sure that in the 25 years since I last saw it the clay will not have washed away in the slightest degree – this is why earthen dams have a clay core. Yet just this afternoon I walked about fifty yards from home and photographed an old mill made of sandstone and the stones were seriously eroded by perhaps a combination of exhaust acids and road salt.

Three hundred yards from home I photographed the corner of a carpark that now has a covering of topsoil and vegetation including small trees and I am sure that if left untouched for just fifty years that carpark would be a miniature forest with a deep layer of rich topsoil, lush vegetation birds and small mammals to provide mute evidence of the Earth being young, and The Flood was real.

The Grand Canyon is one of the most spectacular exhibits of exposed Flood strata held up by the Evoquacks' as an example of millions of years of erosion by their university 'groupthink 'woozle' aka brainwashing' despite it dating closely to 4,350 years ago when the receding waters of The Flood drained off the new continent of North America and set off racing towards the sea eroding the many young layers of probably still soft sediment laid down perhaps as little as a few months previously. Mount St Helens has a new grand canyon too – all washed out in just a few hours overnight while no-one was watching so we have no video of it to show the howling Evoquacks! The video showing the draining of the old Condit Dam reveals exactly how wet sediments slump, flow and form into the stereotype rills, gullies and scree slopes seen in Grand Canyon and so many other canyons formed as The Flood drained away 4,350 years ago. Watch this video to see canyon formation in seconds not millions of years! https://www.youtube.com/watch?v=4LxMHmw3Z-U&t=5s. On a video of US troops fighting Taliban in Afghanistan can be seen a wide flat valley with steep sides of eroded horizontal strata with scree slopes from higher parts. A photo of Olduvai Gorge area shows these sediments from the Flood. Many cowboy films show typical Flood topography. These sediments are seen all over the world. The Sphinx sits in a dry moat with characteristic rills, gullies and scree slopes proving it is Pre-Flood.

Some Evoquack dreamt up a fantasy idea to explain the many layers of recurring strata by compiling the theory of 'Plate Tectonics' which incorporated a lot of the big bang nonsense.

Basically Plate Tectonics means Earth has a loose outer shell separated into pieces that slide about and climb on each other ad infinitum with the climber having a few hundred million years of supremacy before it too begins to sink down and has another plate climb on its back. Simple really. I was saddened last week to hear one of the leading women of local church claim tectonics was real as she had studied it as part of her archaeology degree- but when I corrected her she was really angry and now seems to hate me... Even Aristotle, a thinking man as ever was, fell for the tectonic plates idea along with earlier metaphysical nonsense. Newton's laws of action and reaction make the climbing plates idea sheer stupidity but that has never stopped all Evoquacks and churchgoers proclaiming it as verified fact.

Naturally they gloss over the fact that all mountains have unopened seashells on their peaks...

In the 1600s many rich people started gathering collections of different rocks from around the world and some specialised in rocks showing fossil creatures. Fossils were soon being displayed that were of weird fish, animals, birds and insects that the excavators and exhibitors claim are 'transitional and imperfect forms and the distant ancestors of today's creatures' but it is obvious to anyone with a brain higher than simian that the darned fossils things is always complete and must have been perfectly capable of leading a good life.

Even I own a piece of limestone containing many trilobites and also a piece of lighter limestone material containing a perfectly preserved fish that must have been buried alive as its death throes left an outline of its thrashing fins bigger than their final stilled size. Collecting is addictive as well as illuminating.

The world's museums and stately homes are full of these intriguing fossils and curios.

It became apparent by the Middle Ages that there had once been vastly more species and kinds of creatures on Earth than there is now. And to confuse matters some were of immense size and very peculiar shapes - especially those with huge teeth! Clearly these were not related to the domestic cows, horses, sheep and goats of the civilised eighteenth and nineteenth centuries – someone would have to invent a name for them! Various 'gists' got in on the act and made money by conjuring up all sorts of nonsense about the ages of the rocks and the fossils in them.

Having 'identified' - actually just noticed
one layer is different from the ones below or above,
but not always different from the one above the one above,
or below the one below the one below, these

Evoquacks claim the fossils in the top strata must be millions of years old

and the rocks having fossils in them must also be millions of years old

and the rock below the the upper layer must be more millions of years old

and the fossils in the lower layers must be more millions of years old

and the manmade artefacts found in rocks must be millions of years old

and the manmade artefacts found in coal must be more millions of years old

and the strata under the coal must be more millions of years old

and the lowest strata of all that has no fossils must be thousands of millions of years old.

And when you add up all these wild estimates of millions of years old

you eventually get to thinking the Earth is thousands of millions of years old

but if you don't believe the estimates that Earth is thousands of millions of years old

disputing it will have Evoquacks smirking that Earth really is millions of years old

and you lacks the expensive brainwashing that tells them Earth is

millions of years old,
but keep on denying Earth really is thousands of
millions of years old
because your Bible shows all Creation is not thousands of
millions of years old,
and you will be told you cannot know the Earth is not
millions of years old
because so many Evoquacks say it is
millions of years old
and they is been to universities and get paid lots to say it is
millions of years old
and who are a nobody like you to say Earth is not
millions of years old
as your old fairy tale Bible cannot possibly prove it is not
millions of years old
and then you start to wonder if it is Satan making them say it is
BILLIONS OF YEARS OLD
- and it is because
SATAN IS THE FATHER OF THE LIE!
Jesus said of Satan at John 8:44: "He was a murderer from the beginning,
not holding to the truth, for there is no truth in him.
When he lies, he speaks his native language, for he is a liar and the father of lies"
Every 'gist' or Evoquack who quotes any of these Flood sediments as 'cenes' and truth is a slave to Satan!

As some rocks faces do show quite definite layers of different rock or shales or loose sands and gravels a child might be easily fooled into thinking the rocks all date from different ages but from the very fact that supposedly ancient dinosaur bones lie next to modern animal bones it should be obvious that both creatures wandered Earth at the same time! But no, the Evoquacks just sneer at that and claim dinosaurs featured in Cene 2 and humans in Cene 6 or whatever! Stephen Gould is the chief EVONONSENSER in this regard with his claim that bones dates the rocks the bones is in.

I had to buy a new computer in November 2015 and searching through the available screens I found one of a shot of the Grand Canyon, now each time I switch the computer on I gaze at the vast canyon panorama of flat plateaus with scree slopes cascading down through the innumerable varves to the little river meandering in the bottom of the canyon. It perfectly illustrates The Flood and the way the canyon was formed in perhaps just a few weeks as the floodwaters drained back underground.

It is a superb scenic shot from a cave or overhang quite high up and overlooking a vast area of deeply eroded landscape with hills, plateaus, prominent buttes, mesas and scree slopes – but the overwhelming feature is that all the land is actually very numerous multilayered loose sediments.

Unfortunately the whole picture is a bit too hazy to count the layers in the middle and far distances but right in front of the camera the rocks can be seen to have layers as thin as perhaps 1/16 inch or 2mm. The entire view seems to be flat plateaus with barren crumbled scree but there are some isolated low bushy plants growing on some of the flatter parts at the base of scree slopes. These plants must have grown from wind or bird-borne seeds that managed to lodge in a crack where a bit of dew sparked them to life. Birds or insects might have visited the growing plants to fertilise it but GOD gave plant roots the ability to dissolve and digest rock so they do have some nourishment in the flood sediments of the canyon. Most of the canyon cannot recover real fertility as it is too steep or loose to give a seed a home and there is probably no way in which worms can be introduced and thrive except by expensive and determined effort by the local land management office. The canyon managers could make pockets of fertility by airdropping tons of farmyard manure with its mass of worms and other creatures, or by having all the visitor toilets discharge their sewage down the sides of the canyon, and adding seeds and worms to create patches of lush greenery that would gradually spread across the desolation. This will not happen as the managers push Satan's and the Evo's claim that these innumerable layers and their differing composition and colours is absolute proof of Earth being billions of years old and was born from a big cloud of dust that had given up most of its helium and hydrogen to the greedy sun and the canyon must remain unpolluted by humans.

One website devoted to the Grand Canyon describes the various sediment layers seen along the canyon walls as everything from the fairy tale of BigBang but really they were all laid down during the Flood. As the canyon is very deep it does have many layers that are also very deep but the truth of these layers is not the childish EVONONSENSE learned from watching Planet of Apes movies but simply from the fact that the Flood was the most fantastic catastrophe imaginable – or rather, to the monkey brains – unimaginable! The Evoquacks who lend their name to studies that named and dated the canyon's layers ladle on so much Woozle to make their silly dreams match the evidence of their eyes by emulating Einstein and fudging the truth with claims that some layers are evidence of greats seas bringing in great quantities of sand... This is their list of EVONONSENSE to describe the finally settled sediments and millions of years their monkey brains assign to each layer.

Kaibab Limestone   300 Million Years Old. Permian.
Torpweap Limestone   280 MYA. Permian
Coconino Sandstone 260 MYA. Permian
Hermit Shale 280 MYO. Permian
Supai Group 323 MYO. Pennsylvanian
Redwall Limestone 350 MYO. Mississippian
Temple Butte Limestone 420 MYO. Devonian
Muav Limestone 550 MYO. Cambrian
Bright Angel Shale 550 MYO. Cambrian
TapeatsSandstone 550 MYO. Cambrian
Grand Canyon Supergroup 4.6 Billion Years Old. Precambrian-proterozoic
Vishnu Schist 4.8 BYO. Precambrian-archeozoic
Zoroaster Granite 5.0 BYO. Precambrian-archeozoic

I'm not going into much BigBang cosmology as any Evoquack has a head stuffed full of the subject and no creationist needs to know much of such a fairy tale, but basically cosmology seems to be the religion based on the worship of dust clouds and especially the one that formed Earth. This dust cloud is supposed to have been thrown out to our present place in the universe around the already burning sun and spun itself into the ball we live on.

Evoquacks and 'gists' drill down through all the layers of dust and notice that some is different colour, texture and composition than others and many contain various bits of fossils that were obviously once living creatures or vegetation.   They then compete among themselves to make evermore fantastic claims about the age and origins of the stones and bones they found. Men and women have been finding these bits practically since Noah's children started spreading across the landscape as areas recovered after the Flood Many bones known and now labelled fossils would have been spotted along river banks by Noah and family but the flood being so recent the remains would have been unremarkable and just dismissed as a polluted creature from before the Flood.

From the flood of about 2,400 BC through the Ice Age decades and especially after the Tower of Babel debacle when GOD ordered the land to open and swallow most of the tower just as He had had ordered it to open to swallow Koah, Nathan and Abiram as recorded at Numbers 16:23, there was little need, urge or time to start exploring and wasting time in barren, eroded lands as life for many humans and creature had become a fight for survival in the unforgiving Earth.

Once a stable life with a sufficiency of food and shelter was possible the family groups diversified into farmers of land and animals, fisher men, builders, toolmakers, leather and textile makers and soon some began to have time to use their talents as artists and artisans or were motivated or paid to make objects of clay, stone, wood, fabric, gemstones and metals and these have been turning up in old city excavations for many centuries and were passed about between private individuals and later displayed in museums or universities. British museums have thousands of useful and decorative items made of gold or bronze but they foolishly date some from before The Flood when obviously they must date to the centuries after the Flood when the descendants of Noah followed GOD's decree and spread out from the Ark's landing place. Satan's slaves, the Evoquacks that run the museums, just love to promote his lies at every opportunity!

Various countries have spawned exceptional craftspeople working with naturally available or easily mined material for many centuries. Everyone is familiar with the glories of Chinese porcelains produced since just after Jesus died although some Evoquacks led by Elisabetta Boaretto claim some ceramic fragments found in a Chinese cave are 20,000 years old! Not surprisingly the Smithsonian Institution, one of Satan's earthly homes, printed Boaretto's report on the fragments as gospel truth in 2009.

The Smithsonian's mission statement is 'The The increase and diffusion of knowledge' but it is engraved around the official logo of a Molochian owl to show that The Smithsonian is ruled by Satanists who destroy all trace of anything that confirms The Bible and The Flood. Old newspapers display miners and explorers posed with giant and strange skeletons and artefacts that they 'sent to the Smithsonian' but now the Smithsonian disavows all knowledge of them. Either the Smithsonian is and has always been controlled by Satanic Freemasons and Illuminati or else they have so much stuff stored in dusty boxes no-one has the will or desire to check for important fossils that might confirm anything in the Bible?

No real records exist of anyone in Southern England's Lyme Bay area finding important fossils until about 1800 though many fossils were found and sold as curios and possible good luck or medicinal charms in the 1700's. The finding of an Ichthyosaur skull in Lyme Bay by Joseph Anning in1811 created much interest as a great many people still believed in GOD and His Creation, but having only the scant mention of a Behemoth in the Bible wasn't precise enough to allow insight into the antediluvian dinosaurs, although on reading Job 40:15 Behemoth is obviously something like a Brontosaurus or Diplodocus and fits the carving of similar beasts on the medieval brass surround of Bishop Bell in Carlisle Cathedral. The Bishop died in 1496 and the carvings suggest someone had either seen the live creature or had been inspired to carve a perfect likeness of one. I've just seen an ancient carving of a TRex on Libyan rocks in the middle of the desert that prove a pair were on the Ark!

Joseph Anning's daughter Mary, found a skeleton to match the ichthyosaur skull to make a complete creature but when eventually sold it was labelled as a fossil crocodile as of course for many hundreds of years British sailors and explorers such as Sir Walter Raleigh had been travelling to lands where great crocodiles and alligators were common and it was perhaps an excusable mistake to label the beast as a crocodile. It was finding a fossil skeleton of the Plesiosaur by Mary Anning in 1823 that fossil hunting and interpretation got into gear! She found it in eroding cliffs along the seashore in sedimentary layers laid down 2350 BC during the Flood but given the silly 'cene' name 'Jurassic' by the Evoquacks and dated about 200 million years old. Those cliffs and sediments have only a thin layer of topsoil to show that they are not ancient and obviously date from the Flood.

The Woozle accreted onto the word 'Jurassic' is incalculable and grows daily as each new proto Evoquack sits at the feet of a superior Evoquack in some ivory tower and accepts the woozle spewed. Failure to accept the woozle or argue it is false will ensure the proto Evoquack will be pushed aside and face a future of slinging burgers in a burger bar! No proto Evoquack dare risk that so readily agrees with all the woozle and eventually becomes a qualified Evoquack in one of the 'gists' and is then able to spread more woozle to more proto Evoquacks.

How can we be sure the fossils really are recent when they are in deep and solid sediment? The dear old Coelocanth fish being the prime example of coming back from the grave and like earthworms making fools of all the Evoquacks. Satan's Wikipedia solemnly states with full and unalterable authority that coelecanth came into existence 400 MYA but after being written off as extinct 68 MYA it started being found off Africa and delighted everyone with its ability to swim upside down and oh my goodness – backwards too – just like dragonflies can! 400 million years is an awful long time so why dear old coelecanth hasn't once looked at the African coast and thought about evolving into a monkey or an efelump is a mystery. Perhaps some coelecanths is just old stick-in-the-muds with enough sense to know that it is nice to remain frolicking in the deep water rather than risk climbing out under that sun and being baked alive? Perhaps they are still waiting for some natural sunblock to evolve thanks to Hawking's marvellous gravity? Then of course we have to ask how ceolocanth survived all those meteor impacts and extinctions. They must be tough old things!

Who designed such a creature as coelacanth?! Did it really exist for 400millionyears unchanged? Obviously not! It is just one of the fish that GOD designed about 6,000 years ago!

But 'NO!' argue the Evoquacks – we don't believe Fred Flintstone had a pet TRex because our Hollywood fantasy woozle tells us dinosaurs all died out 68million years ago when a giant meteorite smashed into Earth to create the Gulf of Mexico and its debris either blanketed the Earth in chilling cloud or flew about the sky and eventually made the Moon!

Mary Schweitzer even found a TRex bone with stinking soft tissue that she claims proves the dinos all died 65 MYA! Oops – it's now 68 MYA! She chose 65/68 MYA because she read some woozle about a meteorite striking Earth 68MYA and killing all the dinosaurs so when she finds a dinosaur bone with smelly tissue she remembers some woozle about a meteorite hitting Earth 68 MYA and so naturally she woozles that the TRex was buried 68 MYA. And she knows the sediments are 68 MYO because they contain a stinking TRex bone that died 68 MYA! And TRex tissue must be able to stay soft and smelly for 68 million years because it was found in sediment reeking like a charnelhouse that is 68 MYO because it has 68 MYO TRex bones in it! Mary is lauded and held in high esteem but is nothing but a foolish if not evil quack feeding the public woozle pablum about the TRex being buried 68 MYA. Or is it now 110 MYA!

Desperate to prevent belief in GOD, the evil Satan inspired and still inspires many prominent people to woozle that fossils are millions of years old and therefore proof that life evolved from nothing on a billions of years old bit of rock called Earth.

They gave each of these layers an impressive sounding name, then over the years they bandied these names about and publish their thoughts about them in the academic press or in lectures at universities until the names have accreted a whole mythology with supporting fantasies that make StarWars seem quite sober and Planet of Apes quite believable.

In some areas the range of fossils in a layer are different from the same layer elsewhere which really baffles the Evoquacks though they generally make up some EVONONSENSE to cover up. The excuse will usually be to date the layers differently as according to their thinking a few millions years is sufficient for one creature to evolve from something else so the different fossils must obviously be from something that evolved a bit over time.

To add to the confusion many of the layers of strata are uplifted or sunk as though earthquakes had been a regular feature of prehistory. The Evoquacks resort to the woozle of plate tectonic fairy tales to explain this.

Further confusion comes from some layers having loops and swirls as though it had all been moved while still wet. I know exactly such a tight loop in Scotland where a new road has been cut through a small hill. The wet layers looped up about ten feet over the lower level layer. The lower layer must have been set and dried when the wet top layers flowed over it then the flowing sediment was blocked and piled up in several layers to make the loop. The looped material then dried and hardened before being covered with further material. In Huddersfield, England, I walked across a bit of footpath paved with very old setts – small hard stone blocks about 8 inches square – and was amused to see the strata in it contained one of these amazing swirls! I'll take a pix of it next time I go there.

I am typing this paragraph just a day after seeing a multi-varved piece of rock with this loop in the garden centre about quarter-mile from home! The centre has a good selection of rocks and stones for decorating gardens and some rock is striking red and cream banding with lots of quartz adding sparkle to all. I can date it precisely to 4,350 years old.(Actually not quite so precise as there is some doubt about the exact date of The Flood with 2350 BC being favoured by many researchers and 2450 BC by others so I hedge my bets at 4,350 years ago.)

The loop in the stone and all larger and more massive loops were made during the resettlement phase of the receding flood when the many layers of thickening sediments were sliding downhill as 'real' gravity made them follow the retreating water levels until they slid onto dryer soil and could slide no more and the following soil pushed them up into loops. Watch videos of train wrecks to see how the looping and piling up happens. EUREKA!

What really baffles the Evoquacks is the fact that in some places there really does seem to be some very simple creatures such as Edicara mixed with pollens and such in the very lowest layer of all! These, crow the Evoquacks, are sure proof of life starting in a primordial puddle from sparks between wet rocks!

Ah dear, how these Evoquacks depress me with their Satanic outpourings! If only they would read Genesis Chapter One and use a little of the wonderful brain GOD designed into all humans! But no! It is simpler to let the worms make fool of them! Is there a perfectly logical explanation of why there are edicara and shellfish remains in the very lowest layer? Yes of course – but first you have to accept there really was about 1,600 years between Creation and Flood! Can you do that?

The Dunning-Kruger Inferior cannot accept any Creation as these are people with limited intelligence who will toss this book away at this stage while the D-K Superior with much superior intelligence will have noticed some connections – the real mark of high intelligence - and be eager to learn how worms, edicara and seashells prove Creation. The connection is the 1600 years.

The Bible quite clearly states that GOD first made earth as a water covered ball of rock and dirt that He stirred for sometime with His active force. That is what created the the first layer aka The Pre-Cambrian or the Bedrock. Next He separated the water and made land appear and followed that with commanding all varieties of vegetation to appear. Immediately the new plants pushed up and began to flower they would be releasing pollens and then dropping petals and all the other parts that fall away as the fruit and nuts begin to mature. A good amount of these must have found their way into the new sea especially from plants growing along the water's edge. The sun would be making gentle breezes during the day and the moon would be making gentle tides twice a day to ensure the waters were kept in motion and all the detritus was spread about. The tides also gently flex the granite bedrock and by courtesy of piezo-electricity give Earth a gentle inner heat – not the liquid iron core so beloved by the

Evoquacks. The next day all the sea creatures were brought into being and naturally included everything from the tiniest planktons and 'ediacara biota' as Wikipedia quaintly puts it. These creatures lack the immortality of humans and lived and died their short lives before settling into the new soft sediments gently swirling about the seas from all the springs that flowed out of all the earth. Rivers can flow many miles into the seas but as the exploration of the Titanic wreck shows the total sediment away from land may only be a fraction of an inch per year as the ship is 370 miles from the nearest land. Crabs can be seen creeping about the Titanic and would no doubt have helped recycle the bodies that formerly filled the several pairs of boots seen lying close together. These crabs are sure proof that the Evoquacks are totally wrong about dead creatures slowly drifting down to the seabed and being buried by sediment over millions of years. Dead blue whales – the largest creature on Earth drift to the seabed and provide over a year's food for hordes of horrible slimy hagfish.

On this point we have to face the fact that when deep water is stirred its finest sediments can remain suspended for many hours if not days so it is likely that the first ediacara would die and be buried in the very first sediments – just as they are found nowadays. Here in northern England we have old canals and the dirt in them is so fine that even if a boat hasn't passed by for several days the water will remain cloudy before all the fine stuff settles to make a varve.

Thus the Evoquacks' 'Pre-Cambrian' is actually Creation's first layer and dates back 6,000 years.

The first large concentrations of fossils are found in 'the Cambrian' and the Evoquacks claim this is when a mass explosion of life happened between 545 and 480 million years ago! This is of course a bit more fantasizing as obviously the layer being on top of the 'Pre-Cambrian' means it can date from less that 6,000 years ago and actually it is datable to the first day of The Flood. On that first day many areas of land would have been stripped of all loose mater and stayed bare until the rising water allowed detritus fromdistant parts to arrive and not be washed away.

Maybe in the Evo's world New Year precedes Xmas with Easter being sandwiched between the two except where earthquakes have tipped the days about?

Which script do you prefer to you believe in: GOD's Creation or Satan's Evonsense? Have you accepted one or the other?

What is more to the point is: can you even understand the two beliefs!

Does it matter?

Do you care?

Should you care?

Well, Yes it does matter and Yes you should care as what you believe in has really serious implications that unfortunately the proponents of either side don't spell out properly to aid their believers fix their choice or even switch sides.

I want to be your guide to the Truth of GOD's Creation and exposing the lies and hypocrisy of Satan's EVONONSENSE such as these claims:

Stars are millions of years old!

Stars have existed for hundreds of millions of years!

Stars are so far away that no spaceship could ever set off and get to even the nearest star!

Stars were formed billions of years ago in a big bang!

Stars burn for billions of years and then blow up spectacularly!

Stars are all suns and may have their own solar system with civilisations like Earth!

You will have seen statements like this every time there is a space item on the news. They are all false statements intended to make you think there is no GOD!

I'll repeat that statement again so you grasp its importance: They are all false statements intended to make you think there is no GOD!

Whoa! You say, GOD! What has GOD got to do with the distance to the stars?

Quite simply GOD said He made the stars and started them on their travels about 6000 years ago - teaching them to be billions of years old is therefore calling GOD a liar!

We don't even know what a star is as they are so far away and still speeding faster than we can think of travelling. Evoquacks like to claim stars is giant suns each surrounded by a solar system with probably one or more Earths but for all we know they may just be pretty bulbs plugged into the firmament. Some stars do blow up or perhaps short circuit to provide us with something to look at. Their distance is supposed to be checked via redshifts but that's like trying to determine the colour of a startled octopus.

Evoquacks desperate to flesh out their fantasies have imagined the Oort Cloud and a typical quote about it is this: 'The Oort cloud is a hypothetical spherical cloud of up to a trillion icy objects that is thought to be the source for all long-period comets and to surround the Solar System at roughly around 1light-year, and possibly 2 light-years. It is thought to be composed of comets that were ejected from the inner Solar System by gravitational interactions with the outer planets. Oort cloud objects move very slowly, and can be perturbed by infrequent events, such as collisions, the gravitational effects of a passing star, or the galactic tide, the tidal force exerted by the Milky Way'. This paragraph reveals how Evoquacks imagine something then make it real in their heads by stating 'facts'. A fact based on an imaginary thing must be an imaginary fact?

Alas, so much real history is pushed aside by the Evoquacks that very few people can actually figure out Earth's real origins and age and what GOD intended for it before Satan decided to rebel and successfully tempted Eve. We are now waiting for Jesus to come 'crush his head' and who knows who will survive Armageddon in the rapturous embrace of Jesus and His angels while Earth smokes and shudders below. Anyone lucky enough to be granted eternal life will have the joy of meeting and mixing with newly resurrected youthful people with names from the Old and New Testaments – and perhaps even names from secular history?

Who knows if Adam and Eve will be resurrected from their long lost graves.

On the previous pages is outlined the dichotomy between the reality of Creation and the fantasy of Big Bang and the Evolution of life.

To believe one is to claim the other a lie.

Creation is known from the record of GOD's Word and has been proven true and accurate in all ways while Evolution is a total fabrication of lies, suppositions, postulations and imaginings with few that can be tested let alone proven accurate. Because it directly contradicts GOD we can be sure that came from Satan – and it rightly needs referring to as EVONONSENSE so throughout this book I will refer to all evolution as 'EVONONSENSE' and its advocates as Evoquacks.

Why the dichotomy?

Why such vitriolic outbursts from some defenders of EVONONSENSE? Schopenhauer really knew Evoquacks when he said they would get violent at mention of Creationism: All truth passes through three stages. First, it is ridiculed. Second, it is violently opposed. Third, it is accepted as being self-evident.

What makes Creationists so serenely confident that they know they were created by GOD just a few thousand years ago when the Evoquacks display multitudes of fossils dug from deep solid rock that certainly appears to be ancient and claim they were killed by a meteorite impact 65 million years ago?

Why should the Evoquacks fly into foul mouthed rage at any suggestion that their millions of years beliefs need shrinking by a factor of 10,000 if not more so that a fossil is correctly aged at 5,500 years old and not 55million?

What makes the creationists – including me – certain that the immense deposits of coal, oil and gas beneath our feet is just a few thousand years old and hasn't been forming for several hundred million years?

Why do the Evoquacks refuse to believe that manmade artefacts regularly found in old coal proves the coal is just a few thousand years old?

What makes the Evoquacks so certain they started out as pondslime with later helpful inputs from aliens from distant planets?

Why do Creationists like to speak of 'GOD' as a real entity to whom they owe their very existence when the Evoquacks scorn all idea of 'GOD' and laugh about an old guy with a long white beard up in the clouds as being childish superstitions and claim themselves evolved from a speck of matter like Hawking claims?

Why do all the Evoquacks worship Hawking as a superbrain when he displays sheer stupidity by his 'gravity created everything?' - and believes he evolved from a monkey that evolved from a fish etc?

Why do the fools like Dawkins and all the fools who follow him get the plaudits and Nobel prizes for blaspheming GOD and His Creation and praising the lies that Satan whispers?

Why, why, why!

There must be some process that over the years has resulted in we humans being vastly superior to all other living things – if by superior we mean living in a technological world and not just mooching about like most other creatures seem to do.

Desmond Morris once labelled us 'Naked Apes', and he will have to explain that to GOD one day, but the label stuck and today huge numbers of expensively educated people all round the world are happy to put themselves in a class with monkeys, apes, gorillas, chimpanzees and such. Well, actually most of them get very precious with me saying they is evolved from monkeys and claim that they didn't really evolve from monkeys but from some apey/gorilla proto-human that just happened to spontaneously evolve into existence one or two hundred thousand years ago and was distinguished from all other monkeys and apes by preferring to walk on its back legs and being able to make fire. If pressed they will grudgingly admit that their ancestor was a beetle-browed hulking hairy thing with limited vocabulary and a liking for brute force, wifebeating and living in caves. Asked how a rough unwashed wrestler with prominent bulges over its eyes turned into the slim sleek gorgeousness of the humans in the high street they shrug it off with a sigh of condescension and say 'Just Darwinian microevolution of course!'

When we point out that microevolution cannot happen they just turn Satanic and call us fools.

We creationists are saddened by this adulation of Darwin as his silly theory of evolution has had so many holes poked in it it is more like a tincan that's been used for target practice but we know that nothing we offer as proof will be accepted so solidly is EVONONSENSE concreted into their brains.

We creationists know that we can have, and mostly have readily taken, the gift of Holy Spirit from GOD and one vital seventh of the gift is discernment and that is what ensures a Christian creationist can see through the fog of EVONONSENSE while the Evoquacks cannot even see the fog they are floundering about in! Their condition really can be called 'A Poverty of Intellect' in that they think they is educated but their knowledge is fragile bubble of fairy tales and nonsense with no basis in reality. Alice Roberts is a perfect example of this as despite knowing Darwin proved earthworms make new soil at the rate of one inch per five years she claims Earth is hundreds of millions of years old – so where is all the topsoil, Alice?

It is GOD's gift of discernment that enables a Creationist to see not the invisible edifice – the Emperor's New Coat - labelled EVONONSENSE that the Evoquacks see but just a crumbled weed grown pile of rubbish that no sane person would want to cling to. The fact that if an Evolutionist neglects to weed his garden path it will eventually get covered in grass sod and creeping things as sure proof of a young Earth eludes their ape brains.

To set the scientific background to the Evoquack's discoveries and claims needs a date of the main discovery that facilitates research into fossils: the microscope.

Making the first magnifying glass occurred thousands of years ago when it was noticed that clear glass bottles full of water enlarged objects behind them. Some maker of bottles may have deliberately striven to blow some very clear and round bottles intended to use as magnifying globes but any such person or date is unknowable. Nero is said to have used clear gems to see small detail better. It was Englishman Roger Bacon sometime before 1292AD who first showed carefully produced glass magnifiers made as one piece of good glass polished to a, dare I say it, UFO-like disc with a thicker centre and thin edge such as we can still buy for a small amount in craft or photo stores. These single glasses can be so graduated from thick centre to thin edge to produce a maximum useful enlargement of about 6 times after which the laws of optics operate and no image is formed but Bacon's originals may have only produced 3 times size. Achieving greater closer images took the thinking of Dutchmen Hans and Zacharias Jansen working on making simple microscopes by putting lenses at either end of tubes and thereby getting higher magnifications than Bacon.

Finally Antoni van Leeuwenhoek in the early 1600's who perhaps hearing of the Jansen's work and after considerable labour with the primitive tools of the time did succeed in making a microscope that enlarged 270 times. He then examined all kinds of organic and inorganic items and must have been astounded to see the life in a drop of dirty water, or blood flowing in capillaries, or the bacteria and yeasts that cause so many bad and good effects. He was thus the first modern scientist to systematically enlarge, study and describe the finer details of many living items and dead materials.

Leeuwenhoek's design was developed over the next centuries but remained limited by the basic optical problems of all glass lenses until, perhaps serendipitously, the genius or demon-possessed Nikola Tesla noticed that the inner surface of silvered glass globes gave immense magnifications when certain rays were projected into them. Today's researcher has multitudes of microscopes available with which to study the amazing variety of fossils found by chance or deliberate search. It is a pity that though they can see the finest details they cannot see the big picture that made the fossil – GOD and Noah's Flood.

Temperature has such strong effects on organic and inorganic things that knowing the temperature can help understand the workings, actions and reactions of so many aspects of the natural world and the final production of a reliable thermometer in the 1600s was a great step forwards. In the decades after the basic device was refined and its operating medium was changed to mercury as we still have today. Once it was realised that water at sea level always froze and boiled at definite points it was easy to calibrate the thermometers in first Fahrenheit with its odd points and the better and more logical Centigrade with freezing at zero and boiling at one hundred degrees. GOD used the Third Law of Thermodynamics on Day Four to dry off Earth and set its temperature perfectly for we humans but the science behind this is beyond the minds of 99.99999999% of humans on Earth. EUREKA!

It must have been obvious to everyone right from Adam and Eve that liquids had different properties. Adam showed Eve the thinness of the water that bubbled everywhere in the Garden of Eden and must have noticed that the juices of different fruits or honey from the comb trickled down their chins and fingers at different speeds and seemed to run as raised lines rather than thin films. Over the thousands of years since then so many people must have noticed that liquids could be thinned or thickened but it wasn't until 212BC that Archimedes had an inkling of the density of different materials. About 500AD a workable indicating device was made but following the destruction of Alexandria it took another 1200 years until 1768 when Antoine Baume invented the basic hydrometer in use today.

Understanding the use of the hydrometer would have perhaps prevented EVONONSENSE being set in concrete as its use may have shown that the supposed casual death and burials of creatures and flora didn't follow the logical process of fast decay or scavenging by other creatures but were buried rapidly in a way that demanded quite extraordinarily powerful actions to produce the strata and dinosaur boneyards and the mammoths in Arctic 'muck' we see today.

However no magnifying glass, thermometer or hydrometer is needed to observe the way grass, aided by worms and provided only with moisture and sunlight creeps over all the land to prove the Earth is young and The Flood was real. This observation can be made all over the Earth's temperate regions but why are Evoquacks blind to it! Is it because by denying Creation and espousing EVONONSENSE they really are in Satan's darkness?

As I type this I am watching a documentary on Peenemunde and note that solid grass has crept over a quarter of the solid concrete entrance road in the 70 years (?) since the site was abandoned at end of World war 2. Give the grass say another 100 years and that road would be totally covered and each century after that the grass fed by water, sunshine and the recycling by worms and other creatures would get deeper and deeper – so if Earth really is old and it is say 15,000 years since last imaginary ice age – where is the hundreds of feet of lush topsoil? Obviously the Flood washed all the topsoil off and left the land bare bedrock.

Can any Evoquack formulate a sound argument for why much of the land of the Earth is barren even if it carries a good mass if pine or similar trees? Or why the soil under beech trees is so jet black?

No Evoquacks can understand the matter of the erosion over the dried land due to the new winds, snows, ice and rains cracking and eroding the exposed granites and limestones and sending them down to the settling places to make extra layers of sediment. They poke about these layers and label them Jurassic, Cretinous, Simian and such when really the are all Plastocinian.

Added to these sediments are those created when Earth overbalanced at the height of the Ice age and buried the mammoths in muck and loess. If these later sediments included post-Flood vegetation and small creatures as most likely happened what a conundrum it would be and what silly woozles would be dreamed up to try explain the whole sediment! These Ice Age sediment layers are just Plastocinian!

Go read the chapter on Mammoths to understand why mammoth and dinosaur bones are rarely discovered together and why mammoth carcasses are not fossils and why dead TRexes is!

Chapter Five
Unbelievers.

Before you get into what to most of you will be unacceptable truths I have to point out that you were described many times by GOD's prophets and Jesus himself; for example your Bible says:

Isaiah 6,9
And he said, "Go, and say to this people, "'Keep on hearing, but do not understand; keep on seeing, but do not perceive.' Make the heart of this people dull, and their ears heavy, and blind their eyes; lest they see with their eyes, and hear with their ears, and understand with their hearts, and turn and be healed."

Deuteronomy 29,4
'But to this day the Lord has not given you a mind that understands or eyes that see or ears that hear'

Jeremiah 5,21
'Hear now this, O foolish people, and without understanding; which have eyes, and see not; which have ears, and hear not'

Isaiah 44,18
'They know not, nor do they discern, for he has shut their eyes, so that they cannot see, and their hearts, so that they cannot understand'

Mark 8,18
'Having eyes do you not see, and having ears do you not hear?'

Acts 9,18
When the scales of Satan fall from your eyes you will be able to see the truth.

Romans 11,8
'As it is written, "GOD gave them a spirit of stupor, eyes that would not see and ears that would not hear, down to this very day.'

When Satan seduces a person they stays blind and deaf to GOD's truth!

Chapter Six

Creation

Creation is the whole account of what a Christian should believe about the origins of Earth and themselves as fully detailed in the Bible. Anyone who claims to be a Christian but believes they is evolved from monkeys is fooling themselves and promoting Satan and his lying agenda. Sadly far too many people call themselves Christians but choose to follow Satan.

Our history all started about 6,000 years ago as accurately as we can calculate using the names and dates of the Old Testament patriarchs, and kings, queens and natural and unnatural phenomena and events.

Open your Bible at Genesis 1:1 and easily follow this sequence of events:

6,000 YEARS AGO GOD and Jesus already existed as did myriads of angels and a few archangels. The most glorious angel was named Lucifer and he sparkled like fine jewels.

DAY ONE. 4000 BC. GOD and Jesus created the Earth as a ball of rock covered with water. Just like that! We know this is true because all round the world is a complete layer of bedrock totally devoid of fossils of any description. (Evoquacks call this the Pre-Cambrian as Wales had not yet emerged from the waters.) Despite their being no universal gravity to make the ball spin or keep it in place it did actually just hang there as Job confirms 'Hanging the Earth on nothing.' GOD was quite pleased with the watery ball even though it had no defining features yet and was formless as the water swirled over it. His and Jesus's presence made a lightness around the Earth but it was not the sun, moon or star light we humans are used to. He made the Earth revolve slowly over 24 hours so that one half of it was in the new light while the other was dark. He called the light 'Day' and the dark 'Night' and 6,000 years later we humans still do. Then He rested.

DAY TWO. The waters on Earth were separated to make dry land appear. GOD made a canopy of water high around Earth by 'lifting' some of the water swirling on Earth. As the universe is permeated by hydrogen molecules it seems logical that He did some trick of science to split the new water to make some hydrogen molecules separate from the oxygen ones with much of the hydrogen flying off above the canopy ready to expand out to fill the present universe with the hydrogen molecules science proves fills the universe in every direction. A vast amount of complete water was contained in the canopy despite the quantity stored ready to fill the universe. The split off oxygen made a viable atmosphere on Earth and began to react with the exposed bedrock to start it breaking down to release the minerals used in the recipe GOD had used and turn the crumbling rock into soils – just as frost and rain does 6,000 years later. The climate was mild. Then He rested and perhaps thought how pretty the blue Earth looked through its atmosphere.

DAY THREE. GOD now got really serious and commanded the Earth reshape itself to let dry land rise and encircle the waters in one great sea. 75% or maybe 95% of Earth was land and 5-25% was sea. Immature Evoquacks refer to the landmass as 'Pangea' or 'Gondwanaland' and fantasize it existed between 315 and 335 million years ago – despite their own charts of EVONONSENSE! Their timing doesn't fit into the silly shooting scripts they dream up as shown in Chapter One! They also imagine it was an island surrounded by seas - another of Satan's lies they love as the Bible says Earth surround the 'sea' – singular. It was the sea being in one place i.e a lake, that allowed Noah to walk about the entire Earth warning of the impending Flood! As soon as the land had stabilised and the excess water and mud had run off all manner of vegetation from the lowest grasses to the mightiest trees began springing up from the soil. The grasses rose, set seed and cast the seed to make more grass; fruit bushes and trees of every variety sprouted, grew tall and wide, flowered and set fruit without even the help of insects to pollinate them! All were gorgeous, spineless, non-poisonous and harmless. The land flushed green sprinkled with rainbows of flowers from one shore of the sea to the other and from north pole to south pole! And all in just one 24 hour day of the sun's rotation! Such speedy growth would not be seen again until Aaron's staff went from dry lifelessness to producing ripe almonds overnight as recorded at Numbers Chapter 17, Aaron's Rod. A mist rose daily to water the entire land - if only Darwin had been able to think that short sentence through he would have exclaimed EUREKA! and tossed his whole silly 'Origin of Species' manuscript into the fire where it belonged! Earth's temperature must have been between 32 and 212 Fahrenheit as that is the range in which water is water? GOD watched the new vegetation getting more productive by the minute before resting.

DAY FOUR. GOD created the Third Law of Thermodynamics! The sun was made and set in place at exactly the right distance from Earth to ensure it warmed and lit it perfectly. The moon was set in its place to reflect sunlight onto the darkside of Earth as well as mark the times and season. The moon was bright and white although now we see it covered with dark dust blown out of Earth during the flood. EUREKA! The other planets were made and hung up to dry but the furthest away were a bit too cold. This is why Mercury has dry river beds and others have ice cover or seem completely ice – wait until the Flood chapter to understand this. GOD thought we'd be delighted to have one planet different from the others so He set Uranus on edge although Evoquacks claim it was knocked over by one of those loose cannon meteorites they is so enthralled by. He also made Venus spin slower as a wind blew at its atmosphere – which makes me wonder why Earth's atmosphere stays in place as we supposedly zoom round the sun at 66,000 miles per hour. The stars were lit and sent on their way with many being arranged in beautiful spiral galaxies. They have star power as radiometers show and soon achieved immense speeds across the vacuum of space that GOD stretched out. Their light stretched out too! Today lots of it look to be reddish so some Evoquack dreamed up redshift after listening to steam trains Dopplering by. All these heavenly bodies give off radiations to warm up all those hydrogen atoms and keep the entire universe a couple of degrees warmer than the nothingness beyond by what is called Cosmic Background Radiation. GOD works in intense cold but has a warm heart! Something called 'The First Law of Thermodynamics'? When GOD was satisfied that all the stars were shining and twinkling prettily He rested.

DAY FIVE. Evoquacks mistakenly lump all their cene's, age's and aeon's into this day but in reality it was merely a 24 hours version of their Plastocine Era. GOD was up early and filled the soft sea bed with forminiferas, trilobites, fish and all aquatic life and the sky with birds. This innocuous sentence makes Evoquacks scream and hurl abuse at Christian Creationists as it contradicts everything Satan taught them. Hawking, Dawkins, Krauss and myriads of others so violently refuse to consider the truth of Day Five that they make a spectacle of themselves. Day Five condenses the Evoquack's 3 billion years into just about 24 hours! Yet, the evidence for Day Five's course of creativity is easily verified by spending just a few dollars on a sample of trilobite fossils! My trilobite fossil is barely as big as three fingers but has several complete trilobites embedded in it. They range from small to large but all seem complete creatures perfectly designed to spend their lives roaming the seabed eating up any detritus they find, nothing more. Until The Flood killed them all with electricity, hot water and warm lime they were perfectly happy living their quiet lives among trillions of their fellow creatures. Extensive drilling and core sampling has shown that beneath the trilobite layers and down to barren bedrock are only tiny fossils or just traces of foraminifera creatures which are basically tiny planktonic things with thin shells and large waving antenna things for trapping the most minute detritus the trilobites aren't interested in. Many had the ability to dig pits or holes for themselves though most would be free swimmers just as today's plankton is. They all display perfectly working bodies so just how they are claimed to have evolved is a mystery known only to Evoquacks. Foraminiferas is complete little creatures quite happy to live their lives being foraminiferas and never think about evolving into monkeys or Richard Dawkinses. Day Five saw all manner of other sea creatures came into being with bewildering contradictions of design that has provided ammunition for several generations of Evoquacks! The sea contains creatures that swim, jet, crawl, slide, creep and walk as if to suggest transition to or from land creatures as Evoquacks like to parrot, but then an awful lot of water creatures is happy just to swim or drift about like kids in a paddling pool. Just a little thought would reveal that the means of locomotion are given each creature are selected not just for novelty or confusion but to ensure a creature fully exploits its own niche in GOD's great scheme of Earth. The choice of flexible foot, oozable body, independent legs, waterjets, suction feet or entire bodies, fins or tail given each creature seems a bit random but when their intended food and most appropriate mouth are added to the mix it becomes obvious that each sea creature from smallest to largest will be completely designed to fit one niche for its own benefit, and possibly on dying to provide food for other creatures. For instance the foraminifera wait patiently until microscopic things trigger their antenna nets to retract to check the catch. Octopusses do not swim up fast mountain rivers to find food as do trout but they are designed to walk jet about the seabed picking up large food items as big as any fly a trout may snatch and they can jump out of water to snatch prey if given the chance! Bullsharks are frightening deadly speedsters readily attacking prey and best avoided while the basking shark is a gentle giant needing nothing but plankton and being quite amenable to brief petting from swimmers as are the most huge creatures on Earth – such as the Blue Whales like as what Evoqucks claim evolved from wolf-tapir hybrids or monkeys or bears. Dig into any sand or mud, lift any stone wetted by sea, pull aside straggling sea or pondweeds and there will be seen all manner of creatures hurrying to a new hide as fast as they can. The creatures GOD and Jesus put into the waters on Day Five boggle the mind for variety and certainly boggle the minds of the Evoquacks who desperately try to label and assign each creature to its apparent group despite often obvious nonsensical explanations such as the truly idiotic one concerning the supposed evolution of wolf/tapir hybrids becoming giant sea whales. I've just had one idiot proposing Lumpfish as proof of EVONSENSE as they are able to climb out of ponds and get about on land – I didn't point out that limbless eels is also quite happy to perambulate from pond to pond. We cannot know if the water creatures were then as fecund as they are now, or if they were all vegetarian and happy to eat seaweeds, or if cannibalism was the order of the day as a means of limiting their shoals to match available seaweeds and the steady flow of detritus of leaves, flowers and fruit falling off all the new land vegetation. Whales have a hunting technique that engulf thousands of fish at once but it is hard to decide if this is original or post-Fall behaviour. The days of immense shoals of fish providing a bounty for fishermen are long gone but some old men still remember days when a trawl net could contain so many fish as to threaten the boat with capsizing. Having helped create all the fishes in the sea Jesus knew perfectly well how to display his powers by making his disciples catch so many fish their net was in danger of breaking and the catch had to be dragged to the shore for counting as recorded at John 21. GOD and Jesus also spent Day Five creating vast numbers of kinds of birds far greater than we have today and each kind had its own range of sizes, shapes, colourings, flight characteristics, preferred food and habitat, method of nest building - their own niche in the system of things. Every bird from those tiny humming birds to great pterodctyls were created to fly around eating from or fertilising all the gorgeous flowers. Just to show virtuosity some of the birds were designed with amazingly accurate global positioning systems in their heads to guide them vast distances – perhaps just for the fun of it, perhaps to try out a new and periodic food supply like the American seagulls that make the annual pilgrimage over the mountains to feast on the brine shrimp on Muroc Salt Lake aka Edwards Air Force Base – how can they know its shrimp season? Suffice it to say that at the end of Day Five GOD could see the sea and sky was filled with an amazing array of creatures. Then He rested.

DAY SIX. This day GOD made all the land creatures. A truly astounding mix of furry animals, hairless animals, scaly reptiles, insects, creeping and burrowing things of all sizes were made with each having its own niche from the tiniest rodent to the great Behemoth. There are far too many extant species to mention, and the museums are crammed with fossils of countless others all complete and fully functioning to mock the Evoquacks who concoct elaborate castles in the air filled with creatures evolving from one kind to another despite a total lack of evidence. I have no need to describe any creature as there are masses of other sources of information even though most 'experts', like David Attenborough, will claim that any creature researched indisputably 'evolved from pondslime over X hundred million years.' Take it as irrefutable fact that all creatures were designed just as they are found today or as their bones are found buried in Flood sediments. No evolution is needed or possible despite all Satan's lies through the lips of Attenborough, Sentamu or any of Satan's other human spokespeople. I just read the explanation of the idiotic claim that tapirs evolved backwards into whales! The big fat piggy tapir likes to swim and use its long nose like a snorkel to breath so some idiot with less intelligence that a dollop of pondslime dreamed that the tapir decided it didn't like living on land and so went into the sea and swam about until it became a huge blue whale.

Just recently a sneering Evoquack posted a list of living creatures having features of some fossil creature that supposedly prove EVONONSENSE. The poor guy is quite unable to realise that the live animals or the fossils show that the creature lives or lived quite happily without any urge to evolve to something else. This is how those 'living fossil' ceolocanths makes fools of all Evoquacks? Almost at the end of the day GOD made a human male from dust. Dissolve a human body to its elements and we turn out to have exactly the same chemicals and minerals in us as ordinary soil. The first human was made with a vastly more complex brain than all the other creatures and an erect upright posture with free swinging arms and fully mobile neck. From his first moment he had the power of speech and a vocabulary fully broad enough for his immediate use – and the amazing ability to create new words. GOD showed the man all the creatures and watched him use his imagination and power of speech to invent names for all of them. Then He showed him all the vegetation and told him he and all the other land creatures and birds could eat any of it they like. The Adam's perfect eyes would delight in gazing at those pretty stars and galaxies! The disputed Book of Jubilees says that after several days GOD noticed that Adam felt lonely so He made him sleep, took out a rib's inner material and built it into a slimmer, prettier, intriguingly female version of him and presented her to him. He would have no inkling that her ovaries already carried the eggs for untold thousands of babies. GOD made a wonderful garden where a river flowed from underground water sources, named it Eden and put the man in it with instructions not to eat the fruit of one specific tree but to eat from any other seedbearing vegetation. **The river would carry fallen leaves and fruit and the detritus of the fish and other creatures in it down to the sea where sea life would recycle it.** This sentence is exceedingly important so to ensure you remember it I have bolded it! Adam and Eve would bathe daily and never think that GOD had designed their bodies to grow a new layer of skin each day as part of the eternal life He intended. As they washed the particles of dead skin would come away and float down river until some minute creature caught and ate it to recycle it into fertiliser for waterweeds to feed fish or waterbirds or crocodiles and alligators. Earth had been placed perfectly to receive the ideal amount of heat and light from the sun with the moon supplying soothing light at night as well as a gentle tugging on the bedrock to generate piezoelectricity to ensure there was a constant upward emission of heat to raise the mist that watered the vegetation and kept Adam and Eve's skins moist and soft. In Eden's good and well cultivated soil the plant roots could travel as far as they wished to solidly anchor the plant and also extract as much nutrients and minerals as possible and we can be sure Adam and Eve ate widely of delicious nutritious vegetation, fruit and nuts and looked gorgeously healthy as did all the vegetarian creatures including TRex. After eating their fill Adam, Eve and the creatures would naturally produce excrement but GOD had designed scavenging lower creatures to take care of all waste. Worms especially love to eat excrement and slither about where roots go and change rock to soil in the microbiome, while urine breaks down into nitrogen and ammonia with potassium and phosphate – just like modern general purpose fertilisers! When GOD had finished He pronounced the Earth and its life were 'Very Good!' Then He left Adam and Eve to enjoy getting to know each other and all the creatures and vegetation knowing they would have a lovely life.

DAY SEVEN. God rested.

In just these six days GOD had created the entire universe, a precisely placed and self sustaining Earth complete with starter populations of land, sea and air creatures and two fine humans. He was right to say it was all 'Very Good!' EUREKA!

GOD's Creation was a perfectly designed planet equipped it with everything needed for humans and all manner of creatures to live idyllic lives with the humans at least having the potential to live forever. They were intended to start making babies once Adam and Eve had studied how the animals mated, gave birth and cared for their young. Eve could have used all those eggs in her ovaries to produce babies and continued having babies as often as her daughters and granddaughters did. It was a flawless system of eternal production and fertility – does any Evoquack ponder on the majority of Earth now being barren and unproductive or does they think the barren landscapes of Planet of Apes and StarTrek are the norm?

At this moment all research seems to indicate that approximately 50% of all people around the world believe they is evolved from monkeys that evolved from fish that evolved from pondslime that sparked into life with the aid of Hawking's Gravity! Naturally that means 50% are quite happy with the idea that a supernatural being created the universe and all life from nothing. This at first glance would seem a fair division but when the various ideas of the supernatural being are studied and separated off the True Christian GOD YAHWEH it is obvious that lots of the remaining 'creationists' actually believe they is evolved with the aid of aliens in UFOs or else had previous existence on other planets where a god-being created them or they believe in nonsense like the Muslim idea of a stone idol named 'Allah' in the kabaa in Mecca creating the universe and all its life. A further category of people actually believe in both creation and EVONONSENSE – David Attenborough is a perfect example of such a one, Archbishop Sentamu is another.

Disregard the Muslims whose inbred heads are so filled with muddled garbage that they cannot formulate a sensible thought, and all the reincarnationists and nihilists with their butterfly pasts and futures, and all the hippydippy young people who were perverted by Planet of Apes, StarTrek and devilworshipping popstars such as George Harrison and his Hindu 'My Sweet Lord' paean to Krishna, and it seems that the total number of real Creationists around the world may actually be as low as 300 million of the 7,600 million on Earth! Satan is winning on his way to his pyrrhic victory!

Ignoring all the idolatrous Muslims, Hindus, Buddists and New Agers the stumbling block all educated people have that makes them doubt Creation is the actual time table and evidence for the Six Days. There seems no way of convincing them of Creation being just six 24 hour days.

Why this should be is strange! Just because we humans can do nothing to speed up or slow down time does not mean the Creator of all time cannot do so? The Bible records several instances of altered time apart from the Six Day of Creation. There is the matter of Aaron's dry walking staff managing to send forth buds, shoots, twigs, leaves, flowers and finally almonds in the course of one night. The Bible records this at Numbers 17:7 'So Moses deposited the rods before the LORD in the tent of the testimony. **8** Now on the next day Moses went into the tent of the testimony; and behold, the rod of Aaron for the house of Levi had sprouted and put forth buds and produced blossoms, and it bore ripe almonds.' The staff was quite old and in the prevailing climate of the location would no doubt have been completely dessicated and dead yet overnight it somehow rejuvenated and produced ripe fruit. Almonds normally take 7-8 months to go from bud to ripe fruit which clearly means that some supernatural help was given the staff. Interestingly almonds are described as the best of fruits and chemical analysis shows almonds to have a very wide variety of vitamins, minerals and general nutrients which is why GOD told Adam and Eve they were to eat fruit and nuts.

The speeding up of fruiting of Aaron's staff is exactly the same phenomena that Jesus and others are noted for producing when they instantly reversed all the cell destroying effects of death as they raised various young and old people to death just as GOD reversed the three days of death effects on Jesus's body in the tomb. One of these restorations of life and health is Lazarus's, and his sisters cautioned Jesus about the likely deterioration of the corpse after four days of heat in Bethany near Jerusalem – but upon Jesus calling to him Lazarus's body was instantly restored in its entireity- tissues, blood and brain cells in a way that is a complete mystery to science and medicine. How could that happen? Anyone with faith in Creation knows that on DAY SIX GOD took soil and made a human body and put life into it so it should surely be no problem to Him to resurrect Lazarus and later Jesus despite the effects of several day's death?

Incidentally Lazarus's sister Martha had full hope of being resurrected on Judgement Day when she said: "I know he will rise again in the resurrection at the last day." John 11:24. If Martha knew and stated that the dead are in their graves until Judgement Day why do so many religions promote the pagan idea that the dead fly off to heaven the moment they die as do Catholics, Muslims and most Christians.

The other incident of altered time that is most baffling to practically all but true believers is that of Joshua's Long Day recorded at Joshua 10:12 'So the sun stood still, and the moon stopped, till the people had revenge upon their enemies. Is this not written in the book of Jasher? So the sun stood still in the midst of heaven, and did not hastened to go down for about a whole day. And there has been no day like that, before it or after it, that the Lord heeded a voice of a man; for the Lord fought for Israel.' The account is of Joshua's army fighting the Amorites to gain possession of the Promised Land just as GOD had ordered him to do. The battle that day raged long with the Amorites retreating quickly and Joshua could see darkness would fall before the enemy was defeated. In desperation he called to GOD and asked that the sun and moon be made to stand still to extend the day for almost another full day. This request was duly granted though whether GOD stopped the sun and moon circling Earth or Earth circling the sun is hard to know as there is still considerable doubt as to whether the Earth is at the centre of the universe or the sun is! The account in Joshua is also worth further study as it mentions the Book of Jasher! Jasher is another book that was obviously seen as a genuine record at the time but was later suppressed and not included in the collection we now call the Bible when the list was compiled in Nicea in 325 AD. It is essential reading along with The Book of Enoch for anyone trying to decide between Creation and EVONONSENSE.

A similar perplexing event was the 2 Kings 20:11 'Then the prophet Isaiah called on the Lord, and the Lord made the shadow go back the ten steps it had gone down on the stairway of Ahaz.' The shadow duly went back the ten steps. Was this achieved by stopping Earth or the sun? Atheists and Satanists claim the sun rays were bent by the atmosphere...

One person who sneers at resurrection and altered time and the Bible in general is Donald Simanek of Lock Haven Woozlestitute who says regarding Bible truths: ' *On many occasions a scientific fact or theory comes into conflict with someone's pet literal interpretation of the Bible. These literalists have suffered many defeats... But whenever a scientific result seems to support the Bible, the religion hucksters proclaim that glorious news from every pulpit! That doesn't happen as often as they would like these days, so some have taken to distorting or inventing science to suit their propaganda purposes. Seldom will the gullible believers trouble themselves to check out the facts. Thus outright fabrications can proliferate for years without challenge.'* By calling preachers 'hucksters' Simanek reveals himself a perfect disciple of Satan's, a perfect example of Schopenhauer Stage One, and also as foolish as can be by imagining himself evolved from monkeys – which surely must be the most enthusiastically promoted and proliferated fabrication ever since Darwin cobbled it together from Wallace's demon uttered inspiration ! Simanek's name will not be found in the Book of Life!

There is one mystery of Creation to ponder – the fact that as GOD made Earth as a ball of water and elements, and as NASA has determined there is masses of water on other planets in our solar system, it is possible that GOD created the planets exactly the same way.

The present positions of the planets ensures only Earth is exactly correctly placed to be habitable – but what if GOD originally intended there to be a whole system of planets onto which humans could be transported to fill just as on Earth?

Mercury the closest to the sun is too hot for life but has ice in deep craters. Was that creation water or water from The Flood ? Silly Evoquacks have imagined that all the solar systems water came from the imaginary Oort Cloud. We can be sure there ain't any Oort Cloud just as we can be sure that creating planets as balls of watery elements would free up lots of water if the planets were moved close to the sun.

We can ask Jesus about this on his return.

Chapter Seven
GOD, Jesus, Holy Spirit, Angels.

I earlier wrote that throughout this book I will be quoting various Bible verses to illustrate the points I make.

I hope you will be sensible enough to understand the context I use them and the context they were originally spoken and recorded.

One problem I wish to clear up is the question of GOD Himself!

Who is GOD YHWH?

Is He Jesus ?

Is He the Holy Ghost?

Is He Allah? Buddha? Krishna? Is He any of the 'gods' worshipped by the thousands of cults?

There have been countless books written trying to explain GOD and Jesus ranging all the way from cold clinical examinations to the most asinine mystical abstractions possible and if you are not satisfied with my explanation I recommend looking for some in the bookshops, library or as an ebook on your electronic tablet.

Actually it is impossible to imagine what GOD really is or how He exists or His origins: we just have to accept that the fact that if we do what He says is good for us life would eventually be very good indeed. Anyone who survives Satan's temptations or attacks until death or is still alive and faithful and is raptured off Earth will meet Jesus and angels and GOD and learn the truth.

It is not disrespectful or lax of me to sum up GOD in just a few paragraphs though many will think so.

Take anyone who has any exposure to GOD based religion – not heathens from the jungles and wastelands who have never seen a Bible or missionary or had any contact with the idea of te true GOD, nor the myriads of atheists and agnostics in the 'civilised' world who have deliberately swallowed the devil's lies - and ask them if GOD is real and is also Jesus or is the GOD, Jesus and Holy Ghost 'trinity' just myths based on tales concocted round the first campfires the mythical Cro-Magnons sat around in their caves a million years ago. Ask them all if GOD is Allah or any of the other names people use for their describe idols or supposed deities. Ask them if Satan and his demons or their version of the chief evildoer and his assistants are real. If they claim to worship a female deity such as Mary, Diana, Ishtar or whatever ask them if she is real.

It will soon becomes apparent that the vast majority of people whom western 'Christian' religion has touched however fleetingly, perhaps as high as 65%, actually believe the trinitarian brainwashing they received as children forced to go to churches, chapels, cathedrals, school services and they will strongly defend the concept of GOD being Jesus and being Holy Ghost to make the 'trinity' or 'the GODhead.' Catholics have the added burden of being forced to accept the belief that Ishtar queen of heaven is a greater god than GOD and her idol must be adored and prayed to and they will not discover their mistake until Jesus comes to destroy them.

Maybe 65% of all pagans also believe in both good and bad supreme beings and each with many willing helpers but I'm not going to bother with all that nonsense as a brief bit of study will show them all based on primitives fairytale or the lies and tricks of Satan.

Just looking into the night sky will confirm that GOD must be amazing, powerful and creative to a degree we cannot understand. He made the monstrous sun and the furthest galaxy but takes a direct interest in us on our little blue planet! Only He knows how it was made and set in place and regularity. What fools the Muslims, Catholics, Hindus, Wiccans and all similar idolaters are to think their lumps of stone, plaster- even gold can make anything at all let alone the universe! And who can try explain how only GOD could create the eyes that let us see the stars and the vocal chords that praise their beauty while practically anyone could make the pagan idols that see nothing and remain as dumb as the scraps left from their making!

Researchers say the universe is slowing down so perhaps GOD decided it would reach a certain size and then no more? Einstein is long dead but he passed his fudge making skills on to many lackeys who all choose to follow his recipes perfectly.

As the universe is pretty big and Earth itself weighs trillions of tons but is tiny compared to the sun and some of the other planets and stars the only thing we can know is that GOD must be absolutely awesome and it is wonderful that He is so fond of us humans that He created a perfect home for us and then sent His only son to suffer a human death on the cross as a scapegoat to erase all the sins of those who believe that Jesus is His Son and GOD is very real!

This is so far above the Evoquack's understandings that they will howl like baboons as they read my words!

There are plenty of Bible passages that show the power and character of GOD, His concern for us and His disdain of Satan, ecumenists and all who worship idols, whether literal ones of stone, wood or precious metals or demonic ones appearing as UFO's, aliens or sky signs and phenomena or imaginary ones such as EVONONSENSE.

Apostle Paul stated how to know the truth of GOD and how futile it is to try encapsulate Him in human thinking and handiwork when he wrote the letter to the Romans:

Romans 1:18 "For the wrath of GOD is revealed from heaven against all unGODliness and unrighteousness of men, who by their unrighteousness suppress the truth.

19: For what can be known about GOD is plain to them, because GOD has shown it to them.

20: For his invisible attributes, namely, his eternal power and divine nature, have been clearly perceived, ever since the creation of the world, in the things that have been made. So they are without excuse.

22: Claiming to be wise, they became fools,

23: and exchanged the glory of the immortal GOD for images resembling mortal man and birds and animals and creeping things.

25: Because they exchanged the truth about GOD for a lie and worshipped and served the creature rather than the Creator, who is blessed forever! Amen."

All the writers of the Bible from Moses and his Genesis to John and his Revelation declare GOD is a caring being who feels pain and disappointment, anger and shock. He loves humans more than all creatures. He created us and gave us a perfect home – the 'Very Good' Earth. He watches us and wants to speak to us and tell us things we can never learn on our own. He designed us to enjoy good food and drink, laughter, music and singing, love and sex, the tiredness of work and the refreshment of a good night's sleep. He made us prefer warmth and moist air and the thrill of cold on our skin. He made us able to live on dry land, climb trees and swim in crystal lakes but never fly by our own strength. He made us unable to see or sense Jesus and angels as well as Satan and demons unless any of them chose to reveal themselves although He made asses, cats and dogs able to see spirits. He made us get a new layer of skin every day and a complete new body about every three months – our seasons are as regular as the Earth's! He was angered and disappointed that Eve preferred Satan's lie about that forbidden fruit. He was dismayed that Cain would murder Abel over a minor issue. He was shocked with disbelief that parents would burn their babies on red hot brass idols they called Moloch. He was saddened but resigned to humans having an inclination to sinful harmful things. He was upset that His original plan for us had been ruined by Satan but stoically stepped aside to let Satan attempt to rule Earth and us for 6,000 years but ensured that Satan knew he had limited power and could instantly be thwarted. GOD can block any move of Satan's and humans can make Satan and his demons flee merely by saying GOD or Jesus's name. He was proud – Well pleased! - that His son Jesus was sinless and willing to carry out his task to be a perfect blameless sacrificial offering for sin. He has constantly told humans and Satan that He has determined to have a day of reckoning despite Satan's lies and false teachings and the nonsense of the Evoquacks. If He has a deadline He ignores its passing in the hope all humans will realise they need Him but He knows that many are only fit for slaughter and eternal destruction and one day soon He will have to order Jesus and angels 'Go cleanse The Earth of the sinful!' In these dreadful days He will watch billions of idolaters and sinful people die in awful ways from pestilences, gigantic hail, calamities and venomous creatures with many being burned to dust as were those in Sodom and the other four towns.

On the other hand anyone can speak directly to Him with no need to go to a church or kneel to an idol or put the forehead in the dirt – just speak to Him any where, anytime! He promises to give whatever is asked for if asked in Jesus's name. He will hear requests and decide the best response though it may not be what is hoped for. The response can be instantaneous, as short as 60 seconds or one hour as I can confirm, or many years; or it may not come before the sword has sliced or the bullet struck but in these cases the response will come as a delight at resurrection time.

But is GOD Yahweh really Jesus? No,
GOD is GOD and Jesus is His only-begotten Son!
GOD says that many times!

GOD said 'This Jesus is my Son!' Jesus said 'GOD is my Father' and prayed to Him regularly and all night too; the angels said it, the Devil said it, the demons said it, the people he healed said it, the soldiers who watched him die on the cross said it.

Jesus emphasised he was not GOD when he said only GOD knows when he will be sent to destroy the wicked and cleanse the Earth. Mary knew Jesus was GOD's son!

Yet so many pastors, preachers, ministers and vicars all seem to be blinded, deafened and unable to think straight by their Babylonian trinity brainwashings and insist that GOD is Jesus and then imagine another being called the Holy Spook!

Many verses in the Bible should make clear the relationships of GOD and Jesus and also the angels but the utterings of false prophets and demons have misled the greater number of people and is a stumbling block to the masses despite Hebrews Chapter 1 being as plain and clear as can be! Hebrews was written by the apostle Paul to the Christian congregation in Jerusalem where it would be copied and distributed to all Christians around the Middle East and Europe where Christianity had penetrated by the date of Paul's writing it – probably in about AD65 or just 30 years after Jesus had risen from the dead.

Quote: **Hebrews 1:1** After GOD spoke long ago in various portions and in various ways to our ancestors through the prophets, 2 in these last days he has spoken to us in a son, whom he appointed heir of all things, and through whom he created the world.

3 The Son is the radiance of his glory and the representation of his essence, and he sustains all things by his powerful word, and so when he had accomplished cleansing for sins, he sat down at the right hand of the Majesty on high.

4 Thus he became so far better than the angels as he has inherited a name superior to theirs.

5 For to which of the angels did GOD ever say, "You are my son! Today I have fathered you"? And in another place He says, "I will be his father and he will be my son."

6 But when He again brings his firstborn into the world, He says, "Let all the angels of GOD worship him!"

13 But to which of the angels has He ever said, "Sit at my right hand until I make your enemies a footstool for your feet"?

14 Are they not all ministering spirits, sent out to serve those who will inherit salvation? Here are 26 verses that specifically state Jesus is GOD's son:

1 Corinthians 15:28 When all things are subjected to Him, then the Son Himself also will be subjected to the One who subjected all things to Him, so that GOD may be all in all.

Psalm 2:7 "I will surely tell of the decree of the LORD: He said to Me, 'You are My Son, Today I have begotten You..

Matthew 3:17 and behold, a voice out of the heavens said, "This is My beloved Son, in whom I am well-pleased."

Matthew 17:5 While he was still speaking, a bright cloud overshadowed them, and behold, a voice out of the cloud said, "This is My beloved Son, with whom I am well-pleased; listen to Him!".

John 1:14 And the Word became flesh, and dwelt among us, and we saw His glory, glory as of the only begotten from the Father, full of grace and truth

1 John 4:10 In this is love, not that we loved GOD, but that He loved us and sent His Son to be the propitiation for our sins.

John 14:13 "Whatever you ask in My name, that will I do, so that the Father may be glorified in the Son.

John 5:19 Therefore Jesus answered and was saying to them, "Truly, truly, I say to you, the Son can do nothing of Himself, unless it is something He sees the Father doing; for whatever the Father does, these things the Son also does in like manner.

John 5:26 "For just as the Father has life in Himself, even so He gave to the Son also to have life in Himself;

Mark 5:9 But He kept silent and did not answer Again the high priest was questioning Him, and saying to Him, "Are You the Christ, the Son of the Blessed One?"

Mark 5:19 and shouting with a loud voice, he said, "What business do we have with each other, Jesus, Son of the Most High GOD? I implore You by GOD, do not torment me!"

Acts 13:13 that GOD has fulfilled this promise to our children in that He raised up Jesus, as it is also written in the second Psalm, 'YOU ARE MY SON; TODAY I HAVE BEGOTTEN YOU.'

Mark 9:7 Then a cloud formed, overshadowing them, and a voice came out of the cloud, "This is My beloved Son, listen to Him!"

Luke 3:22 'and the Holy Spirit descended upon Him in bodily form like a dove, and a voice came out of heaven, "You are My beloved Son, in You I am well-pleased."

2 Peter 1:17 For when He received honour and glory from GOD the Father, such an utterance as this was made to Him by the Majestic Glory, "This is My beloved Son with whom I am well-pleased"--

Luke 9:35 Then a voice came out of the cloud, saying, "This is My Son, My Chosen One; listen to Him!"

Romans 1:4 who was declared the Son of GOD with power by the resurrection from the dead, according to the Spirit of holiness, Jesus Christ our Lord,

Matthew 2:15 He remained there until the death of Herod. This was to fullfil what had been spoken by the Lord through the prophet: "OUT OF EGYPT I CALLED MY SON."

1 John 5:09 If we receive the testimony of men, the testimony of GOD is greater; for the testimony of GOD is this, that He has testified concerning His Son.

1 John 5:10 The one who believes in the Son of GOD has the testimony in himself; the one who does not believe GOD has made Him a liar, because he has not believed in the testimony that GOD has given concerning His Son.

Mark 1:11 and a voice came out of the heavens: "You are My beloved Son, in You I am well-pleased."

Surely we can know GOD isn't Jesus by these facts:

GOD can't sit at his own right side.

He wouldn't pray to himself.
He wouldn't say he is doing the will of the one that sent him.
He would know the day and hour that Himself was going to start Armageddon.
He wouldn't say he has forsaken himself.
He wouldn't say "this is my son, the one I have approved.
If He died on the cross, who was it that resurrected him?
The Bible says that GOD exalted Jesus but why would he exalt himself?
Jesus said GOD was greater than he.
Jesus never asked for anyone to worship him.
If Jesus prayed to GOD how can he be GOD?
If Jesus is GOD why did he have to get baptized?
If Jesus is GOD in who's name was he baptized in?
How did Jesus receive the Holy Spirit if he is also GOD & the Holy Spook itself?
If Jesus is GOD who was it that he prayed too to save him from death? Himself?
Did everyone that saw Jesus after he rose on the third day die instantly as GOD said at
Exodus 33:20"You are not able to see my face, because no man may see me and yet live."
And don't forget: GOD is not a GOD of confusion?
Because Jesus is not GOD.

Isn't it amazing how easily the Bible refutes some cherished teachings when they have no Biblical basis? Why did Jesus receive a 'superior' position from the Father if he was GOD? (Phil. 2:9) If the Trinity means 'co-equal' and 'co-eternal' this would imply that GOD, Christ and the Holy Spirit were ALL co-equal....up until this scripture. Now, according to this scripture Jesus has a 'superior' position........more superior to the one he had BEFORE it was 'given' to him. So if the Bible AND the Trinity are both true, this would mean that Jesus is now HIGHER in position in the Trinity than the Father and Holy Spirit - since before this scripture they were ALL 'co-equal'. How could this be possible? Apparently either the Bible is wrong or the Trinity doctrine is wrong....I wonder which one?

GOD is GOD.

Jesus is GOD's Son.

So now: What is Holy Spirit? Is Holy Spirit a real being just like GOD and Jesus? NO!

Is GOD the Holy Ghost or Spirit? No, that's the old Babylonian Trinity idea rewrapped for Christians and Catholics by Satan to ensure they all lie each time they claim Holy Spirit is a separate being to GOD or the third part of some weird 'GODHEAD!'
Holy Spirit is the spiritual mental gift of understanding and is often backed up with direct

help of angels. I can be described as a new level of conscience or sensibility to help us in our days and works. It is the ability to recognise false doctrines when we hear lying preachers preach. It helps us spot lies of EVONONSENSE and speak the truth about Creation. It is the ability to instantly vocalise an appropriate Bible verse of account to clarify queries, misconceptions and situations. It is the driving force that makes us stand up and preach the Gospel no matter what ridicule it brings on us. It is our conscience that says 'No!' when we feel a lie forming on our lips or when we look covetously at someone else's property.

It is not a spook in a white sheet flapping around the throne of GOD.

EUREKA! - There is no holy spook flitting round heaven!

Hawk'n'Dawk and legions of other Evoquacks all base their EVONONSENSE entirely on the physical realm, after all, like the lesser animals they claim to have evolved from they claim they have just five senses, they can touch, smell, hear, see and speak - well, Hawking lost the power of real speech as he only spoke satanic blasphemies and Dawkins can speak but mostly makes a fool of himself when he does so, his facial expressions and body language clearly reveals he knows he speaks blasphemies and nonsense just as his handwriting reveals he has major psychological problems and is quite immature – which he why he has that truculent schoolboy look on his face so often! Carrier and Strauss are just demonically possessed schoolboys who display mental blocks at the mention of GOD and Jesus.

One quote that will flummox every Evolutionist man, woman or child who ever dared claim themselves a scientist is John 20:19 and Luke's account of the same event at Luke 24:36 when Jesus just appeared in a locked room having just come from heaven to encourage the disciples who were frightened of being rounded up by the hordes whipped up by the temple priests. He must have been in spirit mode and able to pass through Earth's solid material but still had the same human body that had been on the cross which he was able to prove by showing the disciples the nail and spear holes. Many people cannot understand these accounts and either pass over them as Jesus just walking in to greet His saddened disciples or else the event is dismissed as just another bit of Bible hyperbole – after all it seems to say Jesus could walk through a locked door!

The disciples were being persecuted by the Jews after Jesus's crucifixion and had fearfully sneaked into an upper room and locked door! They were Jews themselves, painfully circumsized too, but the Pharisees and Sadducees – the official religious leaders – hated them because they were linked to Jesus and proclaiming His message which showed clearly that He thought Himself the Messiah and had come to end the old ways of the Jews and lead them back to the true worship of GOD. Influenced and spurred on by Satan the leaders succeeded in getting the crowd to choose Barabbas over Jesus - it was probably a demon that shouted "Barabbas!"– and they took Jesus off, whipped and generally humiliated him before nailing him to a cross to watch him die.

He did die after an unusually short time and strangely his death was accompanied by a dense darkness and an earthquake that ripped the curtain of the temple to expose the Ark of the Covenant was not there and the chief priest had been performing a sham ceremonial sin offering many years! The torn curtain also emphasized that Jesus's death granted everyone easy direct access to GOD and an end to the previous need to slaughter and burn animals to obtain forgiveness for sin.

In the days after Jesus's resurrection and undeterred by the supernatural signs around the death the leaders whipped up the people into believing the disciples were still preaching lies and were imprisoning and punishing whatever disciple they could find. The people had refused to listen to the disciple's messages and tried to catch them so that day they were gathered and feeling afraid in the upper room all but for Thomas who was elsewhere.

Suddenly Jesus appeared in their midst. Just like that! No flash, bang and puff of smoke, no fanfare or crash of cymbals, He just appeared and for some moments they were frightened and thought Him to be a ghost and spirit! They thought that because as the door was locked how else could anyone suddenly appear amongst them? He must be a ghost or spirit!

Well, obviously No. Being Jesus and the co-maker of the universe He was well aware of the physical laws that bind all matter together as we humans know it but being spirit from a different realm to us and our matter He just walked through solid walls or doors as though they were not there. He had to ask for food and eat a piece of fish before they accepted he had returned with a real human body. Many centuries previously Jesus and two angels had eaten a good meal of tender veal and bread prepared by Abraham's wife Sarah as written at Genesis 18.

He didn't need the food being a spirit but he and angels do eat to show they appreciate the offer of food and prove they can materialise solid human-like bodies.

Will Hawk'n'Dawk'n'Attenborough ever figure out how He could do eat that meal at Abraham's? Will they ever believe it really happened? No, it will ever be above their level of understanding. If pressed they will go into Schopenhauer Stage Two mode.

Let's try to make Hawk'n'Dawk understand the spirit world. Fill a glass jar with plain sand, shake and pack it down hard until no more can be added. You would now say the jar was filled with sand ie solid matter? Now get a jar of water half the size and start pouring the water on the sand. It keeps going into the sand but the sand does not overflow! Obviously the jar full of sand is not really full but eventually will be full of water and sand. The water has crept through the spaces between the grains of sand. This is basically how the spirit world operates. The spirits including Jesus can somehow make their matter pass through solid matter just as the water passes through the 'solid' sand. No human can or need understand how they do it!

Hospital staff pass rays through solid matter all the time to make our Xrays?

John and Luke's accounts of Jesus's visit and His next visit when 'Doubting' Thomas was there also perfectly confirms that the people of that time were well aware of the existence of evil spirits – as why else would they be afraid when Jesus suddenly materialised in their midst?

Far too many scientists dismiss the idea of spirits - good and evil – and condescendingly smirk at anyone claiming to have witnessed the appearance of a spirit being or the acts of that being whether helpful or harmful to a human. Indeed many doctors, psychiatrists and psychologists, and sadly too many religious leaders, denounce anyone who claims to have seen, felt or experienced spirit beings as being delusional, psychotic, deranged, in the grip of religious fervour or just plain crazy!

Holy Spirit is the range of gifts of understanding given to all who confess Jesus is GOD's Son as detailed here:

1 Corinthians 12: Gifts of the Holy Spirit

12 Brothers and sisters, I want you to know about the gifts of the Holy Spirit. 2 You know that at one time you were unbelievers. You were somehow drawn away to worship statues of gods that couldn't even speak. 3 So I want you to know that no one who is speaking with the help of God's Spirit says, "May Jesus be cursed." And without the help of the Holy Spirit no one can say, "Jesus is Lord."

1 Corinthians 12 details how people with the gift of holy spirit can readily see through the fog of EVONSENSE Satan has thrown around the world. We Christians can understand that creatures cannot evolve from one kind to another; we accept the logic of DNA proving all humans descended from Adam and Eve; we accept the reality of Satan and his demons; we easily determine Islam is Satan's own religion and its recipe book is an awful stew of foul ingredients; we know that if need be we can call on the help of the invisible angels who constantly traverse the world watching us and our enemies.

No EVOQUACK can have the gift of holy spirit as by definition they denounce GOD and holy spirit as a figment of Christian idiocy! Without the gift of holy spirit all manner of EVONONSENSE and false religions seem acceptable.

Perhaps the blame lies on colleges of theology and their demand that to earn a degree in theology a candidate must faithfully parrot the established mantras of Babylonian trinitarianism and idolatry and failure to do so will result in examination disgrace and the ignominy of having to go get a job in the coal mines of secular life rather than live amidst the silken pedophilic luxuries of established religion?

This horror of honest toil among the plebs is also what drives all the professors of EVONONSENSE to constantly regurgitate EVONONSENSE to new legions of proto-Evoquacks each year in the universities the way penguins regurgitate half digested fish to their chicks.

Millions possibly billions of people around the world have seen or had direct contact with spirit beings both angels and demons. Shrines and grottoes around the world are dedicated to 'saints' or 'gods' who appeared to people at various times. Angels never lie or mislead while demons all too often are promoting the idea that they were some messenger from GOD and induce the contactee to believe things contrary to GOD's laws and plans for us humans but by seeming sincere the demon can sway the person into accepting their message as real.

But just as Jesus walked through the locked door or wall or ceiling and convinced His disciples he was real and His words could be trusted so do other of GOD's angels make contact with humans as they fly around the world checking on the people. This is why many of us have had conversations with strangers who obviously seemed to know us and what we want, need or are seeking at any moment.

I had exactly such an experience one sunny morning in 2016. The old mill close to my flat was to be demolished and I decided to take some photos of the progress. It was a lovely clear mid-morning without much traffic to spoil the photographing and I walked about across the road from the building to get different views and closeups. After a couple of minutes I became aware of a man pushing a bicycle towards me. I glanced at him and saw an old tramp. I didn't know him and walked back across the road to take some closeups of the eroded stonework. He followed me across the road and asked me if I knew if the library on top of the hill would be open that day. I said probably not as spending cuts demand it only open a couple of times a week. The hill was really long and steep and I only walked it when a bus was not available and as the day really was warm I thought it would be a very tiring walk for the old man and said something to that effect. At first glance he seemed quite old like someone's old granddad. I thought it strange he asked about that library as the road up to it was about quarter mile back the way he had come. He was wearing the most incredibly patched jacket with patches overlapped patches in incredible numbers. His shoes and cap were also battered old things. His bike was old and nondescript and hanging from the handlebars were refuse sacks apparently full of belongings and an empty milk bottle hung from one side. He asked me about the library in the other town but he had also pushed his bike a long way past the road to that one. I did think it strange he knew of the libraries but had walked well past the obvious roads to them. I told him I thought the second library was more likely to be open. Feeling that he was an angel and unsure what to say I was studying him and being unsure what to say he said thanks and walked along the road – away from the direction of the two libraries! I turned to take some more photos and when I turned back he had disappeared although the road was a long straight one! Although the first impression was of a tramp who slept rough in his incredibly patched clothes he had the loveliest shiny healthy complexion and had never slept rough in his life! A minute later I walked about 75 yards to the butcher's shop run by Fred and as usual Fred was leaning on the counter looking out of the window as not many people use traditional butcher's shops these days and as I was always asking him about the area's history. I asked him what he thought about the old man with the bike but though he had been staring out the window for many minutes he swore that no old man with a bike had passed the shop!

Angels do constantly wander round the Earth either on assigned tasks on on general watching duty. Jacob writes of angels ascending and descending between heaven and Earth at Genesis 28:10. The places he dreamed this may be the site of the temple in Jerusalem and foretells of Jesus coming and the New Jerusalem.

Ron Wyatt the man who found the 'lost' Ark of the Covenant recounted how while trying to find the home of a group of believers and getting lost he prayed to GOD for help in finding the meeting place and moments later an old man guided him and then instantly disappeared. I recently read a similar account from preacher Don Latham. An old lady acquaintance of mine had a similar experience when she was wanting to cross a very busy road. She looked all around but no-one was in sight then she thought of GOD and how He helped people and moments later just fifty yards away was a young man who took her hand, stopped the traffic and helped her across before going on his way 'to a distant part of town.' She described him as a clean healthy man of 30-40 years – which is how many people describe their angel contacts though as the two accounts above show they can also be old scruffy men.

Many verses in the Bible should make clear the relationships of GOD and Jesus and also the angels but the utterings of false prophets have misled the greater number of people and is a stumbling block to the masses despite this passage from Hebrews being as plain and clear as can be! Hebrews was written by the apostle Paul to the Christian congregation in Jerusalem where it would be copied and distributed to all Christians around the Middle East and Europe where Christianity had penetrated by the date of Paul's writing it – probably in about AD65 or just 30 years after Jesus had risen from the dead.

Quote: Hebrews 1:1 After GOD spoke long ago in various portions and in various ways to our ancestors through the prophets, 2 in these last days he has spoken to us in a son, whom he appointed heir of all things, and through whom he created the world.

3 The Son is the radiance of his glory and the representation of his essence, and he sustains all things by his powerful word, and so when he had accomplished cleansing for sins, he sat down at the right hand of the Majesty on high.

4 Thus he became so far better than the angels as he has inherited a name superior to theirs.

5 For to which of the angels did GOD ever say, "You are my son! Today I have fathered you"? And in another place he says, "I will be his father and he will be my son."

6 But when he again brings his firstborn into the world, he says, "Let all the angels of GOD worship him!"

13 But to which of the angels has he ever said, "*Sit at my right hand until I make your enemies a footstool for your feet*"?

14 Are they not all ministering spirits, sent out to serve those who will inherit salvation?

This is so far above the Evoquack's understandings that they will howl like baboons as they read my words!

GOD chose to come to Earth to speak to Moses and His presence on Mount Sinai is marked by the rocks being burned black while Moses's face shone with a radiance. Moses had to hide behind a boulder to avoid death. Sceptics sneer at Sinai's blackened top but the proof is easily found with a hammer's blow showing the blackness is only a thin layer unlike volcanic rock that is burned all the way through.

Chapter Eight
The Bible's Big Picture.

The Bible has about one million words depending on which version and translation is considered. Very few human brains can remember all those words even if it is read regularly as many preachers suggest. The Bible is a record of everything that concerns our past and future from the perfect beautiful new earth of Genesis to its present ruin and on to the renewal under Jesus's care and the work of an unknown number of happy healthy saved humans during the forthcoming Millennium which will be followed by a final sifting to remove the last traces of Satan's seed during Judgement Day.

The Bible is filled with scientific facts about Earth, climate, space, human biology and psychology, nutrition, morality, crime and punishment, agriculture and the husbandry of animals and birds, physics and metaphysics. Its scope really is encyclopedic.

In the Bible can be found everything needed to live eternally with a healthy human body nourished with filtered sunlight, super seeds and fruits, delicious sparkling waters, untroubled sleep and the company of a vast variety of friendly creatures all living equally happy contented lives.

Curiously most people do not want this idyllic existence but believe themselves better off in a polluted rat race in crowded crime filled cities with disease and death catching them at very young age.

The Bible faithfully records GOD's making of the first humans and Satan's tricking of Eve, GOD's resultant cursing of her and all Earth with all the subsequent strife and troubles that have afflicted everything over the last 6,000 years. It's last chapter gives fine detail of how awful things will be in the last few years before GOD says "Jesus - Go cleanse the Earth!"

Nothing is spared about human nature, failings, wilful sinning, addictions and perversions. We can read how shocked GOD was at the very idea that some humans would be induced by Satan to burn alive their babies and children on brass statues and idols especially to the bull/man god Moloch!

All of today's sins and crimes are portrayed in the Bible's pages -as are the punishments GOD directs to the sinners – or not as He thinks fit. Though Paul wrote to Timothy: *2*

Timothy 3:16 "All Scripture is God-breathed and is useful for teaching, rebuking, correcting and training in righteousness, 17 so that the servant of God may be thoroughly equipped for every good work" for most people a lot of the Bible can be condensed down to its **Big Picture**.

The Big Picture is quite simple and logical:

Creation: GOD created the entire universe in just six days in a way that no human can ever know. GOD did not use any sort of EVONONSENSE to create the Universe in just six days.

Eve's Fall: GOD told Adam and Eve to eat any nuts and seeds but not from one tree. Eve listened to Satan and ate the fruit. In response GOD cursed all humans and The Earth for a time until eventually Eve's offspring would destroy Satan and his offspring. That time is almost ended and Jesus will be returning in a few years.

Satan's origins: Satan was created as the most wonderful angel but began to think himself important and eventually superior to GOD Himself. He thought it would be very clever to come to Earth and seduce the new humans into sinning and he was successful as the world's war and troubles all prove. He also taught how to create hybrids like lion-woman, horse-man.

The Nephilim: these were babies born to human women after Satan's fellow angels lusted after having sex with the lovely women. The babies grew huge and vicious, built cities with enormous stone blocks and GOD drowned all of them in the Flood.

Earth ruined: Satan, his angels and Nephilim aided by humans ruined all Earth and the creatures until GOD decided a radical step was needed to wipe out the humans and Nephilim.

The Flood: GOD chose to use water as his weapon and brought about the Flood by smashing the Earth's crust to release the vast waters that were part of his underground heating system to make the Earth the perfect temperature and climate for all creatures.

Noah and The Ark: GOD told Noah that the time had come to destroy all the evil humans, Nephilim and hybrid creatures and Noah must build a giant floating box to preserve a selection of pure creatures. Noah stayed 371 days in the Ark until the Earth dried.

The Tower of Babel: After the Flood the humans once again became rebellious and ignored GOD's order to spread over the Earth and chose to gather together to build a great tower at Babel. To make them disperse GOD struck them with an inability to understand each other's speech so they then drifted away in their family groups.

The Prophets and Kings: These books details GOD's dealings with the Israelites with nothing spared about their idolatry, sins and failings as well as the successful battles as they tried to take control of the lands GOD had promised them.

Prophecies about Jesus: The Bible contains hundreds of prophecies about the seed of the woman being born.

Jesus's birth: We are all familiar with his birth with the animals but we can then read of how he was taken to Egypt for safety before returning to Nazareth to grow up, study the scriptures, become a carpenter and apparently support his mother and siblings until they could manage without him. Jesus was well aware he was GOD's son as we can learn from Luke 2:49 when at age twelve he tarried in the temple causing worry to his mother and father. His mother asked. "Your father and I have been anxiously searching for You." 49 "Why were you looking for me?! He asked, " Did you not know that I had to be in Father's house?"

Jesus's anointing: About age thirty Jesus went to see John who was baptising in the Jordan and was baptised with the Spirit of GOD descending on him and GOD calling from high: 'This is my beloved son in whom I am well pleased'.

The Temptation: Immediately Jesus had been baptised Satan, ever lurking, took him – with GOD's permission – into the wilderness to be tempted. First, Satan tried to get the hungry Jesus to make the stones into bread, then he carried him to the pinnacle of the temple and told him to jump off and see if angels would catch him, and finally he took him to an exceedingly high mountain – on the moon – to show and offer all the kingdoms of Earth if he would kneel and worship himself. Jesus declined all three temptations and Satan left him.

Jesus's ministry: He went about gathering a team of disciples to help him in spreading the good news – The Gospel – that He was GOD's son and had come to start the countdown to the destruction of Satan and all the evil on Earth which will happen in the next few years when the Gospel has been preached all over the world.

The Crucifixion: The chief priest of Jerusalem, his father, and all the temple's elite Jews wanted Jesus dead despite him fulfilling all the prophecies in the Torah the Jews used daily.

The urge to kill him was because they all knew Jesus was aware the chief priests for many centuries had been carrying out an empty worthless ceremony of Passover as the holy of holies was empty and the Ark was not there to be sprinkled with the atoning blood of the lamb and therefore the priests had been lying for many years and cheating the Jews of the atonement they rightly expected.

Pilate found Jesus innocent: The Jews dragged Jesus to Pilate the Roman Governor and demanded he be killed. Pilate could not find Jesus guilty of any crime against the Roman Empire and wanted to release him but the Jews howled for execution. Pilate gave the Jews a choice of the innocent Jesus or the guilty Barabbas and they chose Barabbas. Roman soldiers nailed Jesus to the cross while the Jews derided him.

Jewish Blood Guilt: Jews have been persecuted down through the years and claim it is because they are the only pure worshippers of GOD. The real reason is they constantly turned away from GOD to worship the sun and idols of other imaginary gods and plant Ashtoreths trees beside their idols. Today the Jehovah's Witnesses continue this Jewish idolatry by claiming Jesus died on an Ashtoreth stake as a sacrifice to Baal.

The Resurrection: Jesus's body died on the cross and was buried but being half human and half the spirit Son of GOD his spirit was free to go preach to the fallen angels in their chains.

Preaching the Gospel: Three days later Jesus instructed the disciples to go preach to the Jews with the good news that he had fulfilled the prophecies of the scriptures.

The Book of Acts: This book details many issues of the early church and preaching the Gospel but for Christians Gentiles it is perhaps verse 15:29 that encapsulates our conduct:
'That ye abstain from meats offered to idols, and from blood, and from things strangled, and from fornication: from which if ye keep yourselves, ye shall do well. Fare ye well.'

Paul's Books: These are letters Paul wrote to various congregations informing them of developments and problems, clarifying parts of doctrine, exhorting them to keep faith, be diligent against false prophets and doctrines.

Revelation: This book details the sequence of events leading up to and including the falling away, the sinfulness of the world, the rise of the Catholic church, the great plagues on non-believers and the evil people, the Rapture of the Remnant of the Jews and the Remaining Christians who had escaped beheading by the Beast and Dragon who ruled the world a short time before Jesus returned to being the Millennium.

The Millennium: This is the 1,000 Years during which Jesus will rule Earth and renew it to make it into the paradise Adam and Eve knew.

The Final Sifting: After the Millennium Satan and his fallen angels will be released and given a short time to try find followers in the humans before Satan, angels and followers will be destroyed forever in the symbolic lake of fire.

The New Jerusalem: Once the Earth is cleansed a new Jerusalem will come down from heaven.

Any number of aids to Bible study are available to help explain the meaning and connections of the Big Picture.

Satan has constantly lied about the Bible in general and the Big Picture in particular through his own lips, and the lips of willing slaves recruited from the ranks of the worshippers of idols and evil humans.

And no wonder: Satan is the father of the lie and that first lie is recorded for our benefit at

Genesis 3:4 - "You will not certainly die," the serpent said to the woman.

Since then he has never stopped his efforts to lure all humans to eternal death and has constantly whispered lies into the ears of all who will listen. He often finds his slaves while they are performing forbidden activities: seances and occult practices, criminal acts, lusting after sex, drugs and rock'n'roll, pursuing money, aiding and abetting liars and lawbreakers – the list is endless. He has successfully induced billions of humans to murder themselves and each other and be doomed to eternal death by constantly inventing new 'gods' to be worshipped. These include the Muslim's moon god stone 'Allah, the Buddhists 'Buddha', the Mormon's 'Moroni', the Catholic's 'Virgins of...' etc.

Without doubt Satan's greatest trick was catching Alfred Russell Wallace in a mesmerist's meeting and putting 'The Theory of Evolution' in his head! Wallace and Darwin ardently promoted the fairy tale and gained billions of slaves for Satan who will all realise their mistake on Judgement Day.

Satan has had his slaves lie about Creation, Adam and Eve, himself, The Flood and Noah, Jesus – the libraries and internet are stuffed full of books and articles denouncing all Bible facts as lies, myth and legends when in reality it is Satan's Evolution that is The Big Lie!

Refusal to accept Genesis as true effectively undermines the rest of the Big Picture and GOD Himself but all evolutionists cling to EVONONSENSE and denounce GOD and Creation at every opportunity.

Jesus instructed his disciples to walk away from anyone who refused the Gospel message and said their fate was to be worse than befell Sodom! What could that be? What could be worse than being bombarded with deadly incendiary hail made of 98% sulphur and 2% magnesium?

It is a mystery to many but delving deeper into the Bible than the Big Picture reveals that the inhabitants of Sodom and Gomorrah were sinning through ignorance and the lies of their leaders just as surely today there are many sinners in remote isolated communities who never heard and cannot hear the Gospel – they cannot be blamed for their actions. Some of the people of Sodom may be judged and found acceptable to live on the renewed Earth – while all those evolutionists, recidivists, backsliders, atheists, agnostics and devilworshippers who hear and reject the Gospel message have wilfully turned away from GOD and are doomed to eternal destruction on Judgement Day.

.

Chapter Nine

The Kingdom of Heaven? What is it?

What a mystery the Kingdom of Heaven is to so many people – especially the pastors, ministers and bishops who have spent years studying the Bible! They have a responsibility to know the truth and proclaim it in their churches, temples and synagogues but the entire topic is never truthfully spoken about; instead they all preach variations on Satan's Babylonianisms.

Most people who believe or hope that there is life after death or after resurrection have the idea that all dead people fly off to some heaven where GOD lives. And many people think that the truly evil also go to heaven! And they will live in close to GOD in heaven. And that heaven is The Kingdom of Heaven! This is all nonsense.

The dead of all kinds of creatures live and die and all their functions and thoughts perish - so do humans. The spark of life dies and the body then begins to chill as all its systems and senses stop working. At the moment of death the reward for creatures and humans is total oblivion. Good, bad or indifferent humans all die and ultimately pass from memory just as do all the chickens and pigs slaughtered for food every day. Some humans or creatures may have achieved fame or notoriety and are interred in fancy graves or shrines with professional singers, musicians and great ceremony but most dead get nothing but a basic grave or even the ignominy of the garbage dump.

Exactly what is the kingdom of heaven? If good people can be raised to live in it are they to be raised to heaven and live with GOD? No! GOD lives in heaven. The kingdom of heaven is actually the kingdom that heaven, i.e GOD and Jesus, will set up on Earth for all the godly resurrected people to live wonderful eternal lives. It is not a place for resurrected humans to be with GOD nor is it a place on Earth for GOD to live with humans! The Kingdom of Heaven is to be Jesus's earthly kingdom covering all the Earth in which the laws of heaven will apply and all humans will delight in following GOD's laws.

Earth has been Satan's kingdom for the last 6,000 years ever since Eve chose to obey him and defy GOD. We know this because Jesus said so! Jesus referred to him as 'the ruler of this world' at John 12:31. Paul calls him 'the god of this world' at 2 Corinthians 4:4. John makes a further distinction when he says: 'We know that we are of God, and the whole world is in the power of the evil one' at 1 John 5:19.

Satan rules because he convinced Eve and then Adam to follow him. GOD then told him there would come a time when the descendants of Eve would destroy Satan's seed. Later he radicalised 200 angels to join with him in open rebellion against GOD and GOD's laws for angels. GOD thought that creating angels and humans with the power of free will would mean all would choose to obey Him to live in peace and harmony.

Satan apparently challenged GOD that he could get all humans to follow him given time and it seems GOD has decided to give Satan 6,000 years to try get all humans to follow him or his sinfulness which is why the world is such a mess of sex, drugs, greed and blasphemies.

Much of the troubles can be attributed directly to Satan causing strife and wars but his chosen organisation – the Catholic Church – has actively joined Satan in his sinfulness. But while Satan knows his end is approaching quickly the Catholics and other religious leaders have deluded themselves into thinking they can live as sinful a life as they wish and then go off to heaven at death.

When Jesus returns soon he will bind Satan for 1,000 years and destroy all followers of Satan and rule Earth with the full power and authority of GOD – just as he did when he walked here 2,000 years ago – but his kingdom will be the kingdom of heaven!

THE KINGDOM OF HEAVEN IS NOT HEAVEN.

Chapter Ten

The Iridium in the K-T Grey Layer.

We used to hear a lot about 'The Grey Layer' that is found all around the Earth but nowadays it as been renamed 'The K-T Extinction' as all the Evoquacks have banded together and decided that the layer marks the imaginary impact of an imaginary meteorite and caused the imaginary destruction of most of Earth's creatures about 68 million imaginary years ago.

Masses of EVONONSENSE have accreted to the imaginary K-T that it is now as fervently preached by Fundamentalist Evoquacks as good Fundamental Christians preach the Bible. Sadly it is another fairy tale built on Flood sediments.

The K-T layer is a thinnish band of light grey clay and can be seen in rocks buried deep or shallow all around the Earth. The layer was named in 1980 by a team of Evoquacks led by Nobel prize-winning physicist Luis Alvarez, his son, geologist Walter Alvarez, and chemists Frank Asarro and Helen Vaughn Michel. On testing samples these four found the Grey Layer always had iridium dispersed through it though the rocks above and below had practically none. Clearly the Grey Layer was different. As the Planet of Apes movie frenzy was still roaring along in high gear at the time the four researchers had to dream up a suitably ludicrous script to explain the Grey Layer that would be compatible with the events happening to the monkeys on the silver screen and be acceptable to the great unwashed crowd of Gullibles, and Hey Presto! – "The Iridium came aboard a Deadly Giant Meteorite!".

Talvarez, Alcarex, Asarro and Michel - the Foolish Four -then wrote a novella based around a central theme of Earth being a swampy primitive planet filled with vicious carnivorous dinosaurs and apes that were obviously evolving upwards until one day the Evil Giant Stone loaded with iridium zoomed out of deep space and smashed into Earth with such force it killed everything around most of the world – except apparently the leatherback turtles and crocodiles! The leatherbacks were at the mall looking at leather handbags and the crocodiles were basking in warm salty water still laughing at how they had managed to split off from those dastardly alligators – what selfrespecting croc wants labelling a gator? - about 13 million years earlier!

As the imaginary meteorite smashed into Earth and made an imaginary crater and imaginary shock wave it is imagined to have distributed its imaginary load of iridium around Earth – the Fantastic Four must have watched a lot of disaster movies in their teens! The imaginary impact has been imagined as equivalent to millions of atom bombs exploding simultaneously. Now such a blast of atom bombs would cause great destruction and leave its mark around the Earth but as the K-T meteorite is imaginary it is not the cause of the K-T Grey Layer.

So what made the Grey Layer and its iridium? GOD's fist! Or His hand or maybe His breath. Whatever force He used it smashed the Earth so hard the crust shattered like an eggshell struck by a bullet. Today deep drilling reveals some areas of the crust's granite is shattered in fragments permeated with hot salty water while some is cemented together by quartz. I recently picked up such cemented pieces on the beach about ten miles from home on England's east coast. What the Fantastic Four cannot accept is that GOD's impact generated vast shock, heat and pressures as to create great jolts of piezo-electricity that may have created the iridium by Faraday Effect.

Curiously enough although the Grey Layer is rich in iridium as are volcanoes and meteorites there shouldn't be any iridium in the Grey Layer according to Planet of Apes BigBang EVONONSENSE as it bonds with iron. Iridium is very heavy and adding its own weight to that of iron particles should mean that all iridium is in the Earth's core if Earth really is billions of years old and all the heavy material sinks down to what is claimed to be a liquid iron core? Also the creeping texsonik plates should have carried all iridiums under the crust into all that imaginary hot stuff from where Hawking's gravity would have carried it down into Earth's core. The Evoquack's imaginary iridium loaded meteorite is just a total fairy tale.

As iridium is in the Grey Layer close to Earth's surface it must be very new and young and the only process by which it could have been spread thinly all round Earth is during the period when Earth had just one landmass with great masses of water flowing over it and around numerous volcanoes venting new chemical compounds into the atmosphere: in other words during The Flood.

An intriguing characteristic of iridium is that as fine particles it is reactive and flammable – surely a volcano bursting through the crust would burn any iridium or blow it skywards in the eruption plume? The heavy unburnt particles would sink back to Earth quickly and sink through the liquid sediments to make a layer on top of the bedrock of Creation Day 3?

The power of the crust shattering impact is the method by which vast quantities of Earth crust, water, vegetation and soils were blasted into outer space to become the meteorites with their traces of Earth life. These meteorites have been returning to Earth ever since and naturally retain some of the matter in and on the chunks of Earth that became the meteorites.

Iridium isn't from outer space except after being carried there on Flood ejecta.

Chapter Eleven

Schoolgirl's finger amputated and Baked Jellyfish and Trilobites.

October 2009. A 16-year-old girl lost eight of her fingers when they were baked off after becoming set in plaster of Paris during a school art lesson.

The teenager was making a life size sculpture when her hands became stuck in the plaster as it set reaching temperatures of up to 60 degrees Centigrade.

After 12 operations carried out by plastic surgeons, she has been left with no fingers on one hand and just two on the other.

My bathroom hot tap spouts have a ring of lime around that has precipitated out of the hot water.

Countless trillions of jellyfish and trilobites are baked in the chalk and limestone deposits around the world.

My trilobites sample and all the others offered from deposits around the world are set in limestone. My trilobites are all sizes.

The chalk and lime precipitated out of the hot water bursting out of Earth at the start of The Flood 4,350 years ago.   EUREKA!

Chapter Twelve

Your Brain.

To touch on your brain and the dichotomy of intelligence.

The real problem the Evoquacks have is usually a refusal to correctly understand how they fit into the so-called Dunning-Kruger Theory. This Theory posits a cognitive bias wherein relatively unskilled individuals suffer from illusory superiority, i.e: mistakenly assessing their ability to be much higher than is accurate. It can be quickly summed up as: 'If you're incompetent, you can't know you're incompetent.' This clearly explains why so many Evoquacks promote EVONONSENSE despite it completely lacking any verifiable facts to prove it. A person has to be truly incompetent to be unable to see Darwin's research that says worms build topsoil at the rate of one inch per year and therefore if any fossil or sediment was as old as the Evoquacks claim their should be great depths of soil but as there obviously isn't then they have some incompetence in reasoning. Those people are Dunning-Kruger Inferiors.

I have noticed this so often over the years and also realised that many D-K Inferiors cover their inadequacies with feigned superiority, deceit or bombast when I ask a question that needs thinking about or calculations and will instantly make them look away and show how their mind cannot formulate an answer. Practically invariably they will respond with some sneering remark intended as a put down and an ego defence strategy; poor Hermit the Kermit is a prime example of this. (Hermit the Kermit was a person I once tried to reason with about Creation being the truth. Eventually she resorted to sneering that she had a PhD in EVONONSENSE and I didn't and therefore she knew she was evolved from monkeys! What sort of person names themselves 'Hermit'?) I also noticed this very regularly in my work in manufacturing and work study when my suggestions of making a task easier, simpler and more rewarding invariably met with an angry snarl! Everyone knows both heat and cold (what about the super heat and pressures inside the 'singularity' of BigBangEvononsesse?) sterilises matter yet Hawk'n'Dawk go smirkingly on (well Hawk can't smirk thank goodness!) stating that life started all by itself.

Henry Ford noticed 100 years ago that many people prefer mindless repetitive tasks – these days I think maybe they can shut their brains off for extended periods as quite often these people will seem to snap out of a daze if spoken to when switched off. Eavesdropping on their conversations rarely revealed any deep thoughts beyond the minutiae of television soaps and soccer.

Creationists can have, and mostly have readily taken, the gift of Holy Spirit from GOD and one vital seventh of the gift is discernment and that is what ensures a Christian creationist can see through the fog of EVONONSENSE while the Evoquack cannot even see the fog they are floundering about in! Their condition really can be called 'A Poverty of Intellect' in that they think they is educated but their knowledge is nonsense with no basis in reality.

Jesus truly described the Evolutionists when as recorded at Mark 4:12 he repeated the words spoken centuries before by Jeremiah at 5:21: 'they may be ever seeing but never perceiving, and ever hearing but never understanding; otherwise they might turn and be forgiven.'

Evoquacks are universally non-believers and so cannot access the gifts of holy spirit given so freely to all believers as described by Paul at 1 Corinthian 4: There are different kinds of gifts. But they are all given to believers by the same Spirit.

It is this range of gifts that enables a Creationist to see not the solid seductive edifice labelled EVONONSENSE that the Evoquacks see but just a crumbled weed grown pile of rubbish that no sane person would want to cling to. The fact that if an Evoquack neglects to weed his garden path it will eventually get covered in grass sod and is sure proof of a young Earth eludes their apey/primate/Satan worshipping brains.

D-K Effect should really be re-termed Ego-Compensation in Inferior IQer's Effect or ECIIE for short?

Dunning and Kruger attributed the bias to the metacognitive or inability the unskilled to recognize their own ineptitude and evaluate their own ability accurately. David Dunning and Justin Kruger dreamed up their theory back in 1999 but failed to consider they might have been looking through a mirror darkly and have got it wrong! Their research also suggests that conversely, highly skilled individuals may underestimate their relative competence, mistakenly assuming that tasks that are easy for them also are easy for others. The bias was first experimentally observed by David Dunning and Justin Kruger of Cornell University in 1999 although I noticed it any years before that, perhaps 1985, and most vividly at a job interview when all we candidates were sat together after the initial intro and job overview and given a test paper to complete. I completed my paper at my normal speed and looked up to hand it over to see that all the others were still struggling with page 2 or 3! But the maths and other stuff were simple! I have noticed this so often over the years and also realise that many D-K Inferiors cover their inadequacies with feigned superiority, deceit or bombast but a question that needs thinking about or calculations will instantly make them look away and show how their mind cannot formulate an answer. If the question seems to relate to their competence or religious beliefs almost invariably they will respond with some sneering remark intended as a put down and an ego defence strategy. If pressed for an answer and their mind suggests that their beliefs are wrong they invariably go into Schopenahauer Stage One rage mode! I have dealt with thousands of such sad people in my Creation mission over the years and now know any Evoquack is a perfect example of D-K Inferior.

In the Evolutionist v Creationist camp the vast majority of Evoquacks are quite clearly D-K Inferiors but fail to recognise or accept this about themselves due to their having a parrot like ability to remember trivial facts repeatedly woozled to them by their trainers, tutors and professors, aka Evoquacks, who themselves are prime D-K Inferiors who became professors by meekly learning to woozle what their professors taught – ad infinitum. It's called woozling woozle.

I chose to title the incompetents of the first group D-K Inferiors and the second D-K Superiors.

In reality the D-K Effect is correct in manual workers or sportsmen or even beer drinkers in Cornell Uni's bars but it fails to account for the brainwashing effect of much of today's education and media. This means that a D-K Inferior person may think they can be a great race car driver after playing some Nintendo games but when put behind the wheel they quickly find that they lack the coordination and sensory skills needed to drive fast in dangerous situations. Most D-K Inferiors are easily brainwashed and indoctrinated into becoming Evoquacks as the cult religion of EVONONSENSE is peppered with long words and ability to remember them is supposedly a mark of high intelligences. Conversely a D-K Superior might find it quite easy to drive fast and choose to use little words like as what I do.

I had an amusing exchange with an Evoquack yesterday who sneered at me being uneducated and having a low IQ with himself having a superior education and IQ of 140. I stated mine is 161 and he immediately changed topic being unable to process the fact that many things are above his level of understanding apart from his head being chock full of EVONONSENSE.

Chapter Thirteen

Can we Defend The Bible? Would you Adam and Eve it!

As we wait for the pope to invite Satan to sit in the seat of power we are bombarded by ijits who claim the Bible is full of discrepancies, omissions and alterations. Some of these claims are quite true as anyone who tries to read medieval or older manuscripts can understand.

Our daily speech has altered and continues to alter and naturally influences how we write what we want to say. However what all the devilworshipping Bible criticisers forget and probably cannot understand is: Every Bible eve printed includes The Creation, The Fall, The Flood, John the Baptist, Jesus's birth, life, works, crucifixion and resurrection, The Acts of the Apostles and finally The Revelation. Most of the criticisers cannot understand these broad themes yet they feel compelled to nitpick their way through the entire Bible condemning it for tiny substitutions of grammar and punctuation!

We can know the Bible is true as the 'Books' of the Old Testament that Jesus learned from as a child and later read from were actually rolls – scrolls - of lamb or kid skin on which the Israelites meticulously detailed their beliefs, prophecies, teachings, laws and commandments.

The scrolls were written by specially chosen and trained scribes who spent their lives carefully making copies of earlier scrolls for distribution to new synagogues or believers dispersed around the world. The scribe would start with a clean sheet of skin and ink made from oak galls and copy the words on another scroll. Each word would then be checked by the master to ensure the copy was perfect before the scribe could write the next word. Any mistake impossible to erase without defacing the page ruined the sheet and it had to be thrown away. Each sheet was sewn to the other ones with rabbit veins. All this slow work resulted in scrolls that never varied over the centuries and millennia. The completed scroll was rolled around a wooden stave with end pieces and was kept in a protective cover and was unrolled to be read.

This accuracy in copying each scroll is why we can be certain what we read in our Bibles is unaltered since first written despite all the Muslim and non-believers claiming the Bible has been altered and corrupted many times regardless of the only differences over the years being translations to fit the modern language of the country the scroll is intended for. The Bible's Big Picture has not altered.

Thus we read at Luke 4:17 of Jesus entering the temple in his home town of Nazareth where everyone knew him as the son of the carpenter Joseph, asking for the book/scroll of Isaiah, unrolling it until he found the page with the verse and reading to the assembly: 61:1 'The Spirit of the Lord God is upon me; because the Lord hath anointed me to preach good tidings unto the meek; he hath sent me to bind up the brokenhearted, to proclaim liberty to the captives, and the opening of the prison to them that are bound; 2 To proclaim the acceptable year of the Lord.'. The congregation marvelled at his speaking and as they were familiar with Isaiah's words they could know Jesus was claiming to be the Messiah. From Luke 61 we can know that the Book of Isaiah had not been corrupted during its many copyings and it and the entire Bible are true and faithful records of events.

Old films show Nazis rounding up Jews from synagogues during World War 2 and casually throwing the synagogue's scrolls on fires totally unaware of the intrinsic value or that the scrolls prophesied what happens to people who disobey GOD to kill, pillage and story.

The real problem the Evoquacks have is a refusal to correctly understand how they fit into the so-called Dunning-Kruger Theory. David Dunning and Justin Kruger dreamed up aka postulated the theory back in 1999 but failed to consider they were looking through a mirror darkly.

A perfect example of the faulty conclusions of the Dunning-Kruger Theory is provided by Hawking, Dawking, Attenborough, Krauss, Nye, ad nauseum: who all are on record as saying life evolved from nothing despite that idea being firmly and totally disproved by Pasteur's experiments! Yet the 'professors' who filled Hawk, Dawk and the other's head with woozle passed on all their woozle to later 'professors' who blindly and blandly continue to woozle that life began all by itself on a planet made firstly from superheated matter that then came together in the supercold of space's vacuum!

On the previous pages is outlined the dichotomy between the reality of Creation and the fantasy of Big Bang and the Evolution of life and how D-K Inferiors are blinded to the truth of Creation and the utter stupidities of EVONONSENSE.

To believe one is to claim the other a lie. D-K Inferiors have heads stuffed with Planet of Apes nonsense which they fail to discern is nonsense and so they sneer at all who can see that a Planet of Apes scenario cannot happen.

Creation is known from the record of GOD's Word and has been proven true and accurate in all ways while Evolution is a total fabrication of lies, suppositions, postulations and imaginings with few that can be tested let alone proven accurate. Because it directly contradicts GOD we can be sure that came from Satan – and it rightly needs referring to as EVONONSENSE so throughout this book I will refer to all evolution as 'EVONONSENSE' and its advocates as Evoquacks.

Why the dichotomy?

Why such vitriolic outbursts from some defenders of EVONONSENSE?

What makes creationists so serenely confident that they know they were created by GOD just a few thousand years ago when the Evoquacks slam down multitudes of fossils dug from deep solid rock that certainly appears to be ancient and claim they were killd by a meteorite impact 65 million years ago?

Why should the Evoquacks fly into a potty mouthed rage at any suggestion that their millions of years beliefs need shrinking by a factor of 10,000 if not more so that a fossil is correctly aged at 5,500 years old and not 55million?

What makes the creationists – including me – certain that the immense deposits of coal, oil and gas beneath our feet is just a few thousand years old and hasn't been forming for one or two hundred million years?

Why do the Evoquacks refuse to believe that manmade artefacts are regularly found in 'millions of years old' coal?

What makes the Evoquacks so certain they started out as pondslime with later helpful inputs from aliens from distant planets?

Why do creationists like to speak of a 'GOD' as a real entity to whom they owe their very existence when the Evoquacks scorn all idea of 'GOD' and laugh about an old guy with a long white beard up in the clouds as being childish superstitions?

Why do all the D-KP Evoquacks worship Hawking as a superbrain when he states the sheer stupidity of 'gravity creates things?' - and he believes he evolved from a monkey that evolved from a fish etc?

Why do the fools like Dawkins whom make bigger fools of all the fools who follow them get the plaudits and Nobel prizes for stating the obvious about what GOD created but evilly attributes them to Satan?

Why, why, why!

There must be some process that over the years has resulted in we humans being vastly superior to all other living things – if by superior we mean living in a technological world and not just mooching about enjoying themselves like most other creatures seem to do.

Asked how a rough unwashed hirsute gorilla with prominent bulges over its eyes turned into the slim sleek gorgeousness of the humans in the high street they shrug it off with a sigh of condescension and parrot the woozle they sponged up at woozlestitute: 'Just Darwinian macroevolution of course!'

When we point out that macroevolution cannot happen they just release their demonic controller and call us fools.

Mary Schweitzer is a perfect example of a D-K Inferior as despite all evidence to the contrary she will insist that TRex bones can harbour soft tissues for 68million years – what else but low intelligence or 1 Corinthians 2:15 can explain her inability to evaluate and absorb scientific facts? Today I have just read Schweitzer was mightily impressed at age 5 with a gift book about a dinosaur that wanted to be a chicken...talk about brain programming! Hawking, amazingly as it may seems, is also a D-K Inferior as he stated the most foolish statement ever stated when he said 'Gravity can create everything.' He believes nothing can create gravity which will then create everything? He sank to practically the lowest level a D-K Inferior can sink to when he also said aliens might be two billion years more evolved than himself!

GOD made your brain as a wonderful thing. Basically seeming to be a lump of sloppy grey tissues when dead, when alive it is an amazing power house with unlimited capabilities in storing and processing information and making decisions based of a vast number of inputs from all nine senses. It's gravity as what made your brain – according to Hawking. Jehovah's Witnesses art department Satanists altered a picture of a brain to include a subliminal reptile head...

Very few people use even a tiny portion of the power of their brains and cannot begin to understand what the brain's function really is though they abuse it and stuff it full of rubbish every day. This main function is not running the body's multitudinous systems as complicated and demanding as that is as it is all done automatically with no need for input from us – the fact everything hums along nicely when we sleep is proof of that - the real job of the human brain is simply to maintain a full file on perhaps 20 or 30 or maybe 50 billion other people and GOD and Jesus and various angels!

Until The Resurrection is complete the brain cannot exercise that role, but it can demonstrate this special power by the ease and speed with which it can instantly find the filed away memory of practically anything you have ever seen or done! What enormous amount of computer memory would be needed to keep the complete record of your life that your brain is storing?

Try this small test of your brain's power: Think of the first school you attended and an incident there. See how you can recreate the building outside and grounds, the classroom or was it the dining room or the gym or first aid room? And the person involved. Another child or a teacher? He/she was tall, short, old, young, warm, cold, etc. It's been 60 years but I can recreate daily school events in perfect detail.

If it was possible to have filmed a person fifty or sixty years ago and show them the film, stopping it as they step out the door in the morning dressed in a certain style, and ask if they could recognise the place and what they were doing they may be a little vague unless their outfit was really memorable like a wedding outfit, but if it as a fairly ordinary one they might not remember. Show a bit more film of them now either getting into a certain car or walking to a bus stop and their brain will start sorting its files. Show them driving into the carpark of an office, factory, mall or sports centre or getting off the bus with workmates or shopping or socialising friends and the brain will now be pushing scores of memories forwards. If the day was an important occasion it will come flooding back in perfect detail: the clothing was for a wedding or important interview, the car was a beribboned limo, the other people in the memory were the bridegroom or the vicar, others were family and friends; the room was the church or chapel and looked wonderful with bright flowers and sunlight. Maybe the film has stirred memories of a typical day in a dreary office: old, dilapidated building down a sunless alley, worn steps to a battered front door carrying faded names of previous occupants, the corridor with the sagging old chairs and tired magazines, the office with the frosted glass panel door, a dark wood desk with a green manual typewriter and wire baskets of correspondence, lit by a harsh fluorescent tube; already sat at a brown desk is that stern office manager chainsmoking as usual, a little ancient gas fire is making the room stuffy, big olive green filing cabinets stand against most wall. Make the day the one when something unusual occurred – getting a raise, or an accolade, or being involved in a fellow worker's incident and the brain will now be presenting just one of the billions of files it has stored away.

And so on, all the details are there in our minds – we just need the number and date of the file to open it and watch it unfold with perfect colours, textures sounds, feelings and scents. This really is the true function of our brain – not creating nuclear bombs or squandering billions of shooting rockets off to distant planets while half the population are starving, thirsting or dying for lack of medical care.

One simple technique to discover long disused memories is to set off on an imaginary walk down a childhood location, perhaps the walk to school or the sports ground and paint the walk in full detail as you go – the actual pavement or road, the wall or hedges alongside, the traffic humming past, the various houses, shops, businesses – you may not have thought of any of these since moving away from the area many years previously but it is all still there! In recent weeks a favourite television program has been showing favourite personalities who have done much acting being taken back to their childhood homes. Some haven't seen the homes for 40 or 50 years yet as they enter the gate they describe how the garden used to be, they step inside and contrast the present decorations with what they knew, the layout is remembered with affection and memories of sitting in a certain spot looking out the window or lying in bed watching the dawn light sneak in; they walk the garden and once again can taste the fruit of that tree or bush and the berries from the bushes that father planted, or they look in vain for the chickens that used to live in a pen in the corner.

GOD designed our brains to keep all our memories but also a full file on everyone we meet – and everything about them and how we interacted. He originally intended us all to be able to ramble gently round the world to see our many times removed father and mother, Adam and Eve – what tales they would have and how amazing to sit at their feet and listen to them describing their children, Cain and Abel and Seth, and all the others and their children and grandchildren and speaking to GOD and Jesus and angels and then hear our names and family line mentioned! And all this while fat dinosaurs and huge tigers munch grass nearby and pterodactyls swooped through the sky and frolicked with eagles and vultures and over in the seas sleek plesiosaurs cavort with penguins and sealions!

When Jesus has destroyed Satan and all the evil people and made the Earth into the Paradise GOD intended the human brain will come into its own and help keep track of billions of other humans around the world. EUREKA!

Chapter Fourteen

Rich Deems – a special kind of stupid!

If you have never heard of Rich Deems I envy you! He is an American pastor and like so many American pastors he is one of Satan's best asset in these End Times – he even surpasses Hawk'n'Dawk for stupidity! At least they have the excuse of being British and growing up in the grim austerity 1950's with nothing to eat but cheap white bread and lardy bacon while Deems is a Yank surrounded by GOD's bounty and wonderful landscapes. He claims to have become a Christian and posts photos of being baptised but he was baptised with water and not the blood of Jesus!

On reading his thoughts on the Bible it is sad to see Deems utters many false doctrines as is written at 2 Peter 2: 'But there were false prophets also among the people, even as there shall be false teachers among you, who privily shall bring in damnable heresies, even denying the Lord that bought them, and bring upon themselves swift destruction'. Hi words are lauded by his flock – he sold his soul to Satan and now is the absolute epitome of the wolf in sheep's clothing that Jesus said would appear in the End Times!

He denies Creation, GOD's six days of work, The Flood, The Tower of Babel, The dividing of the Earth, Paradise Earth - all core truths on which to base belief in GOD and hope for life with Jesus. He sneers at those of us who know Earth is young but instead follows Hawk'n'Dawk'n'Ross'n'Nye and all the other fools who claim Earth is 4.8 billion years old. I told him worms prove it is young and as expected he sneered so I told him his friend Satan is delighted to have such an ardent promoter.

If you are new to Christianity and are eager to learn about Jesus and how to gain eternal life I beg you don't read Deems as he will add you to all his other sheep he is taking to Satan's slaughterhouse.

Instead of Deems look at Kent Hovind's teachings which are the truth and fully verifiable with any good Protestant Bible.

Chapter Fifteen
A Day with Adam and Eve and TRex.

A Day with Adam and Eve and GOD and TRex – the T between immorality and immortality.

'There you are!'

'I've been here feeding this animal. Isn't he big? What is his name?'

The man dragged his gaze from the lovely face of the woman to that of the great beast towering above her. It was busily munching orange berries off the bunch she was holding up to its mouth.

'It is Tyrannosaurus Rex. It loves those berries.' He reached over and patted the beast's neck.

(He didn't call it TRex actually as he spoke some forgotten language, but the creature was definitely the herbivore we now call TRex.)

'It has a wife and she has a nest full of eggs. We can go look tomorrow. Right now I'm hungry for some of that tasty white grain we ate yesterday.'

'I think I'd like to eat some of those pink berries; shall I gather some for you?"

His smile was all the affirmation she needed and with a last fond pat to Trexxies neck she strode over to the patch of bushes with the pink berries. The man watched her go, utterly mesmerised by the way her hips swung and how her skin seemed to be liquid as the muscles and sinews beneath propelled her.

Their days followed just the same easy rhythm: wake, stretch, quick wash in the nearest bubbling spring, she sitting for him to comb his fingers through her hair, quiet murmurings about the many birds, animals, reptiles and insects that were also starting their days, a walk and leisurely breakfast on whatever caught their eyes, then Adam bid her a fond farewell as he set off to extend and cultivate the Garden just as he had been instructed, she to just enjoy being deliciously alive among so many other creatures all enjoying their own lives.

Adam and Eve loved watching dinosaurs use huge teeth to strip masses of fruit and nuts off the trees or bite swathes of grass to pack fat bellies before joining the cows, sheep, lions, tigers, rhinos and countless other creatures all lying in the sun digesting their day's food.

Eve might have wondered what it would be like to also have pretty little youngsters the way the lions did as she picked up and cuddled the friendly cubs with the she-lion's full permission. Maybe she was amused at the antics of the baby wolves as they flocked after their mothers and pestered for regurgitated soft grasses. Maybe she was drawn to where swans and geese helped their chicks escape their shells and let her pick up the dry inquisitive chirping babies and looked into their guileless eyes. She longed to help the new baby giraffes learn how to control the legs that had been bent double inside the womb but could only watch as they wobbled and fell repeatedly and harmlessly.

The Evoquacks say Creationists can't know what Eve's day was like as she never existed and all ideas about her are just distorted myths and legends of some primitive ape-like Earth Mother filtered through a hundred thousand years or more. That is sheer nonsense woozled back and forth by generation of Satan's followers. True Creationists who believe the Bible can guess Adam and Eve's days from what GOD and His prophets tell us.

The truth is Adam and Eve had to do nothing more that eat, drink, enjoy themselves and all the creatures with Adam doing a little easy work extending the Garden before the darkening day made them seek a comfortable place to settle into a refreshing sleep. And while they slept their brains would continue to run in high gear arranging and storing the memories of the day, just as ours do as we sleep our nights away 6,000 years later.

The TRex, lions and pterodactyls – and great Behemoth! - also slept their nights away with no worries about being harmed, but their simple brains recorded little of the day, perhaps nothing more than where a particularly tasty fruit or grass had been found or of enjoying being petted by those friendly two-legged upright walking creatures with their constant melodious vocalising. Then again your dog or cat's twitching seem proof that animal brains recall their days.

Life in that Paradise seemed as though it would be wonderful forever but alas no, it wasn't to last. Paradise would be lost for want a little thoughtfulness. Later Adam and Eve would look back to their first days with sorrow as everything changed irrevocably and life became a sad, painful, lonely struggle without the friendship of all those lovely creatures and GOD and Jesus and the angels.

One of Eve's days began the usual delightful way and after Adam tore himself away and went to work she wandered about happy to greet and be among the many furry, scaly, feathered, naked creatures that filled her world with sounds of contentment. She had a purpose that day: find the huge creature that had been bellowing while she and Adam had been eating. She had the idea it was in among the giant fern trees and she stepped confidently into the grove with no hesitation. The grass and fallen leaves beneath her feet were a springy carpet. At each step small bright shiny, slippery and furry creatures moved aside before returning to their task of recycling the detritus. Urgent signals low in her body told her yesterday's food had given up its goodness and now needed to exit her body; without hesitation she scraped a little hole in the detritus, squatted down and firm lumps of waste fell from her followed by a hissing stream of urine. GOD's design for her had included a self cleaning anus and a drip free vagina that allowed toileting to be quick clean and easy. She stood, scooped leaves and soil over her droppings and left the worms and creeping things to recycle them back into nature just the way modern research shows - and worms are particularly fond of human excrement! GOD and Jesus really did think of everything!

She stepped between two enormous trees stretching up out of her sight into the clear blue sky through which flocks of birds of all sizes flew and played. She marvelled at the high speed dartings of some birds and the slow lazy flaps of the big black pterodactyl she had seen on previous days and compared their wide wings to the smooth slimness of her own arms; she thought of asking Adam if she could learn to fly like the birds as surely he'd know.

A loud slurping noise in front brought her out of reverie as a really huge beast rolled over in the mud bath it had created where one of the springs had spread out. It looked at her for a few moments before rolling over once again to coat its other side in mud while letting its long neck settle on clean grass close enough to her feet for her to reach down to stroke the tightly overlapped scales under its chin. It responded by waving its huge cedar tree like tail side to side raising waves of mud onto the grass towards the fat coils of snakes playing there. Never having seen those particular snakes before she decided to give them a closer look but her path took her through the soft mud that squished between her toes as she stepped daintily through it even though she trod into the beast's footprints. Glancing back she thought how small her prints were compared to the great beast's. The snakes untangled themselves to present their heads for her inspection. Bright spectrums moved over their skins as they preened and primped for her inspection. Intrigued by how the serpent's slim black tongues constantly flicked in and out she couldn't help contrast it with the plump pink tip of her own as she pushed it out and squinted at it with one eye at a time. She let one of the serpents tickle her arm with its tongue before pulling back and stepping away to continue exploring. The beast in the mud bath now seemed asleep so she tiptoed past it and again left her small footprints in the huge three-clawed ones of the beast. 6,000 years later similar human and dinosaur tracks would be found in various parts of the world but most noticeably in Glen Rose, Texas before Satanic Smithsonian staffers smashed them in their mission to promote Satan and his EVONONSENSE.

Her seventh sense told her something was watching her from the right and she turned to see a pair of bright eyes regarding her. They belonged to a giant lizard beast, practically identical to the one she and Adam had petted earlier but this one seemed a little, well, different, plumper, bigger? Eve stepped over to it and felt a thrill as she saw the animal was sat beside a nest with several eggs and two small perfect replicas of the adult beast. Kneeling beside the mother she gently touched the heads of the youngsters and felt their mouths nibble her finger tips. One of the eggs was cracking widely and she watched spellbound as another baby struggled into the world and lay gasping from its exertions. Other eggs were moving slightly and one began to crack so she took up the oldest baby and let it sit on her lap and explore her arms and chin while its mother looked on unconcerned. Eve looked into the mother's eyes and though they held no fear or malice they didn't speak to her the way Adam's did. The sauropod yawned and showed her the set of huge rakes of teeth so perfect for harvesting the grasses and fruits. It kept its mouth open for her to feel the sharpness of the biggest teeth to compare them with the blunt evenness of her own pearly white ones. It reached down and gently lifted the baby off her lap and deposited it back in the nest with its siblings and settled itself down to watch the other eggs hatch. Eve stayed a while longer then decided to follow some slim pretty deer passing in the direction where Adam expected to find her so with a fond pat of the small head she left the mother beast to her vigil.

'Eve!' Adam called as he saw her approaching through the bushes and his heart flipped as her face lit up at seeing him.

She cast an appreciative eye over him and noted that whatever work he had been doing had honed his muscles. 'Let's go bathe?' she suggested. He smiled and led her to where water gushed from a small mound and they sat down to splash the cool water over their skins, happy and content as only innocents can be, as tiny fish flickered around their feet and big green frogs hopped around making their monotonous croaks.

Drying off in the rosy setting sun and warm evening breeze the couple ate fruit and nuts and greeted the various animals and birds that came by to pay their respects and enjoy the feel of the human's hands stroking them.

A pair of long white and black striped giant cats padded silently towards them with miniature ones running alongside and Eve felt an incredible urge to cuddle the soft bundles to her. Adam didn't notice as he was feeding his leftover leaves to the adult cats.

A great male beast with spiked head came towards them and picked a mouthful of nuts off a tree beside them before causally walking away.; its tracks were huge. Adam yawned and pushed the cats away, took hold of Eve's hand and led her to a patch of soft dry grass and leaves under a wide small tree that would be a perfect shelter for the night and against the morning's dew. They lay close together, perfectly happy and content, listening as the large creatures around them fell silent. Snakes and lizards crept close to enjoy the warmth radiating off the humans, small rodents crept around scavenging up the bits of fruit the couple had left.

The sun faded away, the moon began lighting the eastern sky, owls hooted, night birds piped, the first humans slipped into sleep. GOD, Jesus and all the angels seemed flown to heaven. Off to the side among the trees a pair of eyes glared malevolently at the humans sleeping so peacefully.

It was Satan. EUREKA!

Chapter Sixteen

Genesis 2:6 and 9:12

Open your Bible at Genesis 2:6 and read about mist rising daily to water all the Earth, then go to 9:12 and read about the rainbow against dark clouds and sit awhile thinking about everything you have ever read or watched about climate, rain, erosion, plants and cultivation.

Read what I have written about the daily circling of sun and moon and how they cause tides in the seas and under the Earth and how piezo-electricity is produced by crushing quartz crystals and how people can see sparks rising when earthquakes crack the bedrock.

If you have wisdom and understanding and a little knowledge you may be able to draw the conclusion about the link between these two verses and how they prove Earth is young, The Flood was real and hence the Bible is truth – GOD's TRUTH! EUREKA!

Chapter Seventeen

Lord of the Crippled Flies.

The more research done on EVONONSENSE the more stupidity is uncovered! This short idiotic article about how flying creatures came about is about as childish as most EVONONSENSE but while Evoquacks read it and visualise the actuality of a creature suddenly growing wings after being blown up in the air by the wind Creationists can see it is absolutely infantile – and must surely have been fantasized by a D-K Inferior!

Quote: 'Likewise it is not much of a challenge for insects to be picked up by the wind and so it was not a great evolutionary leap for so many of these small insects to evolve the ability to fly.'

What nonsense is that! How do they make this stuff up! Do they all work for Disney?

Some creeping insect gets blown about and thinks 'I'll have babies that can grow wings and the muscles, nerves, bone attachments, and re-programmed brain to control the wings and muscles and develop all the flight control systems to ensure take off and landing and steering!'

Surely the insect will either die on landing or survive with broken legs and stagger away thinking itself lucky but I'm sure no insect ever thought it would set about growing wings.

It takes two to tango so the blown about insect has to find one of the opposite sex with the same idea and then they have to somehow redesign their DNA and make sure it is downloaded into their eggs and sperm so that their babies will be able to fly!

Trout and similar fish know perfectly well that most insects cannot fly and any that fall towards water are doomed to be speedily eaten?

Pigeon fanciers know that a baby pigeon will not attempt to fly until it is almost as big as its parents and any attempt to make one fly will merely make the youngster crash to the ground and cripple itself – a crippled bird is as useless as a crippled insect.

Flying ants deliberately cripple their flying by shedding their wings after their mating flight – whoever dreamed up the insect instantly learn how to fly fairy tale needs throwing off a tall building to see if they can evolve wings and the ability to fly before they hit th eground and end up as pulped human!

Humans are vastly more gifted than any creature but can a human order it's descendants to have totally different shapes, functions and the nerves and brain program to control the modified body? Of course not! Nor can we humans make our big toes be as useful and versatile as a chimpanzees!

It is hard to credit these Evoquacks with the brainpower of some of the lowlife that they claim evolved up into themselves? Every cell in the human body carries the full instructions for replicating itself – this is how God designed it so we can live forever!   Just as a long soak in a warm soapy bath will cause almost a complete layer of skin to be easily rubbed off our bodies daily so each day invisibly our entire body's cells constantly check its own condition and arrange for its replacement as needed until about every three months we are entirely new creatures – except for our hair and fingernails which are, as it were, dead from the moment they leave their sheaths.

Each cell carries in it a full code of DNA of such astounding complexity and completeness that only a demon possessed fool would say it all came about from wet stones bumping together billions of year sago. Hawking, Dawkins et al are just such fools. Hawking is especially sad as of all the Evoquacks he really could do with the full power of his DNA to repair his body but his denial of God precluded him receiving the benefits of eternal life that God gave to Adam and Eve.

DNA is usually best described like a book:An analogy to the human genome stored on DNA is that of instructions stored in a book:

The book (genome) would contain 23 chapters (chromosomes);

Each chapter contains 48 to 250 million letters (A,C,G,T) without spaces;

Hence, the book contains over 3.2 billion letters total;

The book fits into a cell nucleus the size of a pinpoint;

At least one copy of the book (all 23 chapters) is contained in most cells of our body. The only exception in humans is found in mature red blood cells which become enucleated during development and therefore lack a genome.

EVONONSENSE is Satan's best accomplishment!

Chapter Eighteen

Bling! All that glitters. Eureka!

I just read a research paper stating quite solemnly that Earth's gold was and is formed in supernova explosions and that molecules, protons, atoms, or whatever, of the golden supernova stuff whizzes across the trillions of miles and eventually hits Earth. This may actually seem quite logical to some uneducated Evoquacks as lots of gold is found in the outer crust. But of course as other Evoquacks claim Earth has a core of heavy elements - and gold is as heavy as it gets - then logically the gold in them thar hills must have come from other sources - especially when it is found in veins of really hard quartz or in great nuggets under a couple of feet of topsoil as that old Australian guy found while digging a new lavatory ditch for his wife.

However, what the gold-from-supernovas gang fail to spot is the big error in their fantasy: if the Universe is vastly big in all directions and a supernova is vastly far away wouldn't any gold atom have very little chance of hitting our tiny Earth? I mean, marksmen know perfectly well that the more distant their target the less chance of hitting it? Thus if you care to draw a supernova explosion as happening from a point squidzillions of miles away then any atom initially launched in our direction would have to maintain exact trajectory to hit our tiny 8,000 miles diameter home? Would anyone with a facility for maths care to do the calculation needed to arrive at exactly how few atoms could be expected to hit Earth if launched from the small end of a cone trillions of miles long? Perhaps the gold-from-supernovas gang think gold atoms is intelligent and can steer themselves here? Or there is a lightwave like a conveyor... If all Earth's gold did come from supernovas can we assume that the much bigger planets will have vastly more deposits of gold? Is thar gooold in that thar Pluto!

Deputy Prime Minister Etienne Schnieder of Luxembourg has reportedly committed his country to invest in a scheme to harvest minerals from quite tiny asteroids which logically have less chance of having gold or other elements – did he do the cone to target calculation? I wrote and told him to look at the research into asteroids being balls of fluffy ice with minimum dirt and ask his partners why they think balls off fluffy snow and dust will have viable quantities of minerals. He hasn't replied.

How are an atom of gold supposed to blast out of a supernova and find its way to Earth then trickle down through solid hard quartz to sink down into the core? Yet the fact gold is still mainly be found in the top layer in solid rock or sediments that were once solid rocks make a nonsense of the trickle down fairy tale? This makes the supernova idea seems like the plot of a Planet of Apes movie. The fiction writer who originated the idea resorted to lots of unprovable hypothesises to shore up his silly fantasy including throwing into the childishly idiotic claim and Satanic talisman '200 million years' which is obviously ridiculous as the worms prove Earth is about 6,000 years ago – and if Earth was 200 million years old there should be thousands of feet of lush topsoil everywhere...Ask Alice Roberts where all that soil disappeared to?

Luckily the Bible and real scientists provide the real truth of gold's origins.

Scientists in Ukraine have been experimenting with super powerful jolts of electricity and have found they can create known and unknown heavy elements just by building up then focussing a big electrical discharge onto a target. The flash of the impact is accompanied by the creation of various elements old and new. While they haven't mentioned the creation of gold it may just be a matter of time before they can create gold at will. But how does this verify Earth's gold's origin?

We need to step back to the moment 6,000 years ago when GOD created Earth as a ball of elements covered with water. A minute amount of gold is found in human bodies and it readily dissolves in pressurised hot water. Rich Orientals like to make exhibitions of themselves by having their food balls covered in gold foil! It logically follows that the Earth, far from being a mass of sterile meteorite dust that accreted by chance, was actually designed by GOD to include everything necessary for life and a fine sprinkling of gold was included in the mix or perhaps in the outer crust. Plant roots must be able to dissolve the gold as how else could we humans have gold in our bodies? Er, I just had the thought that as GOD designed our bodies to manufacture Vitamin D from sunlight – is there any way in which golden sunlight can convert something in our skin or blood t make gold? After all a tiny baby has only a baby's amount of gold in its little body while a huge muscled man has lots more – where does it come from?

We know that gold mines are very hot and in test drillings just a few miles deep there is very hot water, and we know that earthquakes create spikes of electricity so it logically follows that the fine gold suffused as dust and fine flakes through the rocks and especially quartz arrived there when the rock was shattered at the start of The Flood 4,350 years ago and scalding water with dissolved gold filtered up through the rock. Any gold that found itself in pockets would naturally form larger nuggets. The huge nuggets found by the old Australian man may have been the result of settling out of a lost vent during The Flood – or perhaps just as likely – may have been the melted remains of some antediluvian idol!

Whatever the real origin of Earth's gold I'm sure we can rule out supernovas as they cannot have come into existence until four days after Earth was created complete with its motherlode of gold!

EUREKA!

Chapter

Stromatolites

Stromatolites is like lumps of fine-grained hard sediments that form in temperate waters. When sawn apart they show very fine varves of sediments just like the Grand Canyon, but unlike the canyon which was formed as the great flood waters drained off Earth, the stromatolites form by successive influxes of sediments being followed by a spurt of blue-green algae aka cyanobacteria growth.

They is fantasised as being made 2.5-4 billion years ago during the 'Archean' era. The dreamer of that fantasy has to explain why Earth that old doesn't have a hundreds of feet of meteorite dust or a million feet of topsoil on it...and why the stromatolites aren't resting on thousands of feet of sediment...

One EVONONSENSE website states: 'Photosynthesis is the only major source of free oxygen gas in the atmosphere. As stromatolites became more common 2.5 billion years ago, they gradually changed the Earth's atmosphere from a carbon dioxide-rich mixture to the present-day oxygen-rich atmosphere.' As usual this is absolute satanic garbage that only fools and Dunning-Kruger Inferiors believe and basically means that Hawkin's wet rocks sparked into life and gravity let them start making oxygen. Obviously there has been no 2.5 billion years as if so the build up of space dust would be miles deep everywhere! Nor is photosynthesis a source of free oxygen except on Planet of Apes as all Earth's oxygen was provided on Day 4 of Creation when GOD stretched out the heavens and separated much of the water into oxygen for Earth and hydrogen for outer space although obviously to create Earth as a water covered ball must have needed lots of oxygen from Day One?

Other Evoquacks imagine the stromatolites as being started during the Pre-Cambrian period – as they is ignorant of the fact that 'Pre-Cambrian' means the bedrock laid down on Day One of Creation as the waters drained into the empty Earth. Nothing is Pre-Cambrian except GOD and Jesus and the angels who shouted with joy when GOD created Earth - but try tell that to an Evoquack!

Johnnie Moore of the University of Montanain Missoula is a textbook Evoquack. He published a paper about stromatolites forming in Flathead Lake just where the Flathead River sediments are swirled round and drop out of suspension to become available for the plankton and diatoms to extract it to build shells and skeletons. These creatures clump together and become structures on which green algae can grow. As the lake level is reduced for safety before the rainy season the process of building is seasonal. Moore was and perhaps is unaware of the source of the calcium carbonate extracted from the river water by the diatoms. Obviously some will come from the soils through which the river flows but a large amount must come from the green vegetation that falls and rots in the water.

Like most Evoquacks who deny Creation he resorts to lots of fantasising in his comic article by using 'must be' once, 'may' twice, 'possibly' twice, 'probably' thrice – 8 fairy tales- and also tosses in the silly 'Pre-Cambrian' and 'Plastocine' Eras! If he hadn't filled his head with Planet of Apes nonsense he would know that on Day 5 of Creation GOD put starter populations of diatoms into the new seas in order to recycle the calcium carbonate He knew would begin to flow into the rivers and seas as the new vegetation dropped its leaves and from the droppings of creatures – TRex, sabretooth tigers, Adam and Eve – which all ate green vegetation or the fruit and seeds of vegetation.

I emailed Johnnie pointing out how Darwin and my research prove Earth is young and as expected he went into into smug Schopenahuer Stage One: Here is how he responded to my pointing out that the Flathead Lake Nodules may have grown very quickly:

January 2015. My 1980 paper on Flathead Lake nodules...after the glaciers retreated during the big Pleistocene...un-interupted sedimentation in the lake from about 15,000 years...also dated wood using C-14 dating...In the Flathead Lake sediment record there is no evidence for a large flood 4400 years ago...your world would be much more interesting if you understood the Earth and embraced its complex evolution through time....

J. N. Moore,

I replied:

Hi Jonnie,

I got as far as 'big Pliestocene' and then thought 'another one who doesn't know of Darwins Research proving Earth is young.'

Rose

Johnnie's 'big Plastocine ' is really Creation and Flood but he prefers to ignore GOD and be in thrall to that EVONONSENSE Satan whispered into J R Wallace's ear when he caught him in that spiritualist's séance. He thinks Planet of Apes fantasies is more interesting than the reality of Creation.

The fact Stromatolites are being formed today in shallow warm water is a good indication that during Creation and after the Flood there were extensive areas of shallow sea.

Stromatolites are claimed to be extremely primitive but as they grow quite happily and trap sediments to make solid matter they aren't so primitive.

No Evoquack could sit in water and create solid matter by filtering water? Nor make oxygen?

Chapter Twenty

Vegetarian vultures - Do scavengers prove non-eternal creatures?

The multiple births of many birds or mammals does lead me to think they had been designed with limited longevity while the amazing fecundity of insects and sea creature does seem to indicate they also were intended for a short life. Jesus did say that GOD noted the death of every sparrow but does that imply sparrows had eternal life?

In times past in the UK and today in rural Africa birds gather into enormous swarms with unwelcome results. I remember starlings gathering in great flocks each evening when I was young and their droppings were a serious problem to health and to the roofs and gutters of buildings but they are no longer to be seen in such quantities – I was actually surprised to hear and see a starling about half a mile from my flat just a few months ago! African farmers fear the gathering of Red-Billed Quelea birds that can strip a field of grain in just a day or two. Left to breed to their utmost those birds would ruin entire crops – is that what GOD intended or was creation originally so fruitful as to provide excess food for unlimited birds, animals and humans? GOD's prophet Amos did prophecy at 9:13 "Behold, days are coming," declares the Lord "When the plowman will overtake the reaper And the treader of grapes him who sows seed; When the mountains will drip sweet wine And all the hills will be dissolved."

Today's starlings, pigeons and queleas prefer to gather into unwelcome flocks but is that their original behaviour or does Amos 9:13 mean they were, like the humans, intended to spread over the entire Earth where their feeding habits would ensure the lush vegetation's seeds would not be wasted and they would have no need to swarm onto isolated grainfields?

The mystery of why many hilltops are presently barren and have been since The Flood is easy to solve and perfectly logical though Darwin's GOD-hate prevented him solving the mystery even though he saw the cause and effect! Jesus said 'They see but cannot understand...'

Checking the longevity of creatures shows some are still juveniles when humans are dying of old age. Scientists say that human bodies are totally renewed every three months which with the daily new skin does seem to show we are intended to be eternal but does the long life attained by some creatures prove they are eternal too? Longevity charts of creatures shows some birds, animals and sea creatures can easily outlive humans. But is it that these creatures have the the lowest birthrate and live the longest to ensure a continuity? If Eve had ignored Satan's whisperings via the serpent she would still be happily alive but would any of the creatures created the same day be still alive? Elephants have human lifespans but giant tortoise last three or four time longer. No-one knows why.

That afterbirth eating kites, ordinary vultures and the specialised bone eating bearded vulture exist alongside hyenas, blowflies and Nicrophorinae - carrion burying beetles - does seem to show that dead creatures and birth products would need to be recycled by scavenging birds, beasts and insects to cleanse the Earth. Of course these flesh eating creatures could have had their diets changed at The Fall and may have all been originally created to recycle unharvested fallen fruit, nuts and vegetation. The kite and its liking for afterbirths is curious though...

Studies have shown that when a large African beast dies the first bird to successfully penetrate the beast's tough hide is the vulture with no feathers on its head and neck. This ensures that once its sharp beak has ripped open the carcass it can reach right inside to feast on the soft organs without fear of all the blood congealing on feathers. This first feeding allows the blood and fluids to drain out of the carcass ready for the next kind of vultures which have feathers on their neck but not their head allowing them to rip out soft flesh and not have their neck feathers soaked stiff with blood. Other vultures and birds finish stripping the carcass and peck out the tissues connecting the bones leaving them ready for the last bird – the bone eater or Lammergeier. This bird cracks and eats the smaller bones before carrying any larger ones to a height and dropped them on rocks to shatter them to swallowable size. GOD must have put that instinct into the bird's head at Creation. The biggest bones, skull and pelvis, are either dragged away by animals with the necessary jaws to deal with them or else they are quickly buried in situ by vegetation which in turn is fertilised by the bones and the droppings of various creatures coming to feed on the bones. Worms can bury large objects given time. One South American bird has been studied and shown to be able to smell buried dead meat from many miles away – surely GOD must have designed that bird's olfactory system and its ability to detect a few molecules of odour over several miles?

This procession of birds, beasts, insects totally consuming a dead beast or a human does seem to prove the creatures were created with limited lifespans compared to humans and plenty of their dead carcasses would need recycling on a daily basis. On the other hand the vulture's lack of head and neck feathers would also be a boon if the birds were eternal and vegetarian and had to breach thick rind to get inside the gigantic calabashes, gourds, melons, pumpkins, squash and similar fruits of the ante-diluvian world!

Perhaps vultures should be labelled vegetarian! EUREKA?

Chapter One
Majerus's Peppered Moth woozlenonsense.

Michael Majerus, a devout Evoquack living in silken luxury at Cambridge University and naturally parroting the very phrases trained parrots can best parrot , was so determined to worship Darwin and Satan that ignoring all the logical arguments put forward about why the 'Peppered Moth as evidence of evolution was EVONONSENSE' he set about doings some research and experiment in his own backyard which confirmed that, quote: 'The new data, coupled with the weight of previously existing data convincingly show that 'industrial melanism in the peppered moth is still one of the clearest and most easily understood examples of Darwinian evolution in action.' He actually hung all his belief in EVONONSENSE on a trivial little moth that quite happily lives its little life exactly as GOD created it!

In other words, due to the Industrial Revolution that swept over Britain between 1800 and 1940 there was so much soot in the air that houses, factories, rocks and tree trunks in many areas became black, so black that in fact a large part of central Britain gained the name 'The Black Country'. In my own area most older people remember how all old buildings such as churches and chapels, mills and factories, town halls and railway structures were a uniform black from all the soot that had stained them over the 100-150 years of their existence. At this very moment I can turn and look at a railway viaduct constructed at great expense in 1875 to carry the railway over a deep valley and stream and in many places it is still black but large parts of it have been washed cleaned by the rain to expose the warm sandstone colour that makes waking and seeing it lit by the rising sun a delight for most of the spring, summer and autumn.

Majerus's research proves British EVONONSENSE as much as black Africans believe eating the skin of spotted or totally white babies born to black parents will bring good luck. This is clear proof of what racist whites and Muslims have been proclaiming about blacks for millennia.

To prove this is fact an experiment can be carried out by Majerus's disciples in which they go to Africa, kidnap or buy a good selection of these spotted or white children and pin them to trees during the day and in the morning check what predation has occurred from local species of black peoples. If security cameras can be installed to monitor the predators and allow them to be identified and traced it should be possible to carry out long term research on them and their children to confirm if in fact ingesting the skin of albino children does cause albinoism in black peoples. We now know that the Muslim ISIS liking for eating human brain is giving them all a variety of Mad Cow Disease and it may be that eating albino children causes changes in black African brains.

But as for eating albinos causing the eater to become albino I am sure it would not and I am just as sure that there is some bias or weighting in Majerus's work that needs further examination and experiment – and there is!

Deflating Majerus' Peppered Moth Woozle might be as simple as realising that birds and bats are not stupid but do learn, and just as blue tits learn that they can flip upside down to land on the seed balls I put out for them but the robins and sparrows cannot learn or perform this trick, so we have to accept that having learned that Majerus was putting moths on the trees in his garden in the green countryside the local birds dutifully hunted them out. I watch bluetits approach my seed holders, land, grab a few and fly off all in perhaps two seconds. This speed of feeding may seriously skew Majerus's research. They ate crumbs from my hand just as fast.

Majerus kept the moths, presumably purchased, indoors and each night put one per tree in a net fixed round a branch. Just before dawn the nets were removed and the position of each moth was recorded. Any moth not there in midmorning was assumed to have been eaten! It beggars belief that all moths will stay in a place they are deliberately enclosed in as logically any creature will move to what it feels is the safest roosting spot and may move out of the sun or wind? And again, the local birds are not as stupid as the average Evolutionist and will note that when a human comes to a tree, fixes a small net around it and puts a moth in it, it signals that shortly a juicy moth will be an easy snack! I put seed on my bird table and within a minute the jackdaws on the old mill are swarming in to gorge themselves – surely Majerus would have known that's how birds operate?

Also of course the experiment was seriously flawed in both its operation and observation: quote: 'Moths absent from resting positions 4 h after sunrise were presumed eaten by predators as they rarely fly away during daylight unless greatly disturbed. Of those that disappeared, approximately 26 per cent were seen being eaten by birds via binocular observations.'

A good number of birds are active well before dawn and large spiders are active all night and naturally any of these creatures will take any moth it comes across.

The speed and darkness of the typical bat is such that I know for a fact many people cannot see them in flight yet I have them whizzing about my yard and up close to the windows most warm nights. When I worked alongside a river I could spot bats whizzing along the bankside vegetation but no matter how I spotted for him a workmate could never see them – so did those bats exist or not?

Dragonflies are also opportunist feeders and will catch, de-wing and eat moths on the wings in a couple of seconds.

Blue tits themselves are quite adept at landing on trees and brickwork – and the corners of my windows! - to take spiders and can land, search, capture and fly away again in just a few seconds. They like the sparrows, wrens, robins and longtailed tits are selective eaters and don't visit bird tables and feeders for weeks on end so again only if moths lost to these birds can be tallied would Majerus's work have any validity.

Again we have to understand that many creatures have sense far better than our human ones and it may be that Majerus's speckled moths were being eaten because they were easier to find perhaps due to their domestication and diet giving them different scents or pheromones than wild moths.

Homosexuals, lesbians and pedophiles follow their noses just as animals, birds, reptiles and insects do too.

To claim the rise and fall of the peppered moth population is proof of EVONONSENSE is as idiotic as claiming eating albino children's skins will make black Africans white. The bones of many stewed missionaries and murdered children cry out the falseness of this belief.

The final nail in Majerus's EVONONSENSE coffin is this quote: "Professor Majerus compiled enough visual sightings of birds eating peppered moths to show that, in rural Cambridgeshire, the black form was significantly more likely to be eaten than the peppered."

That is the equivalent of turning some black people loose in any redneck town and they would be quickly predated upon as Youtube shows, or turning some white missionaries loose in darkest Africa and hear the natives whoop: 'Hotpot tonight!'

Obviously Majerus was unaware of Darwin's research on worms that proved the Earth was created 6,000 years ago and The Flood was real!

Majerus proved nothing other than he worshipped Satan but his claims about moths is now enrolled in the woozle hall of fame and constantly paraded by Evoquacks .

On the other hand – what if Majerus had crippled his moths before he fixed them to the trees?

Chapter Twenty Two

The 65million year ago meteorite that killed TRex woozlenonsense.

Evolutionists parrot the mantra that "68 million years ago Earth was doing well, many shapes of monkeys were evolving from all sorts of fish and then giving birth to attractive humanoids, though some monkeys got fed up of being dry all the time and decided it would be better to go swim out to sea and turn into whales! Proof of this woozle seems to be the macaques that look so cute when they laze and soak in the Japanese warm springs and especially the ones that have learned to hold their breath under water when retrieving the seeds thrown into the water by the dastardly Japanese.

Study a book or a website describing them and showing how incredibly monkey like their skulls and teeth are! You would not want to meet a human with a skull like that when you stepped into your local Dominos Pizza would you! Going to a select new spectacle frames would not be easy either. Claiming macaques became whales is a monstrosity of a woozle!

But anyway these macaques quite enjoyed life and teasing the TRexes and pterodactyls and generally lived quite well. Stephen Hawking had not yet reared his ugly head so all the creatures had no idea they were supposed to be vicious carnivores and just loved munching on grass and fruit.

One day about lunchtime a lot of mammoths were happily gobbling grass when out of nowhere a bright light and thunderclap sounded and they all died – just like that! A massive meteorite had just smashed into Earth and sent a shock wave racing round the planet shocking may creatures to death.

It was accompanied by a blast of heat that wasn't hot enough to burn anything according to the scientists who haven't been on Youtube and seen how instantly many things can burn up. The heatwave was closely followed by vast amounts of absolutely freezing air that poured through the hole the fictional meteorite had torn in the real canopy the Evoquacks say never existed. This air instantly deepfreezed the mammoths in thick sludge. Many did not have time to finish chewing the grass they had just bitten off.

Passing over the mammoths the blast of fictional destruction fictionally blasted countless billions of trees and creatures to death and made the fictional tectonic plates start fictionally slipping and sliding over each other like they is purported to do to this days. Tectonics is a super woozle!

This fictional meteorite was imagined as so big it must have made a big dent in the Earth and made a lot of it fly off and make itself into a beautiful cold sphere up in the sky. The monkeybrained people who worship Satan call this object the Moon. GOD created the moon and woozling otherwise is another of Satan's lies?

The woozle continues with the meteorite making the moon really clever as it just happens to have ended up in the right place and have just the right orbit to be pretty good for marking and ruling the night.

I'm not sure if an Evoquack takes 'ruling' as meaning measuring or ordering or keeping watch – but anyway here is the verse for you to interpret as you wish: 'And GOD made two great lights; the greater light to rule the day, and the lesser light to rule the night: he made the stars also.' This is the record from Genesis 1:16 and is much more precise that the Evo's 68MYA nonsense.

Anyway if we pursue the Eononsense for a bit down to the Yucatan Peninsular and take a aeroplane ride to a great height it is possible to see what appears to be a great circular depression half in the sea and half on land. The tale is that this big meteorite slams into Earth and makes the crust wobble like a soft jelly and eventually stop wobbling and levels itself off.

This sounds pretty reasonable to the typical macaque but when all other large 'meteorite impact craters' are viewed and they all look the same it behooves one to do a little lateral thing and ask ' Er, excuse me, but shouldn't craters all show glancing impacts with elongated shapes?'

And when we think of all the videos we have seen of large vehicles pushing soil about we have to ask 'Shouldn't a meteorite impact site' have a high rim or debris pile at one side and hardly any at the other where the meteoritie was likely to have come zooming in?' And where is the debris? And surely under the rim piles there should be masses of dead trees and TRexes?

Studying the crater walls and the many layers of sediment and the perfect 32degrees scree slopes to the nice smooth bottom it behooves us to think that perhaps what these craters really is is big vents up through which came all the water at the start of The Flood. Up and up at vast speed and pressure came the water, lifting the crust and topsoil and crumbling it to pieces and spreading it far and wide as the massive flow continued day after day. When the Flood waters stopped rising they sloshes about for about a year which is ample time for the craters to fill quite a bit. When the waters started draining off the Earth so Noah could get out of the Ark the craters also dried off and the last muds slurped down and gave the really nice smooth bowl shapes so typical of craters. Many craters filled level and dried out to give a nice low bowl of good grazing.

Logically any of the bigger craters will hold very big reserves of gas, oil and coal from all the vegetation washed into them during the entire Flood.

Chapter Twenty Three

Hell's Bells – Hell Creek Woozle!

Wiki's Hell Creek Formation website is the most ludicrous compilation of EVONONSENSE since Hawking tapped out his original ABHOT!

Hell Creek is a large area of eroded and almost barren land mostly in Montana extending into Wyoming and North and South Dakota. Photos clearly show it to be a perfect example of drain down terrain left in Phases 7,8 and 9 of The Flood 4,350 years ago. Various sections of the area naturally display a different mix and arrangement of sediments and creatures to other sections and so the Evoquacks who root about in the formations have given themselves licence to conjure up new names for beds and streaks of sediments they fondly imagine is old, older, very old or very much older than the layer below it with the result that my chapter on Cenes could be doubled if I felt like doing so – but do I really need to list idiotic Trekkie terms like 'Campanian', 'Danian', 'Puercan' which are all claimed to be parts of the Cretinous era but really are just different patches of sediments laid down in different Phases of the Flood? Despite the fantasised great age of the formation many fossil creatures are found very close to the surface. Also found are extinct and extant trees and plants all juxtaposed by the flood waters and the drain down.

A prominent feature of the area is the 'grey layer' which is the layer of clay material with a measurable amount of the rare mineral iridium that can be found all round the Earth. Scientists know volcanoes discharge iridium and it is also found in meteorites. Dinosaurs are found below, in and above the grey layer. Evoquacks claim the iridium came from meteorite impacts that raised dust clouds that blocked the sun causing Earth to go into an ice age that killed most life except for crocodiles and sea turtles. As the Bible states GOD shook Earth and a hard shake would shatter the crust and release vast quantities of hot water with dissolved iridium the real sequence of events is not 68 MYA meteorite-dust with iridium-blocked sunlight-dinosaur extinction but is GOD's impact-iridum in water-drowned dinosaurs-sediments - all in 371 days about 4,350 years ago.

At the end of The Flood the Hells Creek area had some large lakes and many drain off rivers and streams that helped great masses of soft sediments flow away. Some sediment filled lakes remained and flushed green with algae and pondweeds. Eventually these lakes dried out to be marshes and bogs with typical marsh and bog fauna and flora but a preponderence of angiosperms which are seed bearing plants with magnolias being a typical example. Some of the lakes dried out to make peat and thin coals. Magnolias look lovely as their flowers get ready to open, and spectacular when open, then the seeds develop and are supposedly very nutritious with a lot of fat content! Lucky Adam and Eve!

However the whole Hell's Creek area has a mass of eroded gully and canyon sides and sediment beds in which can be found a truly amazing variety of creatures that all went extinct in the horrific conditions of the Flood. The list of creatures covers all supposed ages and kinds though they all existed in the 1600 years between Creation and Flood. They are mingled in great confusion with many showing that in death they were torn or smashed to pieces. TRex lies beside plesiosaur beside ammonite beside fish beside turtles beside frogs – I think Noah was glad he didn't have to make room for all such kinds on the Ark!

Triceratops are particularly common but rarely is the great skull found with the skeleton – why can this be when the creatures were massive and if they had merely died from hunger and cold in a dark world where meteorite raised dust coated all the grass? Documentaries from drought-stricken lands show large beasts just lie down to die and their corpses are pecked apart by various scavengers – but what scavenger could move a triceratops skull a long way from the skeleton? And how could masses of great bones and skulls be piled up as they often are in the dinosaur graveyards? A meteorite impact could not do it but both the pre-Flood Nephilim and the Flood could create bone heaps.

Mary Schweitzer found a TRex and imagines it lived 68 MYA but she cannot explain why the whole area stinks of rotting meat after such a time when anyone who has handled mummified bodies of humans or animals knows that dried bodied don't have much smell...she is such a good little Evoquack!

There is never a mention of The Great Karoo Bone Beds stinking of rotting meat...not the other bone beds...Mary knows her TRex died 4,350 years ago though she will not admit it.

Evoquacks from the universities of Berkeley, North Carolina, North Dakota, Washington and Montana all contributed bits of Planet of Apes fantasies to whatever fossils and sediments they found during field trips over the whole area of Hell's Creek.

Museums of America also send Evoquacks to Hell's Creek. One of these is Antoine Bercovici a fellow in the Department of Paleobiology at the National Museum of Natural History who equals Hawking in sheer stupidity by stating 'the only large animals that lived during the Late Cretaceous were dinosaurs'. That will be news to GOD and Jesus! So there were no elephants or mammoths around 65 million year ago when Planet of Apes was being filmed?

To his credit one Evoquack: Kirk Johnson, believes that The Flood wiped out 80% of all the plants on Earth! What a delightful delicious beautiful variety of plants Adam and Eve would see as they ambled around The Garden in their brief innocence!

To clarify Hell's Creek: It is merely an area that was scoured by rushing rivers during the early phases of the flood. Most of its creatures and plants died horrific deaths and were buried in sediments containing volcanic and piezo-electrical iridium during Flood Phases 2,3,4,5. Phases 7,8,9 redistributed many of the remains to confuse the Evoquacks.

Hell's Creek seems a natural magnet for the Evoquacks who worship Satan?

Chapter Twenty Four

The 'Strange Prehistoric Creatures That Still Exist ' Woozle

The internet is full of websites compiled by simpletons with pictures, information and factoids about all manner of extant creatures that supposedly became extinct or evolved over millions of years. The idiotic claims of these websites are truly illustrative of how today any kind of woozle is propagated.

Here is just one such website's claims.

The Aardvark is so unique it is the only one of its kind and evolved over 55 Million Years. Elephant shrew evolved about 66 MYA.

Bees thought to have evolved 100 MYA about the time flowers appeared.

Egg laying Echidna evolved out of Platypusses about 50 MYA and both came from fish.

Solenodon. The solenodon diverged from all other living mammals about 78 MYA and is only found on Hispaniola aka Dominican Republic and Haiti.

Goblin shark has been around for 125 MY without any evolving.

Giant Chinese Salamander goes back 170 MY.
Sturgeon have been around for 200 MY.
Tadpole Shrimp show no change over 300 MY.
Cockroaches 360 MY.
Frilled Shark 95 MYA but Evoquacks think maybe 300 MYA as fossils are in deep Flood sediments.
Freshwater Stingray 100 MYA.
Nautilus 500 MYA and survived several mass extinctions!
Sea Turtles 200 MY.
Hag Fish. 360 MY.
Sea Sponge. Oldest found in rock dated 750 MY. Creation and Flood all occurred recently.
Horseshoe Crabs 450 MYA.
Ghost Shark 450 MYA. Thought to be slowest evolving fish around.
Ceolocanth 400 MYA.
Jellyfish 700 MYA.

Obviously new species will be found occasionally simply because the oceans are so extensive and deep and unexplored. Some jungles are still unexplored. Many wild creatures have perfect camouflage colouring and textures and can remain still while humans pass closely by.

Chapter Twenty Five
The Four Billion Years Old Comets woozle.

Comets is pretty things that creep slowly across the sky trailing a long tail that looks lovely in the dawn or night sky. As they orbit the sun they lose a little of their mass until eventually they just fade away. Some may be sucked into the gravitational field of the planets and end their days in a brief splash. The trails of comets are so fine that it is thought one is barely a spoonful of particles. They is giant fluffy balls of snow, ice and dust and were formed from water jetted out at terrific escape velocity when GOD shook Earth to crack its crust to start The Flood 4,350 years ago.

Some Evoquacks have actually stated that comets brought all the water to Earth's dry dust ball billions of years ago! They get more stupid with every passing day! A rough estimate is about 2.5 million cubic miles of water on Earth – just how many fluffy ice comets would be needed to hit Earth to leave that much water! Does no-one teach university students how to use calculators and calculate volumes? Is being cretinous a prerequisite for seeking a PhD in infantile Planet of Apes woozle?

In 1997 a new comet was spotted independently by Alan Hale and Thomas Bopp; it was named for them as Hale-Bopp. I used to see it in the dawn sky as it slowly crept across the sky about 120 million miles away. As the sun is only 93 million the comet was lit by sunlight that warmed it up to cause a constant stream of particles trail after it to make the two tails seen through telescopes. Scientists estimated the comet as being about 50 miles across but composed of nothing but ice, snow and dust! That is remarkable considering the Evoquacks claims of BigBang being a billion degrees hot...

When the Americans smashed a spacecraft into comet Tempel1 it revealed the comet was like old snow that has been partly melted and is crumbly and holed. NASA's website about that comet has great amounts of Woozled EVONONSENSE words such as: Tempel 1 'might have', 'it may', 'probably', 'it is possible', 'could have' and 'may have'. This woozle added to the disappearance of comets as they totally melt away is evidence that they have no idea of comet origins but stick to Planet of Apes fairy tales instead of admitting that the only place known to have expelled masses of water and dust in the universe is Earth during the start of The Flood – just as it does constantly worldwide wherever volcanoes are active.

NASA found Tempel 1 to be very black – like the black grease and carbon that builds up on a barbecue but containing silica dust like fine sand powder and the greasy stuff. They are mystified of the nature of the black goo but a Bible believer can know that the water and powdered sand are evidence of the power of GOD's shaking of Earth and the goo is most likely the remains of the swarming creatures of the sea – fish, plankton or bigger creatures.

Asteroids are the cousins of comets as they are rock blasted out of Earth's crust in the same first moments of The Flood. 20 years after Hale-Bopp disappeared the Luxembergers invested in a scheme to harvest precious metals from asteroids...read the Bling chapter to see how futile this scheme will be.

Countless evoquacks have ridden comets to the Oort Cloud wonderland of riches and the plaudits of other evoquacks as they concoct elaborate fairy tales about the comets somehow being ancient and being spawned out in the Oort Cloud which is like a fantasy place full of bigbang dusts and such – don't ask how comets have lots of water and methane from the superhot superdry bigbang as methane is made from organic matter and organics has to have water!

Here is a selection of quotes from websites including NASA's full of woozle about the Oort fantasy:

'The Oort Cloud is a theorised shell of icy objects...'
'Jan Oort, who first theorised its existence...'
'it is thought to be...'
'the cloud is theorized...'
'the most likely theory...'
'gravity is thought...'
'the cloud is a reservoir of ice left from the bigbang...'

'Oort Cloud is believed to be a thick bubble of icy debris that surrounds our solar system...'
One Evoquacks website states 'FACTS ABOUT THE OORT' then constantly uses theorized, thought, hyopthesized... thus perfectly illustrating what WOOZLE is!

Did you miss Hale-Bopp comet last time round? If you were not born or too young to see it in 1995-97 put the date of its return visit in your diary for 4385 if you believe Jesus is the Son of GOD and was raised from the dead to return to heaven and sit at GOD's right hand.

If you missed it last time and don't believe the Bible you will never see it which will be a multiple shame as it looked quite cute in the morning sky.

I was just looking at photos of Huygens and Hadley high width to depth craters on Mars and Wiki's quote: 'It is supposed that billions of years ago Mars was much warmer and wetter'. That amused me as the universe is only 6,000 years old and the craters were formed by a gigantic fluffy honeycomb snowball asteroids made by water blasting out of Earth at the start of the Flood and expanding in the emptiness of space until they hit Mars like a gigantic custard pie in an old vaudeville comedy routine! When the snowballs hit Mars the impact formed large circular craters and splashed up the rim and the surrounding area. Then the water ran back into the crater making the rills, gullies and scree slopes we see in so many other water formed features on Earth such as the Grand and Bryce Canyons, the Condit Dam draining, the Sphinx moat. The many small low-ratio craters were made by impacts of more solid material from Earth.

The giveaway is the many smaller craters with central peaks!

New craters that appear on planets and Earth are made by snow/ice balls, soil and rock all blasted out of Earth. They definitely are not formed by rocks inbound from the mythical Oort!

Chapter Twenty Six

Sci-fi and the EVONONSENSE Media.

All those sci-fi movies and magazines! So much occult garbage on television! So many websites and books promoting UFOs and aliens! Such floods of 'evolution is true' lies. So many high priests of EVONONSENSE being honoured by royalty and academia! So many queens and kings of the occult being celebrated on chat shows! So many sinners prominent in news room and entertainment! It gets worse each day! It's like a flashback to Sodom and Gomorrah before that firestorm 3,000 years ago!

For 150 years Satan has been using Hollywood and television studios and printed media to foster delusions in the minds of the audience for science fiction, occult and EVONONSENSE movies and television series and as the return of Jesus looms ever closer this outpouring of occult, demonic, Satanic and evil media has increased drastically to the point where some days there will be four or more such programs and films on popular television apart from what is showing at the local cinema or is available in written and illustrated form in books, newspapers and magazines – while today's internet is available practically everywhere and offers everything ever made about Satan and his agenda – except the truth! Jesus said Satan rules the world for the time being and how right He was! Read Matthew

It is this plethora of media that is responsible for so many people around the world being fully convinced they is evolved from monkeys or perhaps the offspring of demon-demon-aliens who came to Earth and had sex with monkeys – there are plenty such garbage scenarios in the cinema archives and video store. The media is full of reports of demon-aliens from UFOs abducting human and animals to obtain samples for experiment in the 'admixture of species' that Jasher wrote about. I will refer to the UFO occupants as demon-aliens as the UFO craft and its occupants are both manifestation of Satan and his demon army.

In August 16 2016, we had video proof of Satanists at work in CERN performing a ritual human sacrifice in front of idol of Shiva – the imaginary destroyer of Hindu devil worship. They claimed it was just fun! Quote about Shiva shows Satanic idiocy: *'Shiva is the god of the yogis, self-controlled and celibate, while at the same time a lover of his spouse (shakti). Lord Shiva is the destroyer of the world,'.* Why has CERN in a Catholic country chosen to erect a pagan idol as its symbol? Are they like the men who attend the secretive Bohemian Grove event in USA who are suspected of performing a real live sacrifice in front of their idol of owl god Moloch. They also claim it is just good fun!

Alfred Russell Wallace was possessed by Satan while in a mesmerists spiritualists meeting and became a medium able to control others before progressing to trying to contact his dead mother - GOD warns against speaking to demons or attempting to contact the dead as Paul warns of how Wallace and so many others were easily fooled by the utterances of demons: 1 Timothy 4:1 "Now the Spirit expressly says that in later times some will depart from the faith by devoting themselves to deceitful spirits and teachings of demons." Wallace gave Darwin Satan's lies about humans evolving from monkeys and Darwin was led by Satan to write and publish his nonsensical tome compelled by his own refusal to accept the Bible truths.

But the leader of the Satanists pushing what can only be labelled 'EVONONSENSE' in living memory of most is Stephen Hawking, aided latterly by Stephen Gould, Richard Dawkins and David Attenborough. This clique absorbed so much media EVONONSENSE in their growing years that they are now Satan's staunchest promoters ever eager to call GOD a liar and foster belief in EVONONSENSE. Just as Satan has legions of demon-aliens at his command Hawk, Gould, Dawk and Attie also have myriads of followers eager to promote their lies all the way to their eternal destruction on Judgement Day! Over 12 million Britons watch Attenbrough's nature documentaries in which he always says every creature evolved from another over tens and hundreds of millions of years. Gould's seminars on youtube videos gather millions of viewers as do Dawkin's Christophobic rants in front of crowds as remarkably rapturous as the crowds Hitler hectored!

I'm sure one of the first Attenborough documentaries I saw was him filming scenes in darkest Africa where villagers danced themselves to a trance and fall down believing themselves possessed by spirits called Dambulowayo – did one of those spirits slip into Attenborough and make him one of Satan's most prominent and persuasive slaves?

Stephen Jay Gould – actually he's a 'professor' but I refuse to honour an idiot with that debased title – displays the nadir of evolution in brainpower when he quite seriously writes that 'bones date rocks!' He really does believe the deeper a bone fossil is buried the older it is! He has obviously never watched a flashflood or tsunami on the news! It gets worse each day! How Satan loves Gould and all who think like him.

The 'Lucy' fossils make a fool of Gould and his deeper-older dating method as in fact the bones were found buried quite shallow in what is obviously perfect topography dating back to the middle and last phases of the Flood of 4,350 years ago. Go look at the Hadar Formations idiotically dated at 3.2 million years old! It bears quite a resemblance to the Grand Canyon with its serpentine courses, plateaus and scree slopes.

Hawking's science fiction novello, A Brief History of Time, ABHOT, was originally planned by him to be stuffed full of fancy equations and calculations all intended to show that Earth is perhaps 4.5 billion years old and the universe considerably older and vastly more enormous than anyone can imagine or describe. Luckily his publishers told him to cut most of that crap out and make the book suitable for the mass of couchpotatoes who had lapped up all the 1950's, 60's and 70's movies pushing demon-aliens, extraterrestrials and monkeymen such as 1949's King of the Rocket Men, 1966's StarTrek, 1968's 2001 Space Odyssey; the 1968, 70, 71, 72 and 73 Planet of Apes series of monkey fests; 1977's Star Wars and its later spinoffs; Men in Black; The Dedi, E.T, ad nauseum. The viewers of such movies were so bewitched by them that some grown men could be parroting 'May the force be with you!' from Space Odyssey while the toy makers made fortunes from offering related toys and books with which to brainwash children into swallowing the false premises of the movies. Sci-Fi is so addictive especially to men who perhaps love how the heroes of sci-fi have gorgeous female assistants in tight uniforms that they are willing to spend large amounts of time and money on living out the fantasy themes. These rabid sci-fi fans pushed Hawking's book onto the best seller chart just as a generation later they would push the demonism filled Harry Potter books into top rankings.

Walt Disney's Bedknobs and Broomsticks film has an image of Baal prominently lit and children chanting spells for Ashtoreth! What goes on in that secretive Club 33 in Disneyland?

It cannot be known how many of Hawking's A Brief History Of Time's were sold to the oldies - the millennials - whose first exposure to demon-aliens was via the old Flash Gordon and the Mud Men films of Saturday matinees that had Flash travelling in a rocket inspired by Nazi V2's; or the almost as old 1951 film, The Day The Earth Stood Still. This film encapsulated all the themes of all the later UFO/alien movies in having a UFO made of impenetrable metal that finally succumbs to some common Earth matter like pencil lead, an interior with large screen and banks of flashing lights, armament of a deadly ray gun, a crew of a robot and one or more unsmiling humanoids. The 1951 film is a good example of the genre highlighting Satan's modus operandi. It featured the handsome Michael Rennie playing the frozen faced alien Klaatu aided by Gort, a lumbering tincan robot with a deathray. Filmmakers seem to have difficulty naming their characters but believe lots of K's and double vowels is suitably extraterrestrial. As with all these UFO films the humanoid Klaatu explains that he has been sent from a distant galaxy to warn Earthlings that messing about with atom bombs is worrying his 'advanced civilisation' who have been closely watching earthlings evolve from monkeys and now are capable of splitting the atom and destroying themselves back to, er, Hawking's barren red hot dusty planet?

This message is actually almost quite true but the 'advanced civilisation' is actually double with GOD, Jesus and the angels on one side battling Satan and his demon-aliens on the other. Both sides have regularly sent messengers to speak to us humans but GOD sends plain angels while Satan sends crossdressed ones pretending to be Jesus's long dead mother Mary, or awesome winged ones claiming to be Gabriel, or grey androgynes in UFOs. Demoniac David Icke says some come as reptiles able to enter humans including The Queen, Obama and other leaders during which the human morphs back and forth between reptilian and human. The Christmas song 'While shepherds watched their flocks by night' explaining about GOD's visiting angels perfectly. Then we remember that Christmas is actually the demonic Saturnalia...

The media constantly shows what are claimed to be UFOs, aliens or occult happenings and I like to amuse myself during coffee beaks glancing at videos and news reports on Youtube. Most of the imagery is too vague or easily dismissed as idiotic fooling about or the efforts of photoshoppers with the equipment and time to alter clips of video to make them more believable but here again they make elementary mistakes in sizes, lighting, locations, that debunk their work. However there are lots of videos which taken with what we know about both angels and demons does highlight the fact that we are surrounded by these invisible beings with angels performing good deeds and demons mostly evil ones unless they follow Satan and seem to do a good deed. This might be why Jesus said the humans who are chosen to live with him will judge 'angels' – are those angels the ones that fell with Satan or after Satan is released from the pit for a last crusade as mentioned in Revelation 20 and as portrayed in the pope's new throne? 20:2 'And he laid hold on the dragon, that old serpent, which is the Devil, and Satan, and bound him a thousand years..3 And cast him into the bottomless pit, and shut him up, and set a seal upon him, that he should deceive the nations no more, till the thousand years should be fulfilled: and after that he must be loosed a little season...7 And when the thousand years are expired, Satan shall be loosed out of his prison,'

Hawking's ABHOT found a ready audience of fanatical believers who quickly boosted it into the best seller charts and elevated Hawking to position of chief BigBanger whose word was absolute truth. I cannot find a recent 2017 tally of its sales but by 2013 it had passed 10 million. Just how many of those 10million purchasers read and agreed with Hawking is not known, nor is the total of further readers to whom the books were passed on or loaned it from libraries but can I assume maybe 30-40 million people have waded into ABHOT? Could it be as high as 100 million? What percentage of ABHOT's readers could understand all Hawking's fantasies is not known either, but I do have friends who refuse to watch Sheldon and friends in BigBang Theory because they think it is full of incomprehensible science! Sheldon admitted 'String theory' was just nonsense. If only Hawk would admit his bigbang is nonsensical!

Thanks to Hawk'n'Dawk the vast majority of people believe the bigbang and it's occurring 8/9/11/14/15/18/22/13.7 0r the now generally accepted 13.8 billion years ago but I haven't the time or patience to sit through Space Odyssey and the various Apes movies to learn what BigBang dates were pushed onto the masses by the makers of these films. I like watching bits of 2001 Odyssey occasionally as the classical music and the nice slow action is very soothing even though the story is ridiculous EVONONSENSE.

I am quite sure that all those who believe in Hawk's bigbang cannot quite grasp that his FAST, HOT and DRY BigBang fantasy is directly contrary to GOD's SLOW, COOL and WET Creation. The contradiction between the two is a gulf like as what only Christians can grasp. It is also a typical Satan lie. And once again it is proof of Jesus's repeating of Isaiah's words at Matthew 13:14 'In them the prophecy of Isaiah is fulfilled: You will be ever hearing but never understanding; you will be ever seeing but never perceiving.' No Evoquack will listen to proof of Creation. Laurence Krauss gets almost apoplectic as he howls that EVONONSENSE is so perfectly proven that there is no need to even mention Creationism or GOD! He must be home to as many demons as the man in the tomb whose many demons

Jesus sent into the pigs!

In about 56 AD Paul wrote First Corinthians and 2:14 'The person without the Spirit does not accept the things that come from the Spirit of GOD but considers them foolishness,and cannot understand them because they are discerned only through the Spirit. 15 The person with the Spirit makes judgments about all things, but such a person is not subject to merely human judgments,' This is why I can state that worms prove BigBangEVONONSENSE is nonsense and judge anyone supporting it to be blind, deaf and a slave to Satan.

Hawking's rise to chief EVONONSENSER and 'the most intelligent man in the world' began back in the 1950's but by 2010 he'd shown himself to be a true monkey-brain by stating: 'Because there is a law such as gravity, the universe can and will create itself from nothing," He also said. "Spontaneous creation is the reason there is something rather than nothing, why the universe exists, why we exist.' Two hundred and forty two years earlier, in 1768, an Italian priest and scientist Lazzaro SpAllahnzani, proved experimentally that heat killed bacteria, and that they do not re-appear if the product is hermetically sealed. Louis Pasteur repeated this scientific fact in 1862 and then stated: 'Never will the doctrine of spontaneous generation recover from the mortal blow of this simple experiment. There is no known circumstance in which it can be confirmed that microscopic beings came into the world without germs, without parents similar to themselves'. Thus Hawking's saying spontaneous creation is possible in 2010 shows him to be a sloppy, lying, deceitful, hypocrite – the offspring of the serpent – who pushes his silly sci-fi and belief in fairy tale woozles. Or maybe the paralysing demon in him has a lying demon accomplice? Dawkins and Attenborough also choose to make fools of themselves over spontaneous creation – and denying SpAllahnzani and Pasteur's sealed flasks and the evidence of worms prove just how stupid these three are!   EUREKA!

Worms plus 6,000 years plus Flood equals Young Earth.

Hawking, Dawkins, Attenborough, Strauss, Nye, Darwin, Haldane, Thomas Huxley, Sagan, Wallace and new kid Robert Carrier – are just a few of the multitude of monkeys who have misled billions with their sci-fi EVONONSENSE as they all parrot the same woozle idiocies. Look online for a more complete list of famous Evoquacks and read the brief blurb about their silly ideas. It makes depressing reading as you try to follow them down their particular wormholes into the giant WormhAllah.

Satan is no doubt delighted to see that second on the list is Dawkins The Smirker. He is followed by legions more devilish acolytes such as: SJ Gould and his 'Punctured Equilibrium Theory' which says that once a creature evolved to a nice place it stopped evolving – this being proven by Ceolocanth's 300 million year stasis. Gould is now stasised in his grave until making a brief appearance before Jesus on Judgement Day as he only managed 61 years of life so I cannot cross swords with him and ask where the 300 million years of nickel rich space dust is.

Dead-ended by Gould's dying I traipsed after Geerat J Vermiej who being blind from age 3 can perhaps be forgiven for dreaming up his 'Escalation-hypothesis'. This basically seems to mean that small and simple shellfish is more likely to have survived the billennia than the bigger molluscs that had to lug a heavy shell about though shellfish on the fishmonger's slab and the masses of closed shells on the highest mountains might disagree. Elephants cannot scale cliffs the way mice can but curiously efelumps live vastly longer than mice! There are some giant snails around today too! I'm sure I once saw mention of a sci-fi film involving giant snails? My fossil sample has several trilobites in it indicating that before The Flood the seafloor was covered in countless trillions perhaps even googles of the little scavengers. They were one of GOD's first creatures and given the task of keeping the new sea beds clean by recycling all the detritus filtering down from the new vegetation and the soon to be created birds and beast. Didn't they have any enemies, Geerat?

Intrigued by his name I next slid down a fetid hole after Theodosius Dobzhansky and found him famous for his science-fiction essay "Nothing in Biology Makes Sense Except in the Light of Evolution' influenced by the paleontologist/priest Pierre Teilhard de Chardin's woozles. Dobzhansky thus blocked his brain against anything not in the pondslime-to-humans comedy script. My perfectly complete trilobites would have easily sorted his thoughts out but he and de Chardin have both entered stasis until Judgement Day. It is a chilling thought to ponder who their demons moved onto when these human hosts breathed their last!

Glancing at de Chardin's biography I found he was a Jesuit Catholic and famous for calling GOD a liar by naming monkey bones 'Peking Man' which inspired him to conjure up 'The Omega Point' which encapsulates 'The Noosphere'. This is like as what means our physical world will one distant day morph into just a plane of spiritual existence in a sort of reverse BigBang – a BigShrink, BigDissolve, BigFadeout? That is surely the ultimate occult science fiction? I'm sure GOD will ensure his 'noosphere' will be null and void as Jesus definitely said idolaters and liars won't enter his kingdom.

De Chardin was only cribbing EVONONSENSE from an earlier 'Noosphericist' - Alfred Wallace, 1823-1913, who displayed his utter stupidity by evolving from believing he had evolved from a simple forminafera lifeform via fish and monkeys into a higher evolved being that had evolved an eternal soul that had evolved from nothing with the miraculous aid of Hawking's evolving gravity! He let Satan seduce him with the idea that though his dead son had rotted in his grave as all the bacteria in his gut and on his flesh had gone into overdrive a spark of life had floated off into the cosmic 'noosphere!' He went to seances and played with ouija boards and received messages from demon-aliens that convinced him something lived on after death in the 'noosphere.' One day a demon filled his head with Satan's 'Theory of Evolution' which he shared with Darwin only to have Darwin plagiarise the theory and receive the adulation of all the foolish. Wallace was the rare-avis Evoquack who worshipped four of Satan's lies: EVONONSENSE, Parakinesis, Spirit Guides and LifeAfterDeath. Most sci-fi does involve aliens though?

Do Hawk'n'Dawk'n'Strauss and all the others also mess with ouija boards? Satan really messed with Wallace but he was a willing slave who had only to read a Bible to learn how GOD's orderly Creation rubbishes EVONONSENSE – and how messing with spirits is the way to eternal death on Judgement Day. Satan has been fooling people ever since getting upset with GOD being proud of Adam and Eve. The prattlings and lies of Hawk'n'Dawk'n'Hitchens'n'Gervais'n'Strass'n'Nye'n'Carrier do seem like nonsense.

To rubbish EVONSENSE is simple. Gather a bucket of fertile soil, put samples under a microscope to check on the multitudes of life wriggling through it and growing on it, sterilise it with just 250 degree heat, bottle it with a perfect seal and watch it for a thousand years – will any creatures or fungi spark into life? Would it even if you stood the bottle outside during a million year's of thunder and lightning? Or if connected up to a sci-fi Dr Frankenstein spark generator?

NEVER!

Most warm evenings between sunset and final darkness I can see bats flitting about as they use their echo location facility to catch flies around the trees and over the little stream over the carpark fence. How on earth was the first bat supposed to have stepped off a tree branch and discovered it could locate and chase insects by chirping and listening to returned chirps -while zooming around doing incredible aerobatics? And in the dark of night! How many gigabytes of information does a small bat brain process in an hour's hunting? Would anyone be foolish enough to dare provide a chart of the circuitry of every part of the bat's sonar system and how it is connected to its flight and navigation system? NEVER! But lots of sci-fi stories have pilotless Ufos zooming around?

Most Evoquacks faithfully emulate Satan and lie about EVONONSENSE by claiming creatures evolve from one kind to another by macroevolution when they know there has never been an instance of MACROEVOLUTION except in the fantasies of sci-fi but we can all see examples of MICROEVOLUTION at any dog show!

Evoquacks idiotically claim a wolf/tapir hybrid thing MACROEVOLVED into a whale though GOD ensured both wolf and tapir DNA will not blend nor can anything be altered at random except by chemicals a la thalidomide or mercury. Careful breeding might make prettier wolves or slimmer tapirs but that is MICROEVOLUTION. The primary Evoquack, Darwin, wrote in the first edition of his EVONONSENSE book that two bears walked into the sea and became whales! He must have had a moment of sanity, or maybe his demon was visiting someone else, as he deleted that sentence in the second edition. Polar bears have been filmed swimming miles across the sea but I'm sure none of them ever thought of becoming a big whale.

However Satan and his fallen angels had been busy mixing the DNA of various creatures just before The Flood but his willing slaves ensured the Council of Nicea kept that information out of reach of most although Jude mentioned Enoch's account of the hybrids and Jasher detailed it. There are far too many reports of UFOs abducting humans and creatures of taking samples of blood and flesh to be dismissed although as UFOs and aliens are spirits they often don't leave tangible traces. If I had a gold coin for every sci-fi story with UFOs taking samples...

Microevolution means the way farmers and nurserymen produce bigger and brighter coloured plants and animals and other creatures by selecting what they see as the best for their needs. A man trying to breed white animals will choose the whitest he can find to start with as it would be foolish to expect black animals to produce white offspring. A grower trying to grow the biggest flowers or vegetables has to select seed from the best samples of his present crop and hope that the average size of the next crop is bigger than previous. My daughter has just bought a microevolved dog: a Shih Tzu-cross-Yorkshire Terrier and it is a frisky ball of fluff. If she crossed it with either a Shih Tzu or Yorkie she could reasonably expect to get some pups resembling Shih Tzu and some Yorkies but it may take several generations careful selection and use of only Shih Tzu or Yorkshire mates before all the pups look like typical Shih Tzu or Yorkies. Our pets most likely have been microevolved far beyond their natural states as has the fruit, vegetables and food in our kitchens and the lovely flowers in the vase on the windowsill. Jesus said 'the lilies of the field are really lovely' – do they need messing with? Snuffling bulldogs would bark affirmative! A brief glance at dog breeding reveals that today's dogs are susceptible to an enormous number of problems due to inbreeding to emphasize a certain characteristic.

No spontaneous evolution and no amount of crossbreeding can ever make one KIND of creature into another yet that is exactly what Krauss, Hawking, Dawkins and all other Evoquacks claim! Their latest desperate cry is that everything absolutely evolved but in such small steps that no evidence survive! Or in other words every different fossil found is a separate species of existing kinds that is still waiting for one of its members to have a baby that is obviously different and more evolved! Maybe those trillions of trilobites in the Pre-Flood seas were all waiting for Trilobite-Plus to be born?

If only the Evoquacks could provide some Missing Lynx! Something like a fossil of the wolf/tapir thing with nostrils migrating up its face to the top of its hairless ambergris filled head with a mouth filled with balleen rather than sharp fangs and proto-flippers instead of feet! Or Darwin's pair of bears morphing their plump hairy bodies into the slim hydrodynamic one of a whale?

Over the years since Dolly the sheep was cloned and died at an early age there have been constant reports of hybridising experiments and plenty of pictures of odd creatures in the media but as usual many of these turn out to be frauds or mistaken identification of decaying or damaged creatures that can easily be explained when some basic knowledge of how bodies bloat, discolour, lose hair or skin and often lose vulnerable soft tissues quickly. But as usual there is a core of evidence that defies explanation and attempts by the authorities to explain it as mistake on the finder's part. This easily spawns allegations of cover-up that then turns into a storm of accusations between authorities, experts and amateurs. Sci-fi films have been pushing such storylines as the market never gets saturated.

Sci-fi's rich history of mutants, aliens and UFOs has obviously made many people susceptible to false identification and especially since the Space race began 50 plus years ago. The weird and wonderful craft and outfits and the fireballs of the departing and returning craft get thoroughly scrambled in the head of some onlookers and maybe are partially recalled whenever an odd light or movement in the sky is seen for a short while. Poor or exceptional eyesight may make this more likely. Video of early lunar orbits quite often show UFOs but were easily suppressed in the early days until onboard live reporting became the norm though it was soon realised that the imagery transmitted to the public had actually been delayed a short while to allow any scenes showing UFOs or aliens could be edited out. One spacewoman is famous for being shocked at the sight of a UFO although the account was quickly rubbished but if she is a devout Catholic her shock was genuine and understandable as of course Catholics have a rich history of seeing aliens. Now Youtube has many examples of what may be genuine astronaut videos with UFOs in frame though most often they are infuriating amorphous blobs even if they perform superb flight abilities. Astronauts prefer to remain tightlipped and dismissive about the UFOs even after they retire. After all, US and Allied aircrew in World War 2 often reported 'Foo fighters' zooming round the aircraft but after being ridiculed and their reports trashed or not acted on the flyers generally decided it best to keep the happenings to themselves although some persisted in making official reports.

The men-in-black in obsolete cars so frequently reported may be demons as there are many reports of unworldly activity involving these men. Their intention may be keeping the lid on truth as long as possible?

Curiously enough the people who most vociferously protest the truth of UFOs are as sincere as those who loudly promote the Occult and those like Krauss and Dawkins who howl EVONONSENSE is true!

Chapter Twenty Seven
Evononsense epitomised!

I thought I had plumbed the deepest concretions of EVONONSENSE over the last couple of years but I was flabbergasted during final revision of this manuscript to come across what must be the absolute epitome, or nadir, of EVONONSENSE ever put into writing.

It was a chance finding while surfing Wiki about the frequency of flints being found in my part of Britain – Yorkshire – and more specifically GOD's Own Country – West Yorkshire, that I came upon a website: Wiki: Geology of Yorkshire, which for sheer EVONONSENSE cannot be beaten.

It is a compilation of claims about all the supposed cenes, epochs, periods, eras, ages and such that have occurred on this area since 'Earth's formation' as displayed in strata in Yorkshire.

Rather than quote it verbatim I decided to merely list every Woozle word or idea and signify them as such with a 'woozle' but just as I found with compiling the BIGBANG timeline some really were so nonsensical I couldn't refrain from adding facetious but truthful remarks!

The curtain rises:

Yorkshire Geological Fantasy, Act One, Cene One, Take whatever:

Carboniferous -Woozle. Fantasy for settling of organic matter during The Flood 4,350 BC.

Permo-Triassic -Woozle. I had to look this up to as 'cene six, take nine' in chapter four.

Jurassic -Woozle. When Hollywood was making madslasher TRex movies.

Cretaceous -Woozle. It ends with loss of 75% of all life including Mary Schweitzer TRex.

Quaternary -Woozle. Inspired by lingering echoes of Qatermass and The Pit

Palaeozoic -Woozle. Old stones or summat.

Ordovician -Woozle. Ordovicia is in Wales, these rocks came here on holiday and stayed.

Avalonia – Woozle 'a landmass from 30 degrees south of the equator ' Wandering plates.

Baltica -Woozle. Wiki: Baltica dates to NO LESS than 1.8 billion years ago! So precise!

Silurian -Woozle. More Welsh rocks on holiday jaunt deciding Yorkshire is very nice.

Laurentia -Woozle. Marked by rusted hulks of Canadian Pontiacs.

Iapetus Ocean -Woozle. Seems to mean The Atlantic.

Devonian -Woozle. Fossilised scones with cream are local product of the rocky cliffs.

Variscan Orogeny -Woozle. The Evoquacks notion of surfing continental land masses crash.

Supercontinent – Woozle. Imaginary, fantasy, mythical continent too big to live together.

Gondwanaland – Woozle. Means imaginary supercontinent.

Pangea -Woozle. Means mythical supercontinent.

Euramerica -Woozle. Means yet another fantasy supercontinent.

Dolomitic limestone -Woozle. Nice hard limestone first made in Italy.

Dolomite Problem -Woozle. Wiki: 'Dolomite Problem' is no-one knows dolomite's origins!

Zechstein Sea -Woozle. Relative of Guadalupians and Lopingians Evoquacks surmise.

Permian -Woozle. Cene 9, take 14. Epic fable leading to loss of 95% of animals and plants.

Triassic -Woozle. Khamsins and flash floods left pebbles for Robin Hood.

Second Triassic Mass Extinction -Woozle. Cataclysm hypothesis destroys 80% of species.

Rhaetian -Woozle. Means 200MYO Triassic sediments defined by chromostratigraphy.

Chromostratigraphy -Woozle. How Evoquacks give silly names to The Flood sediments.

Carnian -Woozle. EVONONSENSE for a sediment from a stage of The Flood day 5?

Corallion Limestone -Woozle. Areas of coral wreckage and limestone mixed during Flood.

Tabular Hill -Woozle. Hills showing flat top due to stirred water central phase of the Flood.

Kimmeridge Clay -Woozle. Flood debris containing dinosaur fossils and their body oil.

Neogen -Woozle. In this cene the British Isles drifted north and made us seem cold people.

Cleveland Dyke -Woozle. Fantasy volcanic action divided British mainland. Eggshells?

Tertiary -Woozle. Posited as period of blowing hot then cold over 10,000 year cycle.

Coal Measures -Woozle. 290-350 million years old but really from 4,350 years ago Flood.

Magnesian limestone -Woozle Supposed giant inland sea evaporating during Permian Age.

Land uplifted 30MYA -Woozle. Evoquackery as the lift occurred in final days of the Flood.

Devensian –Woozle. Fantasy of ancient ice ages but all occurred post-Flood to 1350 BC.
Pre-Archean Eon – Woozle for a long ago and far away. Perhaps 4.4 bya.
Hadean – Woozle. EVONONSENSE about Earth being blazing hot and watery 4.4bya.
Priscoan Period – Woozle. Brian Harland's woozle for Cloud's Eon aka Pre-Archeon Eon.
Oh my poor head! Such EVONONSENSE! But I know the origin of Dolomite and the answer to the Dolomite Problem!

Surfing the 'Devensian' scene I came across a Catherine Delaney and her work. She is a good little Evoquack who obediently perched before an older, more Woozled Evoquack at Manchester's Metropolitan Woozlestitute and absorbed every myth and fantasy uttered. She says the Devensian was warm based and left hilltops showing evidence of 24 ice ages from 120,000 years ago!

I emailed her pointing out that The Bible and worms makes a fool of her but she hasn't replied.

I went to the seaside resort of Hornsea and found many rocks of granite veined with quartz

Next I logged into a website dealing with UK fossiling and read the usual Jurassic, Triassic, Prussic nonsense about shells and bits of bone that obviously came from creatures torn to shreds or buried during the Flood. Naturally, being me, I posted a few comments to that effect and of course received the expected sneering response stating radiometric dating was irrefutable and I was wrong.

I then looked at fossil shops around the world to see their offerings and find they all have masses of trilobites and sharks teeth but most interestingly of all they all offer fish fossils showing the fish with bulging eyes, arched back and wide open mouths – just the way electro-fishing kills them. Everything is labelled 65-300 MYO.

Do these Evoquacks understand how meteorites and Hawaii's volcano have almost identical iridium levels but in 'the grey layer' around Earth there is barely a trace?

I just watched a few minutes of a documentary on Israel's potash mine and there clear as day is the same solid line of petrified jellyfish as in the White Cliffs of Dover – and again over the Israel area it must amount to billions and billions of jellyfish!

To add to the EVONONSENSE the potash is overlain with limestone – and all dated by Evoquacks to 'a 100 million years ago sea that filled the place and then evaporated off leaving all that potash behind!' Wow! That was some sea – needing some rains – and some hills to wash all that limestone off – and billions of years to accomplish it! And surely such rains for such a long time would have fostered the growth of vast amounts of plants and trees which logically would have drifted down the rivers to make great masses of fossil algae and fish in those mega lakes before they evaporated? If EVONONSENSE was a fact there should be vast layers of coal mixed in all that limestone and lots of dirt in it and under it...there ain't – why?

I just read the author of a Wiki EVONONSENSE article display himself to be a perfect Woozling Evoquack because he says some Carboniferous rock lies 'UNCONFORMABLY' on the Silurian! What he means and cannot understand is that the dastardly rocks must have gone for an unauthorised walk to be on top of what they is supposed to be underneath of and thereby upset all the brainwashing he received at some expensive Woozlestitute! Of course the 'Carbonif' lies on top of the 'Siluries' because it is all sediment that all got stirred, stratified and deposited during the flood 4,350 years ago. If I tried to input a correction pointing out the absurdity of these climbing plates I'm sure it would be very quickly deleted by the Wiki Gestapo.

I'll move on and prove that Creation is True and EVONONSENSE is for fools.

Chapter Twenty Eight

Early Evoquacks and Woozlers.
EVONONSENSE merely proves 2 halfwits evolve quarterwits. 2x.5=.25.

A good example of this negative evolution might be Darwin who readily admitted evolution was unlikely and then gave birth metaphorically speaking to the many lesser luminaries who spouted EVONONSENSE who in turn eventually produced offspring like Mary Schweitzer who quite happily states TRex flesh can survive and smell like rotting flesh for 68 million years!

The DNA of every living thing on Earth makes fools of all Evoquacks and many 'gists'.

Every cell in the human body carries the full instructions for replicating itself – this is how GOD designed it so we can live forever. Just a long soak in a warm soapy bath will cause almost a complete layer of skin to be easily rubbed off our bodies daily so each day invisibly our entire body's cells constantly check its own condition and arrange for its replacement as needed and after about every three months we are entirely new creatures – except for our hair and fingernails which are, as it were, dead from the moment they leave their sheaths. I am inclined to think that the DNA of animals, birds, fish and plants does not have eternal life programmed into it as so many of these non-humans have quite short lives.

Each of our cells carries in it a full code of DNA of such astounding complexity and completeness that only a demon possessed fool would say it all came about from hypersterilised wet stones bumping together billions of years ago. Hawking, Dawkins et al are just such fools. Hawking is especially sad as of all the Evoquacks he really could do with the full power of his DNA to repair his body but his denial of GOD precluded him receiving the benefits of eternal life that GOD gave to Adam and Eve. Actually Hawking seems possessed by the same demon that Jesus cast out of the man by the pool as written at John 5:5 'Now a certain man was there who had an infirmity thirty-eight years.' Jesus instantly cured him and could just as easily cure Hawking.

DNA is usually best described like a book: An analogy to the human genome stored on DNA is that of instructions stored in a book:

The book (genome) would contain 23 chapters (chromosomes);

Each chapter contains 48 to 250 million letters (A,C,G,T) without spaces;

Hence, the book contains over 3.2 billion letters total;

The book fits into a cell nucleus the size of a pinpoint;

At least one copy of the book, all 23 chapters, is contained in most cells of our body. The only exception in humans is found in mature red blood cells which become enucleated during development and therefore lack a genome.

Refusal to accept the impossibility of our DNA evolving from nothing by itself is the hallmark of the Evoquack.

It is easy to find lists of names of Evoquacks in the textbooks and annals of EVONONSENSE. Collectively these people sneer at the idea of GOD and Creation which gives me perfect reason to say they have a poverty of intellect and suffer from extreme Woozle!

Worms go merrily on their way making fools of all these Woozlers but no worm ever got a Nobel Prize!

Top of the list of Evoquacks must go Darwin as he is the man responsible for most people thinking they are evolved from monkeys that evolved from fish ad nauseum.

Charles Darwin Evolutionist 1809-1882 Started as Christian but then Satan got him and filled his head with EVONONSENSE that is still accepted as gospel today. He read Paley's watchmaker analogy but then stupidly perverted it to claim that perfect and exquisite design could design itself before it came into being! He was obviously unaware that every baby girl in its other's womb already carries all her life's eggs in her ovaries and each will give a perfectly accurate copy of a human being – barring sadly the problems caused by Thalidomide, mercury and radiation. Knowing nothing of DNA is not an excuse for his idiotic belief and promotion of Satan's agenda.

Stephen Hawking Evolutionist 1942- For forty years he has been crippled by a demon since his teenage years of doubting GOD and experimenting with ESP let demons take possession of him. He invented 'Black Holes' and denies GOD. Latest lunacy is to claim fear of alien invasion as aliens may be TWO BILLION YEARS MORE ADVANCED! He obviously bought into Teilhard de Chadrin's Noosphere fairy tales about humans being able to evolve into pure thought or something? If he wants to get out of that wheelchair he needs to replay the scene of Jesus instantly healing the paralysed man at the Pool of Bethesda.

Aristotle, Evolutionist 384-322BC Believed Earth was subject to recycling itself.

Patrick Matthew,1790-1874 Promoted speciation or macroevolution. He apparently based his beliefs on finding that some fossils were only found in some areas and other fossils in other areas and took this to mean that fossils in one area had evolved differently to the other – or something similar! Living in Scotland with plenty of local rivers displaying the stratification of gravels with each rainstorm and a little travel in mountains and along coasts would have shown him that all Earth was covered in many layers of sediments - apart from that still bare bedrock – and those layers must surely prove the Flood to be a fact.

Robert Chambers Evolutionist 1802-1871 – six fingers and toes! Rare symptoms of lingering Nephilism despite 4,400 years dilution.

William Smith, Evolutionist 1769-1839 Extensive mapping of British Isles strata. Spent many years working or coal mining and railways and naturally saw and collected many fossils in the various excavations. Unfortunately Satan got him and made him believe that the stratification of fossils indicated the passage of millions of years when of course the lack of topsoil shouted loud that Earth was young. Lacking knowledge of stratification he was unable to understand how some small creature's fossil remains would lie lower than a huge creature's in rocks or soils differing widely in composition. He like so many Evoquacks believed that somehow Earth or perhaps space has periodically deposited trillions of tons of fresh soil all over Earth to kill and bury the layered fossils. How he rationalised to himself the vast areas of exposed bedrock and dead shifting sands is not known. His ideas of how the frozen mammoths came to be is not known either.

Jean Baptiste Lamark, Evolutionist 1744-1829 thought moles became blind, animals grew teeth as they came onto land, but paradoxically birds lost teeth and evolved beaks which is about as crazy a notion as Hawking's plaintive whine: 'gravity did it!' Like so many other Evoquacks Lamarck blanked off the Flood as a fairy tale and so was unable to think that the stratification of the fossil shellfish he found might be due to the swirling of water during the phases of The Flood let alone the probable distance the molluscs had been transport during the time the crustal chunks lifted and fell as the burden of water was lifted and then the burden of ice grew.

Carolus Linnaeus, 1707-1778. By looking in a mirror and describing himself he provided us with the definition of GOD's creation - Adam - and named himself 'Homo sapien' meaning 'thinking man' in contrast to all the monkey-like creatures that lack real thinking ability beyond that necessary for bare existence and survival of the fittest in the post-Flood world. He mistakenly thought that the only difference between himself and a monkey was the power of speech! Perhaps he needed spectacles by that date and was unable to spot the monkey's thumbs on hand and feet? Or the monkey's predilection for rear mounting in mating? Just how Linnaeus rationalised his own faculties being so obviously more refined if not totally disconnected from those of all other creatures is a mystery – unless just like Hawk'n'Dawk he was possessed by one or more GOD-denying demons? He was obsessed with fitting all plants and creatures into classes but constantly came up against some that had characteristics he had previously attributed to another class – he was obviously unaware that GOD and Jesus have vastly more imagination and creativity than we humans and had no difficulty creating a creature that seemed of mixed classes – the premier instance being perhaps the duck billed platypus. Linnaeus died 20 years before the first of these odd but perfectly viable creatures were discovered in Australia so he was spared the anguish of trying to assign it to one or other of his creature groupings.

Etienne Saint-Hilaire, Evolutionist 1772-1844 Defender of Lamarks ideas. He further believed that environmental pressures could produce sudden transformations to establish new species instantaneously being ignorant of the fact all females at birth carry the eggs of the next generation identical to themselves!

Richard Dawkins, Evoquack 1946- Christian until Satan got him and now he is virulently anti-GOD and pushes EVONONSENSE with a sad childish lack of commonsense to his devoted claque of equally deluded followers. A favourite of the perverted columnists of the leftist press and the BBC which accepts his denials of GOD as justification for their pushing homosexual and lesbian lifestyles as normal. He is demon possessed since being molested by a pedophile teacher at school and this shows in his handwriting and in the way he violently reacts at mention of GOD and Jesus's names. Dawking's dedication to promoting Satan's agenda and his demonic possession was spectacularly confirmed with what must be the most rabid, blasphemous outburst ever transmitted this side of a Hitler Nuremberg speech when on Ben Stein's program he actually let his demons control his mouth to reel off this amazing and unbelievable sentence:

"GOD is:
Most unpleasant character in Old Testament,
Jealous and proud of it,

Petty, unjust, unforgiving,
Control freak, vindictive,
Bloodthirsty, ethnic cleanser,
Mysogynsist, homophobic,
Racist, infanticidal,
Genocidal, philocidal,
Pestilential, megolomanical, sadomasochistic,
A capriciously malevolent bully".

WOW! Even Satan himself dare not speak of GOD like that! Dawkin's myriads of fans love him and will follow him to the destruction that awaits all who deny GOD.

Bill Nye, Evolutionist 1955- A buffoon from the same mould as Dawkins with the same mental blocks. Had a long running tv program showing how things work but now spends his time promoting Satan's EVONONSENSE. Proof of this is in claiming dinosaurs lived and died millions of years ago while worms quite clearly show that Earth is young and therefore all dinosaur fossil were buried during The Flood about 4,350 years ago.

Laurence Krauss Evolutionist, 1954- A Godless Jewish clone of Buffoon Bill Nye. Argues that teaching Creation to your children is a kind of child abuse – a comment that Satan is surely proud of! Must be possessed by a demon to utter such a blasphemic woozle. Has been worshipped as a 'public intellectual' whatever that is but based on rubbish utterings it seems to mean he is a Dunning-Kruger Low. He has compounded his stupidity, and symbolically married Hawking, with his publicly argued claim that the laws of physics allow for the universe to be created from nothing – with no watchmaker needed – which is the same as Hawking's woozle: 'because gravity exists it can produce watches without a watch designer' nonsense. He – or his demons - gets into amazing spoiled brat rantings confirming his Jewishness when confronted with truth of Jesus sitting at the right hand of GOD exactly as the temple priests ranted at Stephen before they stoned him to death as recorded at Acts 7:8.

Carl Sagan, Evolutionist 1934-1996 Typical smirking Evo who chose to deliberately ignore all evidence of Creation and GOD to push EVONONSENSE while also squandering vast sums searching for ET instead of waiting for Hollywood to bring it forth.

Edwin Walter Mayr, Evolutionist 1904-2005. Dreamed that KINDS of creatures can evolve over time due to isolation or climatic problems and thereby called GOD a liar.

I ask you: Is there anything as dumb as an Evoquack and their Woozles!

Chapter Twenty Nine
Lucifer becomes Satan. Free Will. Watchers Nephilim. Eve's Fall. God's Curse.

The end of Chapter ??? I left Adam and Eve sleeping without a care in the world unaware of Satan glaring at them through the tree leaves. Here is the explanation of how Satan came to be so evil and plunged the humans and Earth into such ruin.

Atheists deny the existence of GOD and believe the tidal wave of EVONONSENSE woozle that says they are just a step up from monkeys. They cannot grasp the reality which is that GOD created the universe and put humans on Earth but before He created the universe He also created millions of angels. They are all spirit beings and so do not need food or water and can travel to wherever they want any time they want. They had nothing more to do than enjoy the company of GOD and Jesus – their life was marvellous.

One of these angels was made more gorgeous than all others. His name was Lucifer and was quite happy when GOD and Jesus decided to create the universe; he even joined all the other angels in rejoicing at each step of Creation and especially when Adam was created. Sadly after Adam arrived Lucifer became jealous of the attention GOD's first human received! Stung by this unwarranted jealousy he started plotting and believing that he could make himself ruler of the universe and show GOD who was king.

He didn't stop to think GOD was unlikely to give him that chance but the more he imagined how glorious it would be to have a throne higher than GOD the more the longing grew in him.

Today men and women around the world constantly scheme about being the biggest and best in imitation of Lucifer by whatever means they can regardless of any harm or injury it may cause – these people really are the seed of Satan! One day unable to resist the urge to act he tempted Eve by using ventriloquism and a friendly serpent to suggest that GOD was keeping her ignorant of many things but if she ate the forbidden fruit she would suddenly gain enlightenment. Does that sound familiar from legions of gurus, 'prophets' and snake oil salesmen and the generations of gullible people who swallow the hippydippy nonsense?

Having been granted free will by GOD Eve thought Lucifer's suggestion perfectly reasonable and so ate the fruit and then told Adam he should eat some too. And as he also had free will he did do. Immediately they felt guilty and hid. Lucifer went off smirking to watch GOD's reaction.

GOD saw the humans had eaten what He had forbidden immediately cursed them and the whole Earth, but His curse was softened by His knowing that Lucifer was the real culprit. He renamed Lucifer 'Satan: the father of the lie!'

GOD and Jesus formulated a 6,000 year plan for Adam and Eve's descendants and He told Satan that he was doomed and would be crushed to destruction by one of Eve's children but that he could have 6,000 years to try rule the world and have all the humans think him the good ruler he had fantasized he could be. During the 6,000 years the angels remaining in heaven would be able to decide if life with GOD in heaven is better than here on Earth with Satan.

Knowing he had been granted just about 6,000 years and his time was short compared to the humans who he knew would be granted eternal life just as GOD originally promised and intended so long as they rejected rebelling, Satan determined to exterminate as many of the humans as possible by every evil trick or practice he could imagine. He seduced one third of the six hundred angels closest to GOD by explaining how by following him they could become gods higher than GOD himself and not just slaves to the real god GOD. He also extolled the delights of life on Earth with the human women – which was a contradiction of his earlier claim but so eager were they to follow him they overlooked the difference between a glorious life in the heavens and a restricted one on Earth. Satan led the gang down to land on Mt Hermon and on to the villages the humans had established. Mount Hermon presently has a temple dedicated to him but named The Lucis Chapel to mislead the unwary.

By materialising handsome bodies and displaying amazing physical and mental powers and revealing secrets that we human do not need to know the angels quickly gained the attention of the women and the admiration of the men. The women were seduced by the angels and began to have their babies – but for some reason – perhaps because GOD had not allowed them to have perfect human genetics in their new bodies - the angel's offspring were oversize and rapidly grew to be giants with appetites and temperament to match. They were the Nephilim. They lacked natural human affection and became evil tyrants as they spread across the Earth grabbing territory and erecting grand monuments and defensive structures against their fellow Nephilim. It is hard to imagine the vast size some of these attained but the size of the stone blocks they built with show that they must have been many times the size of humans. It is possible they had the help of their angel fathers as they cut and placed stones weighing hundreds of tons as easily as today's bricklayers lift and place bricks and blocks.

Their huge size and appetites demanded more than the vegetarian diet GOD had specified for humans and all the creatures and so they turned to killing and eating the animals and birds and then humans – and the slaughter inevitably led the creatures to investigate meat eating until all creatures had corrupted their ways and Earth was rapidly becoming ruined.

Humans were and still are bombarded with Satan's doctrine of free will and eagerly indulge all their carnal appetites regardless of consequences to themselves or others. Drink, drugs, smoking, disease, famine and drought, pestilence, internecine strife and feuds, crime, disaster, small battles and great wars have put a great toll of human lives on Satan's account though he knows perfectly well that each death confirms his own approaching demise. Obviously the humans and their offspring followed their own free will and Satan very quickly learned he could not rule well as he always met opposition from a few who rejected him and preferred to believe in the GOD and Creation they heard about in the old stories handed down around their campfires.

GOD sadly watched Satan lead his angels and their Nephilim offspring into every kind of sin and criminal activity until eventually He decided Satan had overstepped the authority he had been granted. To let him know he is doomed GOD made him and his gang of angels watch their Nephilim children being killed before The Flood drowned all other life except that on the Ark. The Nephilim bodies were human and needed to breathe just as ours do today and so the bodies drowned during The Flood just as so many people die in flashfloods and tsunamis.

However as the Nephilim spirit is angel it cannot be drowned and as the body drowned the spirit emerged and went searching for a new home. On confronting a man possessed by 'legions' possibly thousands of demons Jesus ordered the demons to leave him and enter a herd of pigs. It need surprise no-one that the demon called Jesus the "Son of the Most High God," or that every time Jesus meets with a demon, it knows who he is? Jame 2:19 says, "You believe there is one God. Good! Even the demons believe that. And they tremble!"

After the Nephilim drowned GOD's faithful angels caught and bound all Satan's angels and threw them into a pit where they stay in darkness until being allowed a short period of freedom at the end of the Millennium reign of Jesus. Satan was allowed to remain free as we know from him directly confronting Jesus several times. When Jesus returns he will capture Satan and bind him in the pit with his gang for the Millennium. One thousand years later they will all be released from the pit is shown in the pope's weird throne which has Satan with his long hair fashioned into a subliminal image of a beast's head – just as Jehovah's Witnesses use such subliminals in their devil exalting magazines! Jesus went down to the pit after his body died on the cross to let the fallen angels know their time is short. And Satan's plan had once again misfired.

In the years after the Flood Satan and the unknown number of demons reverted to their practices of deceiving humans by founding new religions based on lies and manifestations of aliens and dead people. This why why Jesus said that one kind of demon would only come out with praying and fasting. Jesus's life, death and resurrection signalled to Satan that his time was shortening and spurred him on into an orgy of persecution of the human race in general and Christians in particular with the result that the years since Jesus ascended to heaven have been one long catalogue of horror that is soon to reach its height with mass deaths and disasters that will wipe out the entire human race except for those who accept Jesus is the son of GOD YHWH. Jesus said at Matthew 24 verse 21 'For at that time there will be great tribulation, unmatched from the beginning of the world until now, and never to be seen again. 22 If those days had not been cut short, nobody would be saved. But for the sake of the elect those days will be shortened.' I constantly preach this Gospel but mostly get sneered at!

Knowing that GOD had planted a need to believe in Himself in the new humans Lucifer perverted this need and established many regional and national false religions based around belief or worship of dumb idols and the sun, moon, planets and stars and especially in the many idols devoted to Baal and its accompanying wood stake Ashtorah. Study the origins of themselves world's religions and notice how many will have been founded as a result of someone having an encounter with a spirit being supposedly representing GOD or a higher power on a distant planet or galaxy. Generally the contactee will be told or promised special favours if they accept the false doctrine. Being the ruler of the Earth Lucifer had access to its wealth and delights and used it all to seduce humans to worship many of these false idols and heavenly bodies by arranging opportunities to steal money and live luxury lifestyles with unbridled and deviant sexual practices, sex orgies or providing 'sacred prostitutes' which are still common today. Secular history records billions of deaths on Satan's records and billions more to come in the near future.

Satan had a major battle and another pyrrhic victory over Jesus and his disciples by succeeding in having all of them crucified or beheaded and countless other followers slaughtered by various awful means. His delight at inciting the Jewish leaders have GOD's son crucified lasted just split seconds before GOD caused a mighty earthquake that not only shattered the rocks and allowed Jesus' blood to fall on the Mercy Seat of the Tabernacle as a perfect offering for sin but also ripped the great thick veil of the sanctuary in the temple and allowed all eyes to see the emptiness within – the Ark was not there! The Law was mocked!

The priests had been lying like Satan himself and breaking a commandment! They were supposed to enter the sanctuary once a year to take in the blood of an innocent lamb offering and using an hyssop stem sprinkle the blood on the mercy seat to obtain forgiveness. For many years they had been carrying out an empty ritual – a lie – and the Jews had not been and could not have been forgiven. No wonder Jesus called the Pharisees 'offspring of vipers!' The priests thought themselves righteous but were hypocrites beyond redemption – and they knew it! They pressured Pilate into ordering Jesus crucified thinking they had solved their problem and could carry on fooling the people but the moment of death on the cross the ripping curtain revealed them as the evil gang they were.

In the twentieth century Satan had the Jehovah's Witness start proclaiming that Jesus hadn't died on a cross as a sinless offering to wipe out the sins of all humans but on a Ashorath or Ashtoreth stake or 'sacred' tree the same as stood beside every Baal altar. This cross or stake dichotomy may seem a semantic triviality of translation until noting the Bible's many references of Israelites and pagans being slaughtered for worshipping at Baal altars with Ashtorath stakes. The Jehovah's Witnesses lying and saying Jesus died on a stake makes him a sacrifice to Baal and reveals their worship of Satan which is why the Jehovahs Witnesses always insert subliminal images of Baal and trees with demon's faces in their publications. They celebrate Jesus dying on a stake with a feast and claim they are really celebrating his successful sacrifice of himself to GOD but really they are celebrating what they see as Satan's victory over Jesus.

History is full of the names and crimes of rulers and generals who displayed gross venal appetites during their rise to their peak and while they stay in power. Their tortures and murders of innocent victims as well as all who cross them is as sickening as today's news images coming out of ISIS controlled territories. Just yesterday I watched a program examining a castle built in England about 900 years ago and one of its lords was described as so barbaric that his favourite pastime was having men hung on hooks and left to die in agony as the hook dug deeper into their bodies; he also liked gouging out the eyes of people who had angered him! He must have possessed by one or more demons but the psychoquacks would just say he needed help controlling his appetite!

2,400 years after the Flood and just as prophesied and promised Jesus was born to Mary. Thirty three years later Satan succeed in killing him only to see GOD resurrect him and eventually welcome him back in heaven. After Jesus ascended to heaven a favourite manifestation of Satan's has been to appear regularly as a lovely woman taken as Mary the mother of Jesus by Catholics everywhere – and sadly, by many Christians who should have known better! Quite recently I heard a Methodist lay preacher say he had been very moved by a visit to the shrine where Satan masquerading as Mary had appeared to the child Bernadette! Very few of these adoring people know that they are worshipping Ishtar and her son Tammuz as they genuflect to idols of Mary.

In 325 AD at Nicea Satan had his minions prevent inclusion of Enoch, Jasher, Jubilees and other books detailing his own origins from the collection of scriptures that would be the Bible. Luckily copies survived to confirm the reference to these books by the Bible writers..

Latterly he tried to keep to keep GOD and Jesus's promises hidden by having his puppet popes execute all those who objected to the Bible being printed in Latin and unreadable by all but a few. Satan's popes chose to pervert the scripture words to introduce worship of the morning star idolised as Queen of Heaven by claiming Jesus's mother had been raised to heaven as a perpetual virgin despite the Bible quite clearly describing her as the mother of several children with Joseph. As it takes sex to make babies we can know that Mary had normal sexual relations with Joseph until he died sometime after Jesus grew to adult and apparently was supporting the family by his carpentry.

Satan cannot rule well at all as he is a natural liar and vandal constantly seeking to destroy all good things and so we have had all the wars, diseases, famines and divisions in what should be one big happy family.

True Christians are those who have used their free will to decide that Satan and all his works is evil and GOD and Jesus offers the true future.

Chapter Thirty

Evoquacks love Satan and his demons!

This worship of the creature – The Universe – is what all Evoquacks do when they claim they has no GOD and religion even as they worship the Big Banged Universe and praise all the high priests who promote ever more ridiculous and magical claims of it and its supposed powers to harm all those who do not respect it. Like most religions they invest heavily in spectacularly costly cathedrals of science to give the idea that they have impressive credentials and are safe guardians of THE TRUTH. They stage impressive public performances where the high priests can make grave pronouncements of EVONONSENSE and promise great benefits to all followers so long as they keep on tithing. At these events the leading high priests are rapturously applauded and rewarded with more fervour than many queens and kings receive! Sadly all the congregations accept the EVONONSENSE as being the truth! Many recipients of the Nobel Prize were promoters of EVONONSENSE as I wrote previously.

GOD sits quietly watching the craziness around the world and one day soon will take action but at the moment is content to let Satan kill as many humans as possible for refusing learn the truth of Him just as for several thousand years He has seen Satan whispering in the ears of his earthly followers that there is no GOD and that all the universe came into existence by itself! Satan thinks every person who dies ignorant of GOD is a success but as GOD intends to resurrect all the dead and will know which died without hearing the Gospel, which turned away from the Gospel and which died wilfully following SATAN the last laugh will be on Satan and his faithful slaves such as Hawk, Dawk, Attie, Nye, Carrier.

The devil is behind all the lies and mistruths put forth by the Evoquacks claiming the universe all started with a Big Bang from a speck of matter as big as the finest cake flour, not even a grain of salt, and somehow all the impossibly intricate life on earth evolved by itself without anyone designing it and definitely there was no need for any GOD when, as Hawking said in his 2010 book: '*Because there is a thing such as gravity, the universe can and will create itself from nothing*," That was as ridiculous as him saying "*Spontaneous creation is the reason there is something rather than nothing, why the universe exists, why we exist.*' How many fools applauded that Satanic outburst! As for his latest one: "*I'm scared aliens may be two billion years more evolved then me*"...the guy is home to a legion of demons!

Yep, Stephen nothing can come from nothing with no inputs of anything – and I'm sure Satan is delighted with you putting that EVONOSENSE in print to indoctrinate the masses.

Who needs GOD when there is gravity so capable of producing anything it can fantasize? GOD created gravity!

Gravity made Newton's apple fall from the tree but it was GOD who created the first apple tree 6,000 years ago. If gravity really was creative it could have made the apple fly into Newton's kitchen and bake itself into an apple pie? Er, hmm, it would have to create the kitchen, the oven, the pie dish and the pastry first but that would be no problem for Hawking's gravity?

Hawking thinks gravity was so clever it wrote and inserted our whole DNA code right into every tiny cell that makes up our body and no GOD was needed. But neither 'Hawk nor Dawk' can explain why any muddy puddle will have very large single cell creatures happily slurping about in the ooze that somehow, despite the highest powered microscopes showing them to possess no muscles, nose, eyes, mouth, brain, heart, guts or nerves, just glide about imbibing specks of food and, unbelievably, they can somehow sense live prey and speed up to grab it! All with just one cell of jelly stuff! Amazing! Surely any Christian would love to ask GOD how these blobs operate? Perhaps Evoquacks believe those jelly blobs are invisibly guided and powered by gravity? Maybe grafting a fe of these single cell blobs into Hawk's head will make him mobile again?

Satan's Big Bang Theory is far from new!

About 1600-1700 years before Jesus was born Job at Job 26:7 explained that the earth was hanging on nothing in space. Darwin would have been aware of this truth from his Bible studies before Satan seduced him to deny GOD and published the fantasy story that we now call EVONONSENSE. (Job also explains how GOD caused The Flood!) Just how Darwin convinced himself that Job was wrong is not known – but then the loony bins are full of Hitlers, Caligulas and Virgin Mary's so perhaps being an Evoquack is a marker of mental health, or more precisely and verifiably, demonic possession a la Hawkins, Dawkins, Strauss, Carrier etc. Satan ensures all the Evoquacks continue to blaspheme GOD to condemn them to death though they may presently be living high on the hog as Satan does currently rule the entire Earth and can reward the adulation of anyone who falls for his lies. Popstars regularly claim to have sold their souls to him and their financial and social success is good proof that Satan can lavish gifts on all who do his bidding.

But to the Evoquacks Job is merely a lucky guess and a joke and they say that Earth is held in place by gravity as it whirls round the sun at 66,000 miles an hour. To them the universe is simply matter that happened to 'come into being all by itself after a big bang some billions of years ago.'

This website has a good picture of the imaginary BigBang of 15 billion years ago: http://www.universetoday.com/54756/what-is-the-big-bang-theory

Similarly, ignoring GOD and hoping to find evidence to prove He doesn't exist has been the sending of messages into the emptiness of the Universe on the SETI project which is a huge long term effort noted for a total lack of any success other than soaking up money that could be best spent elsewhere. Over the 56 years since 1960, Cornell University's astronomer Frank Drake performed the first modern SETI experiment, named "Project Ozma" - after the Queen of Oz and perfectly appropriate – there have been constant updates of telescopes and computers with ever more grandiose aims and claims and outpourings of fatuous hot air.

In October 2015 Stephen Hawking lent his own demonic imagination and support to the SETI dream of other life in the universe, and he, like one of SETI's original promoters, Carl Sagan, is a confirmed atheist albeit Hawking is locked in his wheelchair by a demon so Satan can use him as a reverse SETI to contact and condemn gullible earthlings to eternal death when Jesus returns.

HAWKING IS A REVERSE SETI! Ah, no, we call them gurus.

Chapter Thirty One
Eureka! - Satan loves Wiki and RationalWiki.

In these End Times we can be thankful that the words Daniel the prophet wrote around 550 BC are showing to be true! At Daniels 12:4 we can read of an angel ordering him: "But as for you Daniel conceal these words and seal up the book until the end of time: many will go back and forth, and knowledge will increase".

This was prophesying today when thanks to the printed word, instant news on the media and the internet there is a vast amount of knowledge coming available daily – so much that is a terrible shame that so many people are wilfully ignorant of so many things not least about GOD, Jesus and Creation. Their god is television and its incessant soap operas or movies with the constant satanic themes of sex, drugs, hate, guns and crime.

Wikipedia was set up January 15th 2001 just eight months before the world changed irrevocably when Islam reared its ugly head on 9/11 and showed that Muslims worshipping the meteorite stone idol of the old Arabian moon god they call 'Allah' in Mecca were intent on world domination with forced conversion of every person to worship the idol or else be beheaded. Satan had actually induced his Muslims slaves to an earlier attack on the Twin Towers when they drove a truck bomb in and its explosion shocked the Tower but didn't do serious damage.

Wiki was set up to be a ever-expanding source of knowledge for anyone with access to the internet and a computer. On foundation day large parts of the world were not linked into the internet or the 'worldwide web' as it was then called.

In 2001 the basic aim of Wiki editing was a person could create an article about almost any topic from whatever viewpoint they saw fit, then the article could be edited or supplemented with further verifiable information by other people who felt they had anything to add to the topic. Those original lofty ideals ensured that good and edifying materials predominated but since then so many minority groups and cults have demanded their views, aims, hates and prejudices must be pandered to that Wiki has unfortunately lost its direction allowing Satan and his demons to use it an excellent means of perverting the minds of many susceptible people.

Homosexual, lesbian and pedophile Satanists instantly delete any mention of homosexuality, lesbianism and pedophilia being learned behaviour induced by contact with LGBP material, events or other LGBPs and not born.

Today any Creationist or Christian can search Wiki for information about some aspect of Creation or GOD or Jesus and be dismayed at all the lies and omissions in the articles. Anyone feeling correction or further facts would be welcome or essential can, in theory, use the editing facility to correct some error, mistruth or deliberate lie or try add input or corrections to make the article better sense and provide the truth but in matters of Christianity and Creation it is quickly found that the article's original author is a Satanist who speedily deletes the corrections or additions and sneeringly threatens to blacklist the corrector. Wiki pushes EVONONSENSE and downplays Christianity and Bible truth with insidious sniping or allegations of Bible corruptions.

A typical example of Wiki's Satanism is this about The Flood and Noah: quote "The narrative, one of many flood myths found in human cultures." Lumping The Flood in with other myths makes it unbelievable and faith destroying and takes GOD out of the picture. Satan loves Wiki.

On the Wiki page for Creationism can be found this Satanic lie: "For young Earth creationists, these beliefs are based on a literalist interpretation of the Genesis creation narrative and rejection of the scientific theory of evolution'. Wiki thus implies that Creationism is a fairy tale belief while EVONONSENSE is scientific – as though it had facts to prove it and isn't the biggest fairy tale foisted on the human race since Satan told Eve to try eating the Forbidden Fruit!

Wiki's entry for Young Earth Creationism – ie Bible Truth – has this Satanism: 'Since the mid-20th century, young Earth creationists—starting with Henry Morris (1918–2006)—have devised and promoted a pseudoscientific explanation called "creation science" as a basis for a religious belief in a supernatural, geologically recent creation.[6] Evidence from numerous scientific disciplines contradicts YEC, showing the age of the universe as 13.8 billion years, the formation of the Earth as at least 4.5 billion years ago, and the first appearance of life on Earth as occurring at least 3.5 billion years ago'. The author of that lying paragraph obviously thinks himself evolved from monkeys, fish and pondslime and worships Satan.

Wiki further panders to Satan in its entry for 'homo sapiens' with such EVONONSENSE as 'Homo sapiens are a branch of the great apes family' before continuing to list the attributes GOD blessed we humans with such as abstract reasoning, problem solving, sociality, culture, art and art appreciation, music making, laughter and invention of jokes and comic scenarios, experimentation and wonderment, manual dexterity combined with calculation and creative visualisation, language, inquisitiveness and not least our lack of body hair but ability to grow long and lovely head hair with which to beautify our quite lovely bodies. Homosaps split off before chimps evolved into great apes which must be thought the reason why baby chimps is so amenable to being trained for sitting for tea parties.

Wiki then ladles on the EVONONSENSE by claiming chimps had 95% DNA similarity but had split into chimps and Homosaps between 4 and 8 million years ago. Hmm, see my chapter on worms and ask yourself where is the 4 million years of topsoil...Worms really do make fools of all Evoquacks! EUREKA!

While researching natural gas as an energy company is wanting to carry out some fracking just about three miles from me I chanced onto Berkeley's page on The Miocene Epoch and as expected it was the usual EVONONSENSE about some layers of Flood sediments containing lots of diatoms which are really the first creatures GOD put in the new seas to recycle all the fine detritus He knew would flow from vegetation and creatures. Berkeley dates the miocene to 5 to 23 MYA – which makes nonsense of the Evoquack's claims of Homosaps having evolved from monkeys 4 – 8 MYA! What's a few million years to an Evoquack!

What fools Evoquacks are to insist themselves are mere variations on monkeys, oops, sorry, chimps, oops, great apes!

Another bit of Wiki's Satanism is the entry for Cain and Abel which says: 'Cain and Abel are symbolic rather than real...with the high level of Babylonian myth behind its stories'. Wiki must therefore believe Babylon was the first civilisation when in fact it was the first one after The Flood as described in Genesis 11.

Wiki then further promotes Satanism is that article by deliberately denigrating quite serious Creation facts by prefacing them with 'pseudo-' and underlining and using italics with the words in blue: 'Pseudoscientific branches of creationism include creation science, flood geology, and intelligent design, as well as subsets of pseudohistory, of pseudoarcheology, and pseudolinguistics.'

Yet another example of Wiki's Satanists pandering to Satan and trying to hide the reality of Satan, fallen angels and demons is found in the entry for Demonic Possession which includes: 'In modern medicine, it is now suspected that an underlying cause of what sometimes appears to be demonic possession is actually 'anti-NMDA receptor encephalitis.'

Both possession and this quack diagnosis have the same symptoms though just how a problem with the brain's NMDA levels can produce the enormous strength shown by demoniacs is a matter that I doubt researchers will want to address as they may come to the conclusion that only a demon could give a human the superhuman strength that frequently accompanies possession as in the Bible account of the seven sons of Sceva at Acts 19:16 : 'Then the man who had the evil spirit jumped on them and overpowered them all. He gave them such a beating that they ran out of the house naked and bleeding'. NMDA gives superhuman strength? EUREKA!

It would be pointless for a Christian to try correct any anti-GOD Wiki article with Bible truths as that would be deemed vandalism and deleted by Wiki's Satanists – Satan really does love Wiki!

Another typical online encyclopedia is RationalWiki. Founded in 2007 as a website for a community to post and discuss articles about a range of topics centred around science, skepticism, and critical thinking. Much of the thinking displayed by most contributors is juvenile and shows that they swallowed Satan's EVONONSENSE hook, line and sinker.

Here is a typical quote from the Rationalwiki 'Creationism' portal quote: "Falldidit. The Falldidit is a member of the main trio of arguments used by creationists when it is appropriate to use goddidit due to conflict with Omnibevolence. (The other two arguments are floodidt and satandidit.) Falldidit is the primary rebuttal for any inconvenient medical or biological fact, or even social principle that suggests that God's design is less than perfect."   This is idiotic simplespeak for the young and uneducated and basically teaches the readers to sneer about any evidence put forward by creationists concerning the present deterioration in the human race as being due to the curse GOD put on Eve and Earth at The Fall when she fell from grace by breaking GOD's order not to eat a particular fruit.

More proof of RationalWiki serving Satan – the father of the lie – is this quote about the Genesis account of creation under Young Earth Creationism: quote: "Their belief is derived from a literal interpretation of the two creation myths in the Bible's Genesis."   This clearly proves RationalWiki supports Satan and his agenda of making everyone believe they is evolved from monkeys and fish with no Creator GOD required.

Satan's seed constantly finds fertile soils and willing propagators eager to gain his approval with their blasphemies?

Luckily there are plenty of other websites that offer some truth to give faith in GOD and Creation.

Chapter Thirty Two
DNA of apes almost human is Satanic EVONONSENSE!

Ever since Carl Linnaeus looked at a monkey in a zoo and thought the only difference between himself and it was that he could speak, Evoquacks have been desperately pandering to Satan by claiming this or that monkey as practically the same DNA as humans. It is so strange that Linnaeus was such a meticulous researcher and cataloguer of plants and trees yet couldn't spot any significant difference between himself and the monkeys!

Thankfully we have television and Youtube and millions of videos of every kind of monkey and gorilla and it should be obvious to anyone outside the Institutes of Woozle that not a single one of those creature have anything more than a superficial resemblance to humans! They were all intended to be endearing and amusing playmates for humans.

Obviously during the pre-Flood years when Satan and his gang of angels were creating hybrids by admixing kinds it is possible that an ape-man was one of their experiments – and as we know the Nephilim DNA was still extant after The Flood and some survivors may be the basis of the BigFoot and Yeti reports we cannot totally rubbish the ape-human DNA link: the few giants around the world may have just a trace of hybrid DNA.

In these End Times so many of Satan's disciples are desperately promoting his lies and agenda. The American Museum of Natural History is just such a satanic institution and its website about DNA states: 'Human and chimp DNA is so similar because the two species are so closely related. Humans, chimps and bonobos descended from a single ancestor species that lived six or seven million years ago'. That statement is truly Satanic as it denies GOD, Jesus and Creation, claims EVONONSENSE as truth and declares Earth is ancient! Is it any wonder Americans think themselves evolved from monkeys! Obviously the researchers at the AMNH have never stopped to wonder why an Earth so old does not have a great depth of topsoil but is mostly water, mountains, barren sand and ice!

And why link humans with bonobos when a glance at a photo of them reveals they have masses of differences from humans. One EVONONSENSE is that monkeys have 98% of DNA but monkeys have 10% more or 110% DNA compared to humans 100% - try figure that out! The AMNH DNA article claims there is a 98.8% similarity between humans and monkeys but how much DNA is needed to give monkeys those long fingers and short little unapposable thumb. And how much more DNA is needed to give them the thumb instead of a big toe on their feet! Have researchers found the DNA string that gives all monkeys a hairy body with short hair on their heads while we humans are totally opposite and both sexes grow long head hair constantly? In the photo I am looking at the entire face and head of the bonobo is so unhuman it makes me doubt the sanity of those researchers. The bonobo pictured does display a bit of humanness by looking sad and depressed as well she might in a bare little cage in San Diego Zoo when she and her baby should really be wandering tropical treetops munching fruit and leaves.

Researching bonobos I tumbled into the East African Rift Valley and found myself in a '22-25 million year old crack in the Earth's surface which is claimed to be widening at about 2-3 inches annually until in 10 million years it will split apart and make a new supersea'. The split has its own fairytale entitled 'Large low-shear-velocity provinces'. This is like what means that the area of Africa is being lifted by imaginary superplumes - and the 'hypothesis' fantasy for the plume is that lots of old texcstonics plates jammed together deep in the Earth and began to fester causing hot stuff to rise and lift the Africa. Tekstonics is wonderful. Of course a couple of minutes with a calculator and a basic understanding on Newton's fizzicks would prove how childish the superplume is. Africa is about 11 million square miles and the plumes is supposed to be lifting the top 625 miles and each cubic yard weighs about 1.25 tons – therefore the superplumes is lifting up about, er, a lot. Care to do the calculation? If Africa is so old why is there so much bare sand and rock in the place? My science teacher showed how plumes of potassium permanganate rise in warming water but there ain't no way any plume of hot melted rock can push up that African continent. And - my Bible says GOD created Earth as a cool wet ball...

The AMNH then goes into Planet of Apes mode and claims that the fantasy meteorite that killed all the big armour plated dinosaurs 65 MYA didn't kill the little monkeys! What a super fairy tale! But surely, many dinosaurs including Behemoth in the Bible are described as liking to slop about in swamps while monkeys mess about in the trees so wouldn't a dinosaur with huge bones be unharmed in water while all fragile monkeys would die during a meteorite hit? Soldiers know that hiding in a hole keeps them safe from bomb blast but standing up would certainly kill them? AMNH's Evoquacks just love overturning fizzicks. Woozle, woozle, woozle. They betray Satan though when they say evolved monkeys/bonobos became human about 5-10,000 years ago when they evolved DNA that allowed us to drink and enjoy cow's milk. My Bible says GOD created Adam and Eve 6,000 years ago and gave them domestic animals – such as cows? And didn't GOD promise the Israelites a land overflowing with milk... And He promised 'the hills shall flow with milk', after Jesus has renewed Earth and the New Jerusalem has come down from heaven: Joel 3:18. We humans are the only things on Earth that actively enjoy milk and all the things it can be made into.

Incredibly the AMNH repeats the worn out discredited lie about our coccyx being a useless leftover from our monkey ancestors! The truth is that the coccyx is used for various functions as it anchors and directs tendons, muscles and ligaments. Maybe the director of the museum should insist himself and all staff take up Kent Hovind's offer to finance having their coccyxes removed and see how they manage without them?

They reveal themselves to be satanic Evoquacks by saying: 'Evolution of the primates is written in the fossil record'. As if rocks date fossils that date the rock the fossils is in!

Masses of researchers claim and win prestigious prizes for supposedly revealing the similarities between monkeys and human and all agree on the 96-99% similarity in DNA. Most of this DNA differences seems observable in the laboratory but unfortunately these Evoquacks have feasted and become drunk on Einstein's Fudge until they cannot see the monkey for the DNA and introduce DNA Fudge which is failing to mention that the DNA strings of a monkey are not as complete as those of humans – sort of cut short. If monkeys were cars they would lack one wheel and a battery. The Evoquacks claim to have evidence for 'a stretch of DNA that had changed 18 times faster than expected since the human and chimp dynasties split'. Just how they spot this sequence of changes is a mystery as monkeys supposedly split of from humans a long time before cameras were available to take pictures of each change. If they can spot a stretch of human DNA that is different to monkeys it is because that is how GOD designed Adam.

If monkeys really did have 98% identical DNA to humans they wouldn't be content to run about the trees all day enjoying themselves and they'd all look like humans and not monkeys.

Legions of devilworshipping researchers have produces innumerable fantasies about monkeys and other creatures sharing human DNA but as only the basic body of all monkeys resembles the basic body of humans in a superficial way we have to accept that either DNA researchers are chasing shadows and the similar DNA they discover is only programmed to create the flesh and blood - and the vast number of differences between monkeys and humans is due to some unknown DNA – or maybe the junk DNA they all dismiss?

No monkey needs anything but the monkey DNA GOD programmed into monkeys on Day 6 of Creation.

Chapter Thirty Three

Early Creationists. The Movers and Shakers.

Creation Science's Hall of Fame honours the early and medieval scientists, researchers and explorers who had belief in being created by GOD even if they lacked the evidence of the dinosaurs, geology, fossils and especially worms that we have today.

Some of those honoured actually did point out the true origins of the fossils and phenomena they recorded but few had the ability to set their thoughts straight and in the correct time frame of a 6,000 years ago Creation. Fossils were visible in rocks, erosions, mines and excavations in many parts of the world and could be clearly seen without need of microscopes and were accepted as evidence of Creation by many just as many Evoquacks saw the same fossils as evidence for repeated recycling of earth.

With many brilliant minds accepting the truth of Creation for so many centuries one wonders why the so-called 'brilliant minds' of today choose to be so wilfully blind as to totally deny the existence of GOD and all the mass of solid evidence that has buried EVONONSENSE in a concretion of stupidity.

Johannes Kepler and Sir Francis Bacon are some of those who believed in some form of Creation based on what limited evidence they could examine and place in the correct time frame of a 6,000 years ago Creation but they lived and died before the first 'true' dinosaur was discovered in 1840 a full 20 years after Mary Anning's Plesiosaur.

Leonardo Da Vinci is one who today would fully grasp the totality of Creation and the Flood and be delighted with the mass of scientific and fossil evidence of GOD's work available today.

Roger Bacon sometime before 1292 AD first showed carefully produced glass magnifiers made as one piece of good glass polished to a, dare I say it, UFO-like disc with a thicker centre and thin edge such as we can still buy for a small amount in craft or photo stores. These single glasses can be so graduated from thick centre to thin edge to produce a maximum useful enlargement of about 6 times after which the laws of optics operate and no image is formed but Bacon's originals may have only produced 3 times. He must have been delighted by the detail revealed by his simple lenses while today we can find images that magnify the tiniest part of tiny creatures a million times and always we are astounded to find that there is always further detail to be seen if only can can zoom in another million times.

Achieving greater magnification than Roger Bacon with clearer and closer images took the thinking of Dutchmen Hans and Zacharias Jansen working on making simple microscopes by putting lenses at either end of tubes and thereby getting higher magnifications than Bacon's single lenses.

Finally Antoni van Leeuwenhoek in the early 1600's who perhaps hearing of the Jansen's work and after considerable labour with the primitive tools of the time did succeed in making a microscope that enlarged 270 times. He then examined all kinds of organic and inorganic items and must have been astounded to see the life in a drop of dirty water or blood flowing in capillaries or the bacteria and yeasts that cause so many bad and good effects. He was thus the first modern scientist to systematically enlarge, study and describe the finer details of many living items and dead materials. Leeuwenhoek's design was developed over the next centuries but remained limited by the basic optical problems of all glass lenses until perhaps serendipitously Nikola Tesla noticed that the inner surface of silvered glass globes gave immense magnification when certain rays were projected into them allowing him to see fabulous detail.

Today's researcher has multitudes of microscopes available with which to study the amazing intricacies in a wide variety of fossils found by chance or deliberate search; it is a pity that though they can see the fine detail they cannot see the big picture that made the fossil – Noah's Flood.

Temperature has such strong effects on organic and inorganic things that knowing the temperature can help understand the workings, actions and reactions of so many aspects of the natural world that the first production of a reliable thermometer in the 1600s was a great step forwards. In the decades after the basic device was refined and its operating medium was changed to mercury as we still have today. Once it was realised that water at sea level always froze and boiled at definite points it was easy to calibrate the thermometers in first Fahrenheit with its odd points and the better and more logical Centigrade with freezing at zero and boiling at one hundred degrees. Every Evoquack can afford a thermometer and verify that water is only water between 32 and 212F – so how stupid is BigBang?

It must have been obvious to everyone right from Adam and Eve that liquids had different properties. Adam showed Eve the thinness of the water that bubbled everywhere in the Garden of Eden and must have noticed that the juices of different fruits trickled down their chins and fingers at different speeds and seemed to run as raised lines rather than thin films. Over the thousands of years since then so many people must have noticed that liquids could be thinned or thickened but it wasn't until 212BC that Archimedes had an inkling of the density of different materials and yelled 'Eureka.' About 500AD a workable indicating device was made but following the destruction of Alexandria it took another 1200 years until 1768 when Antoine Baume invented the basic hydrometer in use today.

Understanding the use of the hydrometer would have perhaps prevented EVONONSENSE being set in concrete woozle as its use may have shown that the supposed casual death and burials of creatures and flora didn't follow logic but needed quite extraordinarily powerful actions to produce the mammoth and dinosaur boneyards and the Arctic 'muck' we see today.

As I type this I am watching a documentary on the Nazi Peenemunde rocket base and note that solid grass has crept over a quarter of the solid concrete entrance road in the 70 years (?) since the site was abandoned at end of World war 2. Give the grass say another 70 years and that road would be totally covered and each century after that the grass fed by water, sunshine and the recycling by worms and other creatures would get deeper and deeper. If Earth really is old and it is say 15,000 years since last imaginary ice age, where is the hundreds of feet of lush topsoil? Obviously the Flood washed all the topsoil off and left the land bare bedrock.

Leonardo da Vinci, Christian 1452-1519 Despite his great ability with maths and invention he disputed the total Flood of Genesis but believed basics of sedimentation as the cause of fossilisation of dead creatures. His biological notes of many creatures revealed that intelligent design had been applied to all facets of life.

John Ray, Christian 1627-1705 declared 'There is for a free man no occupation more worthy and delightful than to contemplate the beauteous works of nature and honour the infinite wisdom and goodness of GOD.' Wrote: 'The Wisdom of GOD Manifested in the Works of the

Creation in 1691.' He can be forgiven for not totally believing The Flood as evidence of fossils and like Da Vinci thought many had originated in the original full ocean of Day One.

William Paley, Christian 1743-1805 He saw the perfection of microscopic design in creatures as evidence of GOD and Jesus's handiwork. Quote: "The marks of design are too strong to be got over. Design must have had a designer. That designer must have been a person. That person is GOD." The hinges in the wings of an earwig, and the joints of its antennae, are as highly wrought, as if the Creator had nothing else to finish. We see no signs of diminution of care by multiplicity of objects, or of distraction of thought by variety. We have no reason to fear, therefore, our being forgotten, or overlooked, or neglected. We have to think that Hawk and Dawk have never studied earwig antennae and admitted to themselves that such fine and perfect design must surely have been designed by a master designer able to think through all possible movements and sensations the antennae would encounter during the earwig's life?

Martin Luther, Christian Creationist 1483-1546 Believed Moses wrote Genesis and it was true – this is in stark contrast to Satan's False Prophet- the pope- who is now teaching Genesis is fiction and EVONONSENSE as fact! Luther's Catholic upbringing naturally saddled him with a great burden of lies, superstitions and heresies and addled his thoughts with Mariolatry and Ishtarism but once he had found time to study the Bible he realised just how wrong and rotten the Catholic Church was. He pushed aside the lies and nonsense of Catholicism as he discovered and believed Bible truths such as the dead all resting in their graves until Resurrection Day – which even today most Christians deny no matter how much more evidence and research has been made available since his day! Curiously though, his study of the Bible did not lead him to the inescapable truth that there is no Trinity nor will the dead be raised to go to heaven – but today the churches I attend here in England hold to these same false doctrines despite all I can say otherwise! He told the truth of Islam being satanic and the Catholic church being the antichrist's Scarlet Beast. His essays and conflicts over many doctrines can make interesting reading if many months of spare time are available. He wrote many hymns and even helped a group of nuns escape a convent and married one of them for what seems to have been 21 years of happy productive marriage!

John Wycliffe, Christian 1330-1384 He was born in Yorkshire, England and so was privileged from birth. He believed that GOD's grace was available to all who accept Jesus is GOD's son and that the Bible was the sole authority for a Christian regardless of the claims of popes. He lived through the the Black Death which accounted for the death of at least one-third of the European and local Yorkshire population between 1347 and 1353 and most certainly would have seen the suffering and dead everywhere and felt depressed at the future of the human race - which is strange for a Bible literate person to be and perhaps shows that unless GOD opens the eyes a person cannot understand what they read? He became strongly anti-Catholic and especially hated all the monks and friars then infesting the monasteries and streets of Europe. In his Treatise 254 denounced the friar's sodomy and spiritual impurity as being worse than that of Sodom as unlike the Sodomites the friars were aware they were sinning. He was well aware they stole children to induct them into Catholicism but whether or not he knew of the rampant pedophilia of the Catholics is not known. He advocated dissolution of the monasteries as being hotbeds of sin and heresies but this was not completed until 150 years later when Henry the Eighth tired of the rule of the pope and the moneygrubbing tricks of the friars. He rightly denied that during Eucharist the bread and wine become Jesus's literal body – doesn't the Bible say to keep away from blood? He took a lead in translating the New Testament into English for all to read and see the lies of the Catholics. He quite clearly believed the Creation account and the reality of GOD keeping a Book of Life to be opened at Judgement Day. He died of a stroke at age 64. 30 years later the popes labelled him an heretic and ten years later Pope Martin V succeeded in having his bones dug up and burned! Such evil vindictiveness! .

Thomas Burnet, Christian 1635-1715 He was one of the first modern Creationists! He accepted the Genesis account of inner Earth being the source of some of the water during Flood! He agreed with me that pre-Flood the world climate was a perpetual spring though he doesn't seem to have spotted the link between Genesis 2:6 and 9:12, nor noticed how worms create the topsoil that proves Earth is as young as the Bible states. He rejected Newton's idea that pre-flood the days had been longer and stated GOD was the clockmaker of an accurate clock. Today we know that Earth has a longer year than the perfect 360 day year of Creation which ensured all people would easily determine the times and season by noting which stars and constellations were in the night sky as Earth spun through its perfect months.

John Woodward, Christian 1665-1728 – HE KNEW THE TRUTH! He experimented with hydroponics for growing mint and found it grew better in natural water than distilled though at that date there was no way of analysing the actions of the microscopic tips of roots and how they can extract minerals from water – 180 years later Darwin was similarly unable to understand the action of the gravity weights in root tips and how the roots extract minerals but I stumbled across this when looking at videos of explorations of Hitler's concrete bunkers as I detail in my worms chapter. In his 1695 book he correctly dated The Flood at 4,000 years before his time or about the same 4,350 that I accept and use. He was aware that seashells are found on the highest mountains. He studied the strata of cliffs, mines and quarries and was convinced that all the layers we see had been laid down during The Flood just as we can now see in multitudes of images of the Grand Canyon and cliffs all around the world. He realised flints are dead creatures and not mere accumulations of silica in chance holes as so many believe even today – though he had no idea that flints are baked jellyfish and had no conception of how to bake a jellyfish until millennia later it is seen as a mere piece of stone! He couldn't understand why some shells and fossils were exceedingly common and others scarce, and why some resembled existing creatures while others were only found as fossils. He studied the specific gravity of shells and sediments and noted that some light shells are only found in lightweight chalk. He found it incredulous that learned people of his day refused to accept the reality of The Flood despite fossil sea creatures, seashells and sediments being seen all over the world! He would be shocked to find that 320 years after publishing these facts the vast majority of people including supposedly educated people like Hawk, Dawk, Attie, Sagan and Krauss still refuse to believe GOD once flooded Earth to a great depth!

William Whiston, Christian 1667-1752 He believed, as I do, in Arianism contrary to the near mandatory Trinitarianism foisted on the public by the Catholics, yet, curiously he seems to have believed that the gifts of holy spirit freely given all Christians may actually be demonic possession while also holding to the truth of prophecy and miracles of Christian belief! He lectured on solar eclipses but may not have known that his calculations used the post-Flood 365 day year and not the ante-diluvian year of 360 days which would ensure eclipses were very regular and usable for timing anniversaries just as GOD had intended when He put the sun and moon in the sky as recorded at Genesis 1:14 'And God said, Let there be lights in the firmament of the heaven to divide the day from the night; and let them be for signs, and for seasons, and for days, and years.' Unfortunately while trying to arrive at a cause for the Flood he believed really had occurred, and other catastrophes, and being unable to decipher the Bible account of the bursting of the fountains of the deep, he invented the fairy tales of comet impact that is now held in unbreakable grip by countless Evoquacks. He lacked understanding of comet's origins and so thought them heavenly bodies existing since Creation and not being mere snowballs from water jetted into space as a consequence of the Flood itself! Like many false prophets his prophecy that the world would end by yet another comet impact on 16 October 1736 failed proving that despite the many hours he devoted to study of the Bible and many related books he could not understand The Sermon on the Mount, nor grasp the conclusion that GOD had handed rule of the Earth to Satan for 6,000 years or one thousand years for a day to equal the time that GOD had taken to create and make the Universe 'Very good!' He correctly traced the path of some of the wandering lost tribes of Israel through the Tartars of Mongolia who today's science reveals have distinct DNA and are not descended from today's Russians. He was undoubtedly a Creationist but his errors about comets and prophecy may just show a lack of knowledge rather than deliberate discredit of the Bible as confirmed by Daniel 12:4 'But thou, O Daniel, shut up the words, and seal the book, even to the time of the end: many shall run to and fro, and knowledge shall be increased'.

Joseph Henry, Christian 1797-1878 First Secretary of Smithsonian - he would turn in his grave to know that it has been routinely hiding or destroying fossils that clearly prove Creation! Charter member of National Academy of Science and again Henry would be aghast at how the Academy promotes Satan's Agenda.

Michael Faraday, Christian 1791-1867 early electrical experiments named after him. Thanks to him we know that GOD caused the Flood by shattering the Earth's producing monstrous surges of electrical energy which simultaneously killed trillions of sea creatures and created numerous atomic elements GOD never intended us to suffer from or misuse. Interestingly Ukranian scientists are now demonstrating how Faraday's Law explains the origins of all the dangerous and poisonous radioactive elements on Earth as being due to the tremendous spike of electricity created in the granite/quartz crust at the start of The Flood.

Benjamin Silliman, Christian+Evolutionist 1779-1864 Christian American Journal of Science. He knew that all present sediments rest upon a global bedrock and noted that fossil fish exist in global shoals under various sediments as though all killed at once by some unknown deadly means. He had no inkling that these fish died from piezo-electrical electrocution in the first few minutes of The Flood. He noted that in all the fossil vegetation there were practically none of today's trees but only trees that lived in a warm and moist climate such as Adam and Eve enjoyed briefly. He firmly believed humans had only existed a few thousand years as in Genesis but thought animals had existed much long because he did not find any human and animals fossils together– just like so many Evoquacks in fact. Sadly he is an example of the angel's warning to Daniel to 'seal up the book until knowledge increased in the End Times'. He .didn't realise the reason why human and animal fossil bones are never found together is simply that GOD told Noah 'I am going to wipe from the face of the earth the human race I have created'. Human bodies are quite buoyant and easily washed away as many videos show while the larger beast are more likely to snag on outcrops which explain the paucity of human fossils and the masses of dinosaur bones. Also of course it may be that GOD ordered His angels to sweep over the raging waters of the flood and direct human bodies to where they would sink deeper than can ever be excavated by wind, rain or human excavators.

Louis Agassiz, Christian 1807-1873 Believed glaciation was post-Flood. He denied EVONONSENSE but promoted polygenism or humans having sprung from several sets of Adam and Eves. He first specialised in the study of fossil fish and noted they like my own fossil sample consist of teeth scales and fins with all soft flesh having disappeared. He actually honoured Mary Anning for her finding of fossils. His studies of the Bible led him to think that the author of Genesis knew only of localised flooding in the old Mesopotamian area referred to as Tigris and Euphrates. One hundred and eighty years ago he proposed that Earth had suffered an Ice Age over the northern hemisphere in contrast to the extended glaciations favoured by the Evoquacks.   While believing in Creation and not EVONONSENSE he questioned how fish of the same species live in lakes well separated with no joining waterway, concluding they were created at both locations. He held that the intelligent adaptation of creatures to their environments testified to an intelligent plan. Failing to understand the Bible he was unable to accept that a worldwide Flood would have pushed fish about and left some stranded in what became isolated lakes or river systems. When he extended the idea of the intelligent designer purposely designing some humans for one area and different races for other areas he became labelled a racist; he could have avoided this by acknowledging that all humans can interbreed and produce offspring that may be perfect copies of one or other parent or have no resemblance except to previous generations. Knowing nothing of DNA he was unable to understand how selective breeding of closed societies will reinforce some characteristics of body shape, face and hair and colouring until the society seems different to ones living, say, at the other side of a mountain or on an isolated island.

Charles Babbage, Christian 1791-1871. He believed in Uniformitarianism rather that the Planet of Apes catastrophism favoured by the Evoquacks unable to understand the evidence of the strata but he was however unable to accept Genesis as literal despite holding to its basic chronology. He rejected the Athanasian Creed's trinitarianism as contradictory – as do I – and concluded that nature demanded an intelligent creator - and that was the GOD of the Bible. He defended the reality of divine miracles as recorded so often in the Bible and in the secular world. He used his GOD designed brain to delve into many scientific realms but is best remembered from designing several version of early mechanical computers.

Mary Anning, Christian 1799-1847 Found the first plesiosaur fossil and many smaller fossils in a small area of eroding low cliffs by the beach at Lyme Regis, England. The Evoquack who composed the Mary Anning article in Wiki stupidly referred to the low cliffs as Jurassic meaning dating between 195-210 MYA. This date and name is EVONONSENSE as in fact the cliff perfectly displays many layers of sediments laid down during the year long Flood about 4,350 years ago – and also if they were a minimum of 195 million years old there surely ought to be a great depth of topsoil on the sediments – but there ain't! Ergo the cliffs are not Jurassic but Plastocine and arrived at their present conformation during the settlement of lands after The Flood and perhaps after the post-flood ice age. The cliffs are visibly composed of many layers of sediments arriving over the many days of the three phases of the Flood GOD ordered and as such they are very unstable as seasonal rains dissolve strata leading to regular erosion and serious landslips as the one that killed Mary's dog. The many kinds of fossils Anning found are not evidence of constant turmoil of Earth and creation of new creatures but merely the settling of creatures drowned and washed to their final resting place at some point during the phases of The Flood 4,350 years ago in this Plastocine Age we live in. Mary's life was a serious of highs and lows no different from many working class people of her time with recurrent family and financial crises and she finally died from a most awful woman's complaint – breast cancer.

Chapter Thirty Four
Must Christians be killjoys?

Much hatred of GOD, the Gospel and Christianity in general is due to the perception that somehow Christianity is a joyless religion that bans practically everything pleasurable or desirable. This is partly true as today what is seen as desirable is what GOD calls detestable. Sex, drugs and rock'n'roll pastimes seems be a major focus of many modern lives as are fast flashy cars, extravagant homes, overpriced fashion and jewellery and a debauched lifestyle that glorifies criminal and unsocial activities and personalities. Equally many people are denied much joy due to poverty or evil religious edicts, and again many more people have been brainwashed to think that life must be a boring joyless existence of uninspiring work, junk food and hours spent staring at depressing and unsuitable garbage of television. None of this is GOD's fault as His intention was humans should have an eternal life of light work, silvery sunshine, good food and the happy stimulating company of countless other humans and all manner of friendly creatures. Only Adam and Eve has managed to achieve that idyll for a very short while before Satan ruined everything.

But like all accusations against GOD and the Bible this one about GOD being a miserable killjoy can be laid at the feet of Satan and his lying nature. Having designed every iota of us GOD actually approves of humans enjoying lots of sex and designed our bodies to make procreation a delight – which shows how Satanic Islam is as the Muslims carve up girl's most pleasurable part! He never said 'No sex' to humans though He has ordered temporary abstinances for religious reasons by various people. He made women to be beautiful with soft delicious flesh under smooth skin and men to be taller and stronger and both to delight in the other if marriage has been their intention. He promises a renewed Earth with grapevines so productive they will drip with wine. He enjoys music and is serenaded with charming singing and angels who delight in anything He creates. He dressed Satan in finery with many precious stones just as He instructed Moses to make the Israelite priests' clothing special and top quality with gold and jewel adornments and to make an Ark to hold the stone tablets of the Ten Commandments of fine mothproof wood covered with beaten gold and adorned with two cast gold seraphims.

It is curious that all Evoquacks claim ascent from various monkeys all of which have limited sexual opportunity most of the year and then only the briefest basic and unimaginative sex, no real musical skills and no ability to turn grapes into wine! I wonder which bit of our DNA they think evolved enough to enable all these behaviours and pleasures! No doubt they'll quote some dumb stupid Evoquack's biased researches or like Hawking will claim 'Gravity did it!'

GODs plan for the Millennium is for humans to have their own homes with no restriction on style, design or size but I am sure the renewed perfect climate will need simple constructions in harmony with the Earth and built from eco-friendly materials not grandiose palaces made of expensive materials sourced regardless of damage to the environment or land. I doubt if any sort of currency will be seen on the renewed Earth!

The people raptured off Earth during The Tribulation will watch as the entire Earth is burnt off with great fires that must reach deep underground to incinerate bunkers and nuclear stockpiles, the evil CERN and similar instAllahtions. Early warning systems and missiles silos will attract deep blasting flames. Secret military and political shelters, gold and precious stone vaults will crumble to dust no matter what solid granite they have been excavated under. All concrete and tarmac roads and the mines, pits and production facilities will burn as will nuclear, oil, gas and chemical refineries, pipelines, storage and distribution systems and all the waste dumps and contaminated lands they caused. Every hut, igloo, house, mansion and palace will shatter and disappear. Railways, canals, ports, airports, boats, cars, trucks, trains and airplanes will flee from existence. Fertile field and forest, barren desert and mountain will smoke and crumble away. Rivers, lakes and the seas will steam, shrink and dry up to reveal all the rubbish and man-made things in them that will then burst into flame. Rusting, shattered wrecks of famous ships and long forgotten ones will appear fleetingly before melting away. Leaking rotten barrels of chemicals and poison gases and boxes of discarded ammunition and weaponry, nuclear and biological warfare artefacts of every type, secret submarine detection and tracing systems will be briefly exposed before flaring up to leave no trace. Earth will resemble a smaller, much dirtier sun for a short time as fires blanket its surface to destroy all trace of its 6,000 tainted years under Satan's rule.

Take a look at Sodom and Gomorrah to see what hot fire GOD can cast down on Earth!

And exactly as the hail of fire destroyed every perverted devilworshipping human in Sodomia so the final fires of Jesus's returns will destroy all the evil people who follow strange gods and worship idols.

For a brief while many will mourn the loss of the evil people including relatives but then the delight of living safely under Jesus's rule and he guidance of the 144,000 redeemed will erase all sadness and life will become a daily delight of easy work, good food and the stimulating company of vast numbers of other people and for many will come the opportunity to find love and marriage with the opposite sex and bring the marriage to fruition with babies perfect in limb and mind. During that time – The Millennium – women will learn that in their ovaries they hold enough eggs to birth a thousand babies. What a shame it would be to leave all those eggs unused as so many have done since Eve forfeited her chance to utilise all her eggs!

During the Millennia masses of babies will be born to build the Earth's population to perhaps 50 billion.

Chapter Thirty Five

Woman's Childbirth

Women's childbirth v animals

Childbirth?

Four women giving birth.

Which scene is as GOD intended?

WOMAN ONE.   She is pacing around a modern maternity unit. A trained midwife is with her and has lots of elaborate expensive medical equipment around room monitoring her and her baby and is ready in case there are problems in what should be the most natural event every woman should delight in.

After some minutes or hours she climbs onto the bed and her baby begins forcing its way into the world. The midwife guides the woman towards pushing the baby in time with the contractions of the muscles.

It seems hard work as racked with pain, her face contorted with effort she pushes and heaves with all her might with great deep breaths and cries ranging from anguish to agony and blasphemy. She demands painkillers and soothing gas and perhaps wishes she was one of the many women who now choose a ceasarian section rather than the indignity of natural childbirth. Obesity and lack of exercise is resulting in far too many of these ceasarian operations.

Her family and midwife offer soothing hands and words but the woman is uniquely fighting her own battle with her body and this baby.

The midwife checks the baby is coping well and not showing signs of stress.

Encouragement comes from all sides as the baby's head appears and the midwife urges her through the final moments as the baby comes out of the woman wet, slippery, bloodied but hopefully breathing and yelling in indignation at all the stress of leaving its warm, quiet, friendly cocoon to come into this bright, noisy world.

The woman forgets the pains as she is handed the baby and can see and feel its tiny fingers and its eyes that struggle to focus and make sense of this thing called life.

For some minutes the mother lies still holding the baby and passing it round to family but soon she feels another stirring in her womb and pushes out the empty afterbirth sac that held her baby for nine long months.   The midwife checks the afterbirth was complete and showed no problems.

Hopefully mother and baby will be fine, healthy and able to go home in a few hours or a day or two.

WOMAN TWO.   This woman, girl, is sad and apprehensive and trapped in a marriage to a pedophile who regardless of the fact her young vagina was mutilated by FGM – female genital muilation   and her birth canal and womb are not matured enough for sex let alone nurturing and eventually passing a baby out - is typical of backward societies and especially those Muslim countries where pedophilic marriage is the norm and the child brides are made pregnant by older, often much older, rapists. FGM is evidence of Satan's hatred of Eve and all women and his contradiction of GOD making vaginas as delicate sensitive organs designed to make sex delightful. Muslim's insistence on FGM is due to a combination of repressed pedophilia, homosexuality and Satanic teachings.

When her time comes this girl/woman will struggle to force the baby out but the sheer mechanics of sizes and strengths of materials may mean that the mother and child may die during the process but if delivered of a healthy baby the mother may find herself condemned to a life of painful disgusting problems and possibly disease and death due to her vagina and rectum walls being torn. Sex will be a nightmare rather than the delight Eve and Jesus's mother, Mary, could look forward to.

Muslim clerics and leaders all over the world are upholding their pedophile's right to rape little girls – the Grand Mufti of Saudi Arabia, Sheikh Abdul Aziz Al-Sheikh says: Good upbringing makes a 12 year old girl ready to perform all marital duties at that age. Good upbringing presumably means browbeating the child that she has to let a man rape her when she should be at school learning.

Skeikh Mohamed Ibn Abderrahmane Al-Maghraoui says: A nine-year-old girl has the same sexual capacities like a woman of twenty and over. Does he realise he is too stupid to know that a nine year old girl's entire vagina and womb are too small and having a baby may damage the girl internally if not actually kill her? Al-Maghraoui excuses himself by saying is it the will of the stone idol in Mecca – in other words he follows Satan's vicious punishment of all women.

When was last mention of old king of Saudi Arabia only being able to get aroused at the prospect of raping a six year old girl? It's been a long while since the media dare condone such disgusting pedophile rape !

We westerners should perhaps be grateful for the lack of media reporting in these backward Muslim and GOD-less countries as if we knew the full extent and details of each case we would be truly sickened.

WOMAN THREE. She has walked with obvious fatigue and despair to a church and sits outside in the overflow seating.

She is called into the church by the preacher and she walks awkwardly forward with everything about her showing tiredness, despair and hopelessness. The preacher tells her to sit and tells her she is in difficulty because doctors told her the baby is in the wrong position and she hasn't felt it move for some days and fears it is dead and getting it out may kill her.

The preacher tells her not to worry and in the name of Jesus tells the baby to come out. Within seconds the woman feels the baby kick, then she has to grip the chair arms as she feels her womb and its contents being manipulated.

Another few seconds and the preacher and the woman and her friends can see her stomach moving as the baby is rearranged and begins to be born.

The woman has to fight the urge to stand and rush but the preacher bids her wait but seconds later making desperate attempts to keep herself modest but allow her waters to break the woman rises from the chair and steps behind it seeking privacy but her baby will not let her.

It forces her to squat down and raise her dress and seconds later a full size baby shoots out of her and lands on the floor.

Her friends rush to scoop up the baby while seconds later the full afterbirth slides out and lands on the floor to be quickly scooped up and taken away.

As soon as she can stand safely the mother is helped to get cleaned up

Total time that passed from the mother being called in to safely delivering her baby was less than three minutes and she never uttered a yell of pain or anguish or even had to gasp for breath.

Not too many minutes later the mother walks back and shows everyone the baby.

WOMAN FOUR. She feels her muscles moving and knows her baby is about to be born.

She is happy and comfortable with her husband and mother holding her steady as she sinks to a semi-crouch to allow her waters to break and drain then within seconds her baby slips out of her without struggle, stress or pain and is caught by the husband who cradles it carefully and watches the umbilical pulsing for a few seconds before stopping then the woman's mother takes the baby while the husband quickly nips and pulls the cord apart.

The woman reaches to touch the tiny fingers of the baby squirming quietly in its grandmother's arms.

Just a few seconds later the new mother tenses herself, opens her legs and crouches to allow the complete afterbirth to slip out of her to be caught and wrapped up by the husband.

A bowl of warm clean water is waiting and the woman quickly washes her thighs and hands, dries them and rearranges her clothing before sitting in a comfortable chair and reaching for the baby that has now lost its dark birth colour and is relaxed and pink and looking around.

As she cuddles it close its eyes seek hers. She whispers sweet words to it it reaches to touch her and then moves its face to the warm swell of her breast seeking its first meal.

Which of these four scene is as GOD intended?

Woman One is typical of those who every minute around the world have their babies in modern maternity hospitals or under the supervision of trained midwives. The mother-to-be is regularly checked through her pregnancy to ensure her health is optimised and problems addressed with the aim of delivering a healthy baby. And after a certain length of time and perhaps the aid of midwives and painkillers and even surgeons the birth should be successful but the worries and pain of the delivery are not what GOD intended women should bear. Many of today's women have various post-natal complications.

Woman Two is quite often Girl Three has the misfortune to be born into a Muslim, Catholic or other Satanic society where females are treated like possessions or cattle and often have no choice in what man they will marry and have to face the daily misery of having someone loathsome invading her body while constantly subjecting her to physical or mental abuse.

Woman Three is typical of the woman who has a complication at her due date that may prove fatal to her or her baby but luckily is saved by the power of prayer calling on GOD's help.

This happens regularly in many Third World countries where medical care is either too expensive or too distant to be accessed by the pregnant woman. Examples of these and other miraculous births can be found on the internet though as ever detractors claim the mother wasn't really in difficulty or the videos may have been faked. Wrongly set babies can be turned to the correct position, tight birth canals can be miraculously loosed, stillborn babies can be given life.

As GOD and Jesus designed the human body, and angels can carry out any miracles they are ordered to, there is no reason to doubt that GOD heard the prayers. Obviously non-Christians will be unable to understand this.

No doubt the mothers are grateful for GOD's help and the baby but this intervention is again, not what GOD intended.

Woman Four does have her baby in the way GOD intended! It as conceived by a Christian couple after Jesus's Second Coming when the Earth was cleansed. Thanks to the loving relationship between its mother and father the baby spends its nine months serenely and blissfully safe from medical or physical harm while being nourished to optimum size and condition by the pure and wholesome vegetarian diet of the mother.

These babies suffer no shortage of vitamins, minerals, carbohydrates or proteins as the mother-to-be has a vast choice of herbs, grains, fruits and nuts growing all around her. She doesn't suffer from denatured, adulterated food, nor a contrived shortage engineered by men desperate for money, nor does she suffer any dietary shortage from lack of money.

Just like Eve this Woman Four can walk around a beautiful and productive Paradise eating anything at all with no worries that the food maybe forbidden or dangerous.

In western societies most babies are born healthy but sadly there is so much ignorance and sheer stupidity that far too many women abuse their bodies and their unborn child by poor hygiene, diet and exercise with the result that some women's bodies cannot deliver their babies without surgery and some women die during childbirth.

When Jesus has returned all women lifted off Earth during The Rapture will have easy childbirth and healthy babies as they fulfill GOD's original mandate to Adam and Eve: 'Be fruitful and fill the Earth!' EUREKA!

Chapter Thirty Six

Cain and Abel. Cain's Wife?

Incest was the plan! EUREKA!

Whenever Evoquacks or atheists try to ridicule Creation one of their chief woozles is the matter of Cain's wives and where did she come from. They hate having to acknowledge they are the result of ancient incest! Maybe this is why they prefer to claim evolution from unknown and countless monkeys, fish and pondlife or innumerable seedings by meteorites and UFO aliens?

Fools like Dawkins and Strauss sneer at the Creation account but so do many church leaders who should believe the Bible. They and the vast majority of common people who are asked about the woman's origins will say either they don't know or more likely she was from another tribe that was on the Earth at the same time! After all, they say, if there was only Adam and Eve and Eve only had two males babies then Cain's wife must be from another tribe. Maybe this is how they rationalise all the different colour that humans manifest?

I have in front of me two letters – replies to my letters – one from John Sentamu, Archbishop of York, the other from David Attenborough the nature documentary maker. Both writers sneer at me for being naïve enough to believe Creation! Attenborough claims he is a

Christian and he evolved from monkeys. That is bad enough in Attenborough but when the archbishop for northern England calls GOD and Jesus liars it shows what a hold Satan has on the UK churches and media in these Last Days.

Just today I read an article of the GODandscience.org website by a writer who refuses to accept the truth of The Flood being a worldwide catastrophe and says it was just one of a series of local floods that the present Mesopotamian area is subject to. By aligning himself with the ranks of the liars the man is able to escape having to think the correct answer to the question of Cain's wife by thinking that there must have been other tribes on Earth as well as Adam and Eve! I was astounded a year ago to hear the devout wife of a Methodist lay preacher make the same claim! Satan has so many allies spreading his lies!

Their thinking is flawed and their weapon is nothing but hot air as the woman's origins are easy to understand by anyone with a brain: 1 Corinthians 12:8 'To one there is given through the Spirit the message of wisdom, to another the message of knowledge by the same Spirit,'

Go back to GOD telling Adam and Eve to be fruitful and fill the Earth – what a simple command and delightful task they were given! Life was intended to be a constant happy cycle of eat, drink, work, make love, sleep and have babies. A veritable Paradise few have ever experienced but it could have been Adam and Eve's, yours and mine! If she hadn't eaten that fruit she and Adam would now be about 6,000 years old, still be fit and healthy with all their hair and teeth and glowing skins and be happily acknowledged as dear forebears by billions of humans all equally happy and content. But she ate the fruit and the two of them were chased out of the Garden of Eden into a wilderness where Adam first consummated his place as husband. In due time Eve went through the agonising birthing common to all but a very few women and produced a baby man – 'I have acquired a man from the Lord' she cried and named him Cain. She produced another man child and named him Abel.

When they grew up Abel became a keeper of sheep while Cain became a grower of grain. Having cultivated a large area and produced a large crop of grain Cain took some as an offering to GOD. We don't know where or when this became a law or obligation. But as Abel took some of his best lambs it seems there was an arranged day for the offering. GOD was happy with Abel's lamb but not with Cain's produce. Cain became angry at what appeared to be a snub and resolved to kill Abel. This he duly did. GOD discovered the murder and ordered him to be a fugitive but first Cain took a relative to be his wife. GOD put a mark on Cain then told him he would have to go be a fugitive away from Eden in the Land of Nod. Cain went and settled there, founding a city in defiance of GOD sentencing him to be a wanderer. The Book of Jubilees – the book Satan kept out of the collection that at Nicea in 325 AD was declared 'The Bible' – tells of Cain being killed when the roof of his house fell on him. The roof fall may have been divine punishment as similar Bible deaths are shown to be, for Cain's disobedience in deciding to settle in a house rather than wander the Earth and live in temporary shelters of a tent. We have to use some human intelligence and GOD's words found elsewhere to fill in all the gaps.

We are not told how old Cain and Abel were when they started their farming enterprises. Shall we assume that they were mature adults of say 21-25 years? Most of today's people can be thought mature by these ages and capable of starting enterprises. Take Cain as being 21 and Abel something younger: shall we say 20 or 19 or 18? Or shall we let Cain be perhaps 25 and Abel 24, 23, 22? Or did these first children not mature and feel the urge to seek partners until several decades old? Let's settle on Cain at 25 and Abel at 23.

We have to ask why would Cain feel there was a need – a market – for lots of grain and Abel a need for lots of wool or possibly sheep milk and cheeses? After all even though the parents had been chased out of the Garden of Eden the land was covered with vegetation as Genesis tells us that on the third day of Creation GOD had caused all manner of plants and trees to emerge from the land and we can be certain that many more than today's selection of fruit and nut bushes and trees and grain plants were to be found all over the land. So why was there a demand for wool and grain?

We have to consider the fecundity of humans when denied any kind of contraception except abstinence. One baby every 11 months perhaps even ten months is the result and I know a woman who did just that as her husband was desperate for sex just four weeks after the birth of the first child and he said 'it would be safe.' He was very upset when she told him she was pregnant again – and of course it was her fault! Both Adam and Eve still being perfect physical specimens would also have no difficulty making a baby every eleven months for many years.

Or is the old wive's tale of pregnancy being impossible while still nursing a baby actually true and GOD's plan to ensure both mothers and infants can have a good long healthy stressless bonding until weaning? Based on present situations we can believe Cain was the very first baby and Abel the second with less than a year between them. Then what babies are born to Eve? We are not told but must infer that many, perhaps all, were single babies but possibly some were twin or even multiple births. By the time Cain was 25 he could have had a minimum of 24 siblings and possibly many more. If the third child born to Eve was a girl then on Cain's 25th birthday she could have been an attractive woman of marriageable age. The other siblings would have been a mix of boys and girls. On the other hand if Cain didn't feel any need to marry until he had built his farming enterprise by age 40 or 50 he could have been a witness at marriages between his younger siblings and the children of such would have been his nieces and good prospects for wives, as Cain having come from one of Eve's original pure eggs would presumably have been in excellent shape with the prospect of a long life, and a young wife just out of puberty would have been perfectly right and logical. If he had waited to 75 or 100 years old he could have had thousands of relatives of marriageable age.

After murdering Abel and being cursed by GOD Cain was banished from the vicinity of Eden where Adam and Eve lingered and went to roam the wild of uncultivated land to the east of Eden. The area was referred to as 'Nod' meaning wild, desolate, uncultivated and would have made life hard. Until their children were born he and his wife would be a solitary pair without the companionship of others.

Lots of Evoquacks, atheists and agnostics sneer at this account and claim that Cain was indulging in incest yet that is exactly what GOD had ordained! As second generation humans Cain and his wife's DNA or genes would be as perfect as Adam and Eve's would be because at birth a girl carries all the eggs already formed in her ovaries and so unless GOD had specifically cursed the eggs in Eve's ovaries then her children would still be perfect with little risk of birth or later problems. The Earth and its vegetation was cursed and it may be that some of the food the first humans all ate after the Fall was beginning to contain some of the toxins that we now know can cause birth deformities or later life shortening and that their parents. The Book of Enoch – kept out of the Bible by Satan at Nicea - tells us the fallen angels revealed to the women how certain roots could be eaten to abort babies or induce infertility – just as today plant extracts are in legal and illegal abortion pills.

Many legends and fairy tales accreted to Cain but most can be dismissed as Jewish or Muslim nonsense as can the claim that Abel is buried in a mosque outside Damascus which is obviously ridiculous as perhaps 1500 years Abel's death the Flood would certainly have completely washed away any small building like a mosque.

Cain's family tree eventually extended to a great number of tribes but just as he perished when a house he was building fell on him so they all perished in The Flood leaving the human race to be carried on through his brother Seth.

Led by Satan and ever ready to pervert the Bible and GOD's words the Muslim, Said Ibn al-Musyib is responsible for the idea that Cain and Abel were not Adam's sons but two other Israelis as the Muslims claim descent from Cain and are thus rooted in incest! This is the reason most Muslims sneer at me and other Christians who uphold the truth of Cain and Seth being incestuous with their sisters! I like to remind them that they are the result of incest which despite their claims otherwise shows just how Satanic and idiotic is Muslim thinking as Muslims insist on the practically incestuous first cousin marriage with all its genetic problems being visible in Muslim populations with its exceedingly high ratio of stillbirths, deformed babies and imbeciles with IQs many points lower than normal which is evident from the sullen ape-like incomprehension seen on many Muslim faces especially imams when asked to explain Islamic stupidities!

Likewise the British royal family is a good example of how bad inbreeding can be as just one twig of the family produced three retarded children who had to be exiled to remote mansions before news of them leaked out as secrets is wont to do !

You and I owe our existence to our lineage from incest between Adam and Eve's children – get over it!

Chapter Thirty Seven

Adam's descendants.

Adam and Eve	Created Year 0		4026BC?	
Adam	age	130	fathered	Seth and lived until 3100BC
Seth		105		Enosh
Enosh		90		Kenan
Kenan		70		Mahalalel
Mahalalel		65		Jared

Jared	162	Enoch
Enoch	65	Methusaleh
Methusaleh	187	Lamech

Adam died aged 930 when Lamech was 56.

Therefore Adam knew 8 generations of his descendants.

All descendants could have had first hand accounts of Adam and Eve's Creation and speaking with GOD and Eve's being tricked by Satan. Adam would have given Enoch first hand accounts of the first short time of paradisical life in the Garden of Eden with the friendly beasts and dinosaurs, and of Satan tempting Eve and of the pair being banished from the Garden of Eden followed by Cain killing Abel. Enoch could not know such a blissful existence but perhaps the human pair could tell him of the sorrow at what they lost and the troubles and pains that life became after The Fall.

Genesis 5:28 tells us: 'When Lamech had lived 182 years, he had a son. 29. He named him Noah and said, "He will comfort us in the labour and painful toil of our hands caused by the ground the Lord has cursed." Thus Adam had been dead 126 years when Noah was born but there were plenty of others alive who had known Adam and Eve personally had heard of GOD cursing the Earth and these would have been able to give Noah very accurate descriptions and recollections of Adam and Eve's later years and their accounts of being suddenly brought into existence by GOD and Jesus and the joys of their short time in the Garden of Eden!

The numerous verses that include 'It is written' let us know that some form of writing was known pre-Flood though none of Adam's line's writings have been preserved compared to the carved stones used by the pagans and Nephilim.

Chapter Thirty Eight

Earth's population now 50 billion healthy happy people!   EUREKA!

I'm actually about 1,000 years too early as that total will not be reached until Jesus has renewed the Earth during The Millennium.

Huge numbers of people are making silly claims that various governments and secret societies and groups are actually implementing or planning mass deaths, or experimenting with ways and means of controlling and eradicating up to 90% of us because the world population is too high.

The trouble with all these claims is that no concrete evidence can be found and like all myths it grows and spreads among simple people who lack understanding. They do all sound like the old UFO and Alien Invasion movie plots from decades gone by in which aliens come to warn Earth is getting overcrowded, depleted, polluted etc.

And sadly, just this morning I was clicking all the channels trying to find something watch on telly when I found a black ad white 1950's film on the 'crashed spaceship with weird aliens' scenario. The only problem was the aliens were so dumb they didn't know about electricity and that proved their Achilles' Heel!

The methods proposed for depopulation range from secret toxic ingredients in food and packaging, genetically modified crops and foodstuffs, drugs and vaccinations that will sterilise the reproductive systems or weaken our bodies' immune systems so we can be slaughtered en mass by artificially created hybrid viruses such as influenza. The apparent increase in ADHD and autism-type problems in the young are cited as proof that some covert campaign of genocide is underway.

For many years they governments of the world have been working on all sorts of toxic weapons and at the same time any backyard chemist can brew up barrels of deadly poisons, pollutants and germs to be sprayed along the streets or added to food production lines. Add to all these are the supposed brain altering electrical effects and paranoia-inducing drugs available to induce mass deaths.

With all these easily available and well tested killing methods and products available it is naïve to think depopulation is impossible – it is easily achievable as GOD dramatically demonstrated during The Flood when He reduced the population to just eight humans and a boatload of creatures! Satan wishes he had the power to do the same as his agenda is to ensure as few humans as possible survive either his own schemes or GOD's apocalypse.

The secret societies have no need to resort to poisons or disease – they can continue to kill billions very efficiently just as they have done for thousands of years by subtly promoting EVONONSENSE, atheism, agnosticism, false gods and idolatries on everyone.

Earth's population is presently about 7.6 billions.

Christians: 900,000,000
Catholics: 1,200,00,000
Muslims: 1,600,000,000
Chinese folk religionists: 400,000,000
Primal religionists: 400,000,000
Buddhists: 375,000,000
Sikhs: 24,000,000
Jews: 14,500,000
Baha'is: 7,400,000
Jains: 4,300,000
Shintoists: 4,000,000
Taoism: 2,700,000
Athesis, Agnostic: 1,000,000,000

As these figures do not total 7.6 billion it reveals how difficult it is to establish a person's religious thinking.

Nevertheless if 900,000,000 Christians actively believe in GOD and Jesus they are all nominally likely to survive Armageddon while all the other groups do not. That will be a reduction in population of approximate 87%!

Actually, although at this moment 6.7 billion people would die if GOD brought Armageddon today the actual tally of Armageddon- Apocalypse might be more or less than 87% as during the Last Days many more Christians will have fallen away and many of the non-Christians will have turned to GOD when they see the great signs of the End Times.

Instead of worrying about mythical depopulation focus on believing Jesus is the risen son of GOD and be sure of eternal life.

Chapter Thirty Nine
The Dead know nothing.

It amuses me to know that the original composer of 'The Fairy Tale of EVONONSENSE' was not Darwin nor Alfred Russell Wallace but actually Satan! That's right – all evolutionists believe a lie concocted by Satan and passed onto Wallace when they caught him delving into spiritism. That is how Satan and his demons so often gain control over the ones they choose to use as mouthpieces. Smith the Mormon, Mohammed the Muslim, Russell the Jehovah's Witness, Disney and his pedophile mission, David Icke the reptilianist, King Saul and the Witch of Endor - all were snared into Satan's slavery while open to possession in meditations, spiritualism and secret ceremonies.

When young Wallace had been reading much scientific and philosophical works the subject of mesmerism caught his attention enough to lure him to exhibitions of mesmerism or hypnosis during which was noted: *'effects which have been designated by the names of clairvoyance; intuition; internal prevision; or when it produces great changes in the physical economy, such as insensibility; a sudden and considerable increase of strength;'* or *'phenomena such as the trance,'* These are perfect descriptions of the demonic possession so often met by Jesus and his disciples! The *'physical economy effect'* can also be negative and include Stephen Hawking's 40 years of paralysis just like that paralysed man by the pool whom Jesus cured! I say Hawking could get out of that wheelchair if only he would accept GOD, Jesus, Creation and The Bible are the truth – but he prefers his paralysis.

During the demonstration Wallace had allowed himself to be hypnotised or possessed by a demon just as surely are all other 'mediums' who use evil spirits. He then found he could successfully hypnotise -mesmerise–allow demons to possess his young students when he was employed at Leicester University. Once the demons had him hooked they drip fed him all sorts of anti-GOD and anti-Creation ideas exactly as did the demons who possessed Darwin, Smith, Mohammed, Buddha, Hawking, Dawkins, Icke, Krauss et al.

Wallace's studies then led him to make all manner of false conclusions about life, religion, Earth and the universe. At various times he investigated hybridisations, extinctions, isolations, unique kinds and species of all manner of creatures. He posited his 'Wallace Line Theory' concerning the isolation of Malay Strait monkeys but was unable to understand that this is due to and has only occurred since the Earth's crust was shattered at the start of The Flood and Earth was still trying to achieve an equilibrium many years later during which the Malay Straits opened up isolating one kind of monkey on one side and one on the other. Monkeys hate swimming so have no way to reaching the opposite side of the strait. If his demons had not blocked him thinking in terms of Creation and The Flood for the success or failure of each creature he studied he would have easily understood the separation. He could have solved the mystery of the monkeys if he had opened a Bible and read 1 Chronicles 1:19 "Two sons were born to Eber: One was named Peleg, because in his time the earth was divided;"

To his credit Wallace was intelligent, researched widely, made many detailed studies and published excellent contributions to many legitimate sciences, political and technical matters. He was what Kant would later describe as an INTP and like all INTPs, realised that a person's impressive position, title or the adulation of sycophants could cover serious factual, scientific and doctrinal errors: In 1893 he wrote: *'The whole history of science shows us that whenever the educated and scientific men of any age have denied the facts of other investigators on a priori grounds of absurdity or impossibility, the deniers have always been wrong.'* I concur.

Unfortunately he was looking in a mirror darkly through eyes clouded by demons and was describing himself, Darwin, Dawkins, Krauss, Attenborough, Nye, Tyson, Mary Schweitzer, The Leakeys and myriads of other Evoquacks!

It is really amusing how all these Evoquacks base their entire belief system and faith on the whisperings of demons and Satan!

Satan's EVONONSENSE declares that humans are merely evolved monkeys evolved from fish evolved from pondslime in some primordial soup that just happened to arrange its inert super-sterilised molecules into DNA with life! This is in direct contradiction to many verses of the Bible that clearly state Earth and all life was created by GOD in just six days and is a perfect example of how constant a liar Satan has been since speaking that first ever lie to Eve!

During the Dark Ages after the deaths of the original apostles who knew, or in Paul's case had met Jesus the leadership of the Christian churches scattered around the Mediterranean was usurped by bishops about 150AD and located the new organisation to Rome. The Bible and Gospel was perverted to expunge all mention of the origins of Satan and to introduce false teachings including making the day the pagan Catholics worship the sun – Sunday – the Sabbath contrary to GOD's orders – an example of the typical Satanism of the Catholics, as well as demanding idolatry of Mary and other dead who the popes claimed could intercede for humans with GOD. Bibles were only written in Latin and church services were only spoken in Latin which obviously the lay people could not understand with the result that false doctrines could easily be inserted into the beliefs surrounding religion. The masses were easily indoctrinated into believing that on death every good soul flew off to heaven to be with GOD and Jesus and The Holy Spook while souls of the evil were shipped off to a 'purgatory' in which they were to linger until Judgement Day unless a relative could pay an exorbitant price for an 'indulgence' or certificate declaring that the soul had been ransomed and proceeded off to heaven. The popes became so rich they could afford the finest scarlet robes, golden decorations and costliest jewellery. Not without cause does Jesus call the Catholic church 'The Scarlet Beast!'

The worst sin of the Catholics was removing GOD's name YHWH or YAHWEH from the Bible and replacing it with 'Lord' to have everyone confuse GOD with Jesus. They then claimed the holy spirit is a real being equal to GOD and Jesus and thereby ensured continued worship and belief in the Babylonian Trinity. I preach GOD the Father, Jesus His Son and holy spirit is one or other of the gifts given to all believers – often backed up with the physical help of angels.

Until about 150 AD the true Christian message was that the dead became totally lifeless and went to their grave to return to the soil with all their thoughts and actions ceased. The Rome bishops and later the popes introduced various pagan beliefs to satisfy Satan with the main lie being that at death the soul does not die but has two possible destinations: heaven or hell.

Sadly this false doctrine is still preached across the 'Christian' world by practically all denominations and cults although Jehovah's Witnesses correctly proclaim the souls of all humans are dead and waiting to be recreated on Judgement Day. I also preach this death state but I don't know any of the Christians I have met who also believe it so pervasive has been Satan's lies! Once the lie that the soul did not die with the body became a central belief of all churches it was easy for Satan to introduce the idea that the dead could actually still interact with humans and perform physical works. GOD had said the dead know nothing but as only Catholic priests could read the Bible in Latin a great many people could not know this truth and were easily convinced that as Jesus had taken the thief on the adjacent cross off to 'paradise' maybe it should be possible to contact the 'spirit' of the dead by praying to them or asking the spirit details of the previous life and life in paradise. Demons have access to human's history back to before The Flood and can easily flit around Earth constantly entering homes to eavesdrop on conversations or watch illicit acts and so can make quite convincing statements with genuine facts during seances or occult consultations. Huge numbers of people have been convinced of life after death by these mediums but in every case the medium is merely repeating lies Satan and demons put in their ears. This became and still is a major business bringing billions of dollars annually to all who practice clairvoyance, fortune telling, automatic writing, mesmerism, seances and all similar occult things.

So it was that the 22 year old Wallace attended a mesmerist's meeting and by showing he believed what he saw he was easily possessed by a demon who for the rest of Wallace's life led him into all manner of occult research, studies and occult practices until Satan put the false Theory of Evolution into his head and ultimately made him convinced the spirits of the dead were alive and able to contact the living – totally contradicting GOD's word! This begs the question of just how many other Evoquacks also believe they will fly off to heaven or hell when they die! Attenborough is clearly one such and despite all his sadly comical Satanic posturings Dawkins also hopes to meet GOD on his death!

Many charlatans have spotted the wealth available from the occult and have set themselves up as prosperity preachers with a central message that wealth including gold and paper money can be easily obtained from GOD if only a person will give their money to the charlatan preacher! Satan and his demons can make this message seem plausible by being able to materialise gold dust or paper money out of thin air – which is a cynical copying of how GOD's angels will respond to calls of distress from penniless people. But whereas the angels will provide 'this day's daily bread' Satan's false preachers demand vast amounts of money so they can live in great luxury. The angels are proving GOD hears every genuine prayer and provides sufficient for the day with no attempt to appear or sound like dead relatives or saints.

If anyone tells you they have a message from 'the other side' or from a dead relative tell them they have been tricked by demons!

True Christians such as I can be sure that the dead know nothing – although the Bible does indicate a few saints were raised to life at the same time as Jesus, and as Ron Wyatt found their empty tombs it may well be that these are in the same place in which Enoch and Elijah were taken to. However the vast majority of the dead and all those who will die when Jesus returns will die totally until Judgement Day. The billions of marked and unmarked graves around the world along with the seas and oceans will give up their dead on Judgement Day – and not a single one of those people will be able anything but the moment they died!

Chapter Forty
Earth Ruined. GOD's Curse on Earth, Adam and Eve and Satan.

Today it is almost impossible for most people to imagine a world where there are no poisonous snakes, insects, plants, fish, reptiles or vegetation or all animals are friendly and harmless or where the air is pure and clean and the land is an unbroken vista of lush greenery with food in abundance just for the picking! It seems any excursion outside home is fraught with the possibility of being bitten, stung or harmed by teeth, claws, stingers, spines and prickles on all manner of things. Luckily here in Yorkshire I don't have to contend with much more than wasps on hot summer days, midges on damp warm evenings and self-inflicted scratches when I indulge my passion for fresh picked wild blackberries. Around the world the list of deadly creatures and plants seems incalculable. I did prick my finger while arranging roses last week and had a sore swelling for several hours but it was gone by morning as my body dealt with whatever germ had entered my skin.

It was not always so. When Adam took Eve's hand that first time they met he could lead her through soft grass and flowers without fear her fresh soft skin would be scratched or scraped, he could let her peer closely to see bees and other insects gathering honey with no fear of one stinging her nose, he could let her reach to stroke a delightful flower or leaf or pluck a fruit and know nothing would be lurking on it ready to inject venom; he knew any of the animals and birds she came across would obligingly lower its head to be stroked and its sleek hide or hard scales admired, he knew TRexes and other dinosaurs could be coaxed to stop browsing on soft leaves and fruit to display their impressive teeth so perfect for gathering the huge quantities needed to fill their big round bellies, he knew huge lions would allow her to pick up their cubs, he knew she could safely wrap bejewelled snakes around her arms - nothing had any way or thought of harming either human.

Maybe it was that first walk together when Adam showed Eve the tree of Knowledge of Good and Evil in the centre of the Garden and told her under no circumstances must she eat it as GOD had commanded as to eat it would bring her death. With so many new fruits, flowers, trees and creatures to admire on all sides the tree would perhaps have no special appeal unless it really was the most beautiful fruit possible. Even today we have a wonderful choice of fruit and nuts available so the forbidden fruit must have really looked very desirable.

Eve dutifully allowed Adam to lead her away from it and perhaps show her yet another lovely vista of soft fresh greenery with bubbling springs everywhere for refreshment and tasty fruit or nuts to delight her taste buds and share with any of the tame animals, birds and reptiles.

But sadly the Garden of Eden was already tainted – not by stinging plants or biting creatures but by an unimagined deadly being – Lucifer the rebellious vengeful angel who lurked invisible close by them listening to Adam's words of advice and instruction and Eve's murmurs of appreciation and praise.

After GOD had shown pleasure at the perfection of the first human pair Lucifer became jealous! It is strange to think that an angel could feel jealous as they are so much more gifted than us earth-bound humans: Satan himself could go to GOD anytime he wanted! He could think himself at the edge of the universe and be instantly there among the stars! He needed neither food, drink or sleep. Yet, he felt wronged or slighted because of Adam and Eve and the more he thought about it the more angry he became. He started fantasising inflating himself to a higher position – to even being above GOD! Is this what psychiatrists refer to as megalomania? He began to dream how to achieve his aims but first concocted a plan to ruin the special relationship between GOD and the human pair! To that end he devised the first lie as Jesus would confirm four thousand years later when he said Satan was the father of the lie. Choosing a moment when GOD's attention was elsewhere he sped down to Earth and crept up to where Eve guilelessly wandering around the garden waiting while Adam finished working in a distant place had arrived by the Tree of Knowledge of Good and Evil. Eve was happily enjoying the company of the birds and creatures – or as Balaam's ass showed - perhaps all the creatures had sensed Satan's invisible presence and fled? Regardless, apart from a friendly serpent she was alone and vulnerable and as there had been no previous evil acts or speech she may be excused for not analysing what she heard as the serpent seemed to speak directly to her.

Using words as honeyed as any uttered by conman, pedophile or insurance salesman Satan said "Did God really say to not from every tree in the garden?' Eve innocently replied GOD had indeed said one tree's fruit was not to be eaten on pain of death. She may have been unable to know what death was as to that date no death had been seen anywhere. She plucked and ate the fruit and gave some to Adam.

GOD saw what they had done and thundered His curse on them and Earth and made the serpent travel about on its belly ever after. He cursed Lucifer and said his name would henceforth be Satan. He also gave Satan 6,000 years to try rule Earth to see if he could make it as 'Very good!' as He himself had made it.

Satan was not barred from heaven and was able to travel there and infect a third of the serving angels with the same rebelliousness by coercion or appeal to their egos and vanities until they swore an oath and agreed to do as he said. Until Lucifer's rebellion none of the angels had any need or thought of using leaving their privileged position in heaven or discovering their ability to materialise human bodies but once Satan had canvassed and polluted them with his dream of establishing his own kingdom on Earth one third of them eagerly pledged their allegiance and came to Earth.

A parallel with those angels doing what was so strange and using their free will is to be found at Jeremiah 32:35 which records how GOD became furious 'Because the Jews built the high places of Baal in the Valley of the Son of Hinnom, to offer up their sons and daughters to Molech, though I did not command them, nor did it enter into my mind, that they should do this abomination, to cause Judah to sin.' The Jews did something so utterly awful that GOD had not ever considered: they burned their babies on idols of Moloch the bull headed god so beloved of today's Bohemian Grovers and Jehovah's Witnesses who insert Moloch's into as many pictures in Watchtower as possible! Today's Talmudic Jews still teach it is acceptable to burn babies in sacrifice to Moloch – but only another person's baby – not their own! And the Jews wonder why they are hated and despised around the world and are the walking dead!

Actually, regarding the possible number of fallen angels the Bible refers to 'myriads of myriads of angels' and a myriad is 6-10,000 so 'myriads of myriads' might mean not just 36 million but hundreds of millions! One third of that number causing mischief on Earth would surely explain all the present troubles?

Alternatively, were the fallen angels the smaller number recorded as descending to Mt Hermon and only one third of the 200 close to GOD? Even those 66 could easily raise vast numbers of Nephilim in the hundreds or thousand years leading up to The Flood.

These 'fallen' angels were the watchers mentioned in The Book of Enoch. The Letter of Jude mentions Enoch and his prophecies at v:14 which means they were accepted as genuine utterances when Jesus was on Earth. Later, the church that was under Satan's spell saw fit to declare that The Book of Enoch was 'non-scriptural' in the years after Jesus ascended to heaven and Satan was collecting all the world religions under his umbrella organisation just as his chief lieutenant on Earth now, the Pope, is scheming to have all the world gathered into one umbrella New World Organisation that will pass laws aimed at promoting worship and obedience of Satan.

Once down here Satan's angels found they could materialise whatever shape they wished and have since had great fun and caused great misery by appearing as humans, animals, UFOs and aliens, demons, sprites, dead people and poltergeists. Fallen angels watched the delights of human lovemaking and determined to try it so started materialising handsome bodies and seducing the young women. The bodies were so complete that they were able to fertilise the women's eggs but their sperm was defective enough to cause the birth of the giant babies that we know as the Nephilim.

As we rush up to the The End Time Satan's gang's favourite trick is masquerading as UFOs and aliens with great ability to press intricate designs into fields of corn – England seems especially favoured for this crop circle activity.

At this moment clicking on Youtube and searching for Mozart's Lacrimosa will bring up a recording of the piece playing but the accompanying image will show a marble statue of an angel romancing a woman. Here is a link to the statue and while technically it may be a fine piece of carving it is to be seen as confirmation of Genesis 6:2: "the sons of GOD saw that the daughters of men were beautiful; and they took wives for themselves, whomever they chose".

The sculptor claims the angel is 'comforting' her but of course the truth is that the angel is either Satan or one of his gang of angels who believed his lies and left heaven with him to come to Earth to try the sex they had been watching the humans enjoying with such rapture and he is enticing the woman into desiring sex with him. Most Evoquacks and church leaders sneer at the idea of fallen angels being the antediluvian Nephilim or the post deluge demons just as they smirk at the idea of a worldwide catastrophic flood but then most have now swept Satan under the carpet and preach that 'Satan' is a metaphor for a person's will to sin. Was it Kant or Freud or some similar fool who invented the idea of everyone having a good and bad side in perpetual struggle?

We are not spared the details of how the fallen angels tempted the lovely young women who were all descended from Eve and no doubt had the same sublime natural beauty Eve had possessed in her first years before the hardships of life outside the Garden of Eden started making its insidious mark, but these angels urged the women to apply black makeup to make their eyes look more dramatic which is undoubtedly the case as any celebrity picture will prove.

When these women got pregnant the angels also revealed the poisonous roots that would induce abortions and so enable to the women to be free of pregnancies and remain slim and desirable and able to have sex continually. Those roots must have become poisonous after the Fall and Satan and the fallen angels must have been aware of them. Abortion is a totally Satanic act of murder that even today is seen as a crime and sin even if many governments allow it. The Catholic church's ban on abortion is not a sign of adherence and reverence for GOD's laws but more a way of burdening women in line with the Muslim's Satanic belief that 'a man's evil works are more pleasing than a woman's good works' while simultaneously burdening the pregnant woman with health issues and the guilt of her denying her man the sex he craves.

Unfortunately the result of this unnatural angel-woman sex was babies far bigger than normal and it is likely that the women suffered the same damage during delivery as do the young girls 'married' off to old pedophiles in Muslim and similar societies in Africa, the Middle East and Pakistan. The baby Nephilim naturally had huge appetites and demanded more food which also allowed their gigantism to flourish. They were also unruly children who resented parental authority. As they grew they followed their father's natures and resorted to every trick and sin. They began to fight amongst themselves using weapons revealed by their devilish fathers. As they grew beyond baby food and desired solid food they soon exhausted the meagre resources of the typical ante-diluvian family even if the land was still fruitful under the canopy that provided the daily dew and they turned to eating the birds and animals. Maybe they were burning the great forests in order to flush game as hunters do today? Their game may have extended to humans to feed their giant Satanic appetites just as in these last few years the Muslims in the Middle East have been killing and butchering Christians the way westerners butcher hogs. (One thing these Muslims do not understand is that eating human brain brings on mad cow disease. Maybe this explains how crazy many Muslims are?)

If we accept, as we must, the fact of there being 50-60 foot tall giants 'as big as cedar trees' and there are footprints about 6foot long to show even bigger giants were real, as well as old newspaper clippings of gigantic human bones being forwarded to the Smithsonian, then we have to imagine just how much dead meat such a being could eat during a meal or day! They could easily have overpowered many of the smaller slower humans and animals and the fish of the rivers and lakes. Toothmarks on TRex bones might actually be Nephilim's! The oft-reported but never fully verified Bigfoot, Sasquatch, Abominable Snowman, Yeti may be a direct descendant of the Nephilim – one needs catching to check.

The Nephilim are responsible for the megalithic and other unusual ruins around the world such as the base of the temple at Baalbek and the other great stones in the quarry there. Similar huge stones are seen in parts of Russia and South America. One trick of the Nephilim was to make walls from odd shaped stones that have multiple angles and lengths and fit absolutely precisely together – to this day this defies replication even with the latest computer guided cutting machines. The weapons these giants used keep being found in excavations and eroded places and show that they must have had very big hands and strength in order to wield the tools. While some of these tools seem to be chipped from hard rock and flints it would be foolish to think the Nephilim were stuck at such a stone age level as of course they were guided and taught by their fallen angel fathers who knew the secrets and the power of the universe.

There is no direct record of how many centuries the fallen angels and their Nephilim sons lived, built cities or battled each other. Eventually they were totally violent, lawless and greedy as their fathers, the fallen angels, had led them into trying every evil act that the Earth was ruined. Following the lead of Satan and the Nephilim all the humans did as they pleased. Sin was the norm, idolatry was seen in every home. They carved or cast idols of any beast or heavenly body they wished and worshipped it with acts of lewdness and blasphemy. The stars were eagerly consulted along with every other trick of divination and necromancy until the world and its people seemed as polluted, ruined and evil as it does today as we wait for Jesus's return.

Deeply angered GOD decided to destroy all life except Noah. At Genesis 6:3 GOD sadly declared "My spirit shall not always strive with man, for that he also is flesh: yet his days shall be an hundred and twenty years". He told Noah He was going to bring a Flood upon the Earth to kill all creatures except for Noah and his sons. He gave Noah 120 years to travel round the world warning all humans to stop their evil life and return to a simple honest GOD-fearing life. After 100 years Noah had preached the impending Flood and returned home to start construction of The Ark.

Although the world had heard the warning and the Bible says the Flood killed all except those on the ark it is possible that some may have heard Noah's warning, had cleansed their lives and had died with hope of resurrection in the years before the flood but if so we are not told of any person doing that and the inference is that from GOD's warning to the flood all except Noah and family died.

Chapter Forty One

Noah's Last Day of Warning.

In the temple courtyard above the cawing of ravens on the columns and the small crackles of burning logs from the sacred altar fire was heard the wailing of a baby as the gold robed priest raised an ornate knife to the carved idol behind the altar as he prepared to sacrifice yet another infant to the lifeless brass bullheaded bull with the fire in its belly. Either side of the altar beautiful temple prostitutes stood imperious in revealing gowns, a soft bodied eunuch held a woman-lion creature on a tight chain. An infant less than a year old lay in a carved bowl before the altar; it squirmed in its tight bindings and its eyes sought those of its mother begging she provide every baby's right to warm hugs and soothing words but the mother's face was one of hundreds of rapt, joyous, inhuman faces clustered around the altar, the idol, the priest and the sacrifice: her baby.

The priests' last invocation to the stone idol echoed off the temple's fine marble walls momentarily alarming the black birds sat high waiting for what they knew would soon be warm but tasty burnt titbits on the altar. The congregation held its breath for the downward slash of the knife that would slash the throat, spill the blood into the bowl on the altar and be a signal to them all that they could start throwing off their clothes and begin daubing the blood on each other's bodies in what would be another mass orgy of indiscriminate sexual congress.

The knife began its downward stroke.

'Look!' The yell shattered the expectant silence. Startled, a beautiful altar maid dropped the fine alabaster bowl of wine the priest would have soon drunk; its shattering scared many out of their wits. Startled ravens flapped away. Again, 'Look!' The priest stopped his arm, the baby stared at his face, the throng regained their nerves and turned angrily on the shouter who had spoiled their excitement.

'Look!' Again the person exclaimed and pointed a hard accusing finger across the town square to where a tall man strode directly towards them. All turned hard eyes to scrutinise this stranger – yes it must be a stranger to not know that today was monthly sacrifice day – who this stranger was. Annoyed and frustrated to be delayed in their expected sexual pursuits they questioned one another about what fool could possible interrupt the day's main event. Rough and smooth male hands sought dagger handles, fine and coarse female hands pulled bodices close over beating hearts and suddenly slackened breasts; the man strode on directly towards the priest.

Recognising authority when he saw it the priest licked his fat lips and tried to draw in his fat belly to make his well fed body seem important. The crowd opened wide before the man's deliberate progress then flowed back behind him like wet mud flows around striding feet. He halted, barely a cubit separating his nose from the priest's that now showed perspiration.

Lifting high his staff he whirled it round encompassing the priest, the crowd, the altar with its bright fiery idol, the bound child and the gold and glory of the idol's possessions. 'I come to warn you! The great GOD Jahweh has set a day when all who perform and take a part in this vile worship of Satan will perish! Set this child free! Go home! Stop your idol worship! Shatter the brass faced idol! Tear down this temple and go home and start worshipping the one true GOD, Yahweh!'

His words re-inflamed the crowd's blood and they pressed forward angrily calling down the wrath of their gods, and thunder and lightning upon him. He stood firm and once again swung his staff over their heads and the priests who now buoyed up and given bravado by the crowd's support thrust his own oily face at the man and sneered 'Get away from here before we sacrifice you to the great Moloch!'

The man flicked a glance of disdain at him and once again turned to address the crowd whose mutterings subsided to a resentful silence. 'I come in the name of my GOD Yahweh. He warns you that you must cease this offering the blood of innocent children to idols now or face destruction by a mighty flood!'

'Be off with you, crazy man!' was one of many catcalls thrown from the crowd. 'Take yourself and your floods and water the deserts!' This brought gales of scornful laughter from the crowd and small stones began hitting the man. The priest gained bravado. 'Get away from here before I call on the mighty idol to smite you with pestilence and make you our next burnt offering!'

More stones and dirt and fruit began hitting the man. He reached over the priest's shoulder and contemptuously smacked the idol's face with his staff. 'I give you one last warning! Yours is the last city I will pass through before I begin my journey home. The true GOD Yahweh has told me to warn every city that He has set a day when all who disobey Him and ignore His warnings will perish in a great flood. Today I have warned you and you have rejected my warning. Turn now to your stone idol, sacrifice your children by fire and make merry as soon you will die by water!'

To raucous jeers and hateful laughter the man turned and straight backed walked away off the temple platform, across the square and out through the deeply carved gate onto the road across the desert. He did not look back. The priest, the crowd, the ravens all ignored his departure. The priest's knife flashed down and silenced the baby. Crimson splashed the altar.

As night drew near the man's pace slowed and tiredness weighted his bones. He approached some rocks and wondered about finding a safe niche to sleep the night away. A heavenly being suddenly grew visible stood on one of the rocks, bathed in bright comforting light and beckoned him. 'Come here, Noah, you can spend the night here.' 'Yes, I have walked far today. My belly grumbles and my mouth is dry.' 'Pass through these rocks.' replied the angel. 'I have prepared a sleeping place, food and water for you.'

Noah stepped between the rocks and saw dry grasses laid deep and smooth, fresh water trickling from a split rock; ripe corn stood one side of the water, a small bush weighed down with lush fruit grew the other side. 'Wash, eat, drink and sleep safely for tomorrow you have the long journey home. I will guard you through the night.'

Noah, for he was the one who one hundred years before had been given the task by GOD of telling the world that the devil worship and despoilings had become abominable and He had decided to cleanse the earth, laid down his staff, untied his sandals and cloak and knelt to slake his thirst and wash his face. The angel looked on with sympathy. Refreshed, Noah took fruit, blessed it, ate it without washing it knowing it would be pure and unblemished. He pulled ears of corn, rubbed them between his palms to loose the chaff, blew it away, and chewed the grains happy at their sweetness. The angel motioned to the dry grass. Noah went to it stretched upon and it and fell instantly into the deep sleep of a good day's work.

The night grew cold. A lion crept forward and lay behind his legs warming them with its huge black mane; a lamb tiptoed from the other direction and curled against his arms. All three slept in the pale glow of the angel and the pale glow of the moon.

Morning light stole across the desert and found him alone behind the rocks. He awoke and looked around but his guardian and the animals were gone. He washed and tidied himself, ate more fruit and corn, drank deeply and filled his skin bag with the trickling water then stood to leave. Immediately the water stopped flowing, the corn stalks collapsed and crumbled, the fruit bush shrivelled and died the fruit dropping as dry pebbles on the now dry water course. He looked where he had lain but rats were already stealing the fragrant dry grass. By the time he had passed through the rocks onto the road the niche was dry, bare and inhospitable. He strode off to the task of saving his immediate family and as many creatures as GOD saw fit.

Chapter Forty Two

The Ark building.

'Make the ark 300 cubits long, fifty cubits wide and thirty cubits high' said GOD to Noah.

We assume Noah was about the same size as large working men today and taking a cubit as the length from elbow to finger gives us the modern size of the vessel as 450 feet long, 75 wide and 75 high. Such a huge vessel was not to be seen for another 4,200 years!

Scorn is poured on the Ark's size and design by the vast majority of Evoquacks and even shipbuilders who should know better. Denying the Flood was a unique and worldwide event of catastrophic water upheavals and downpours the sceptics sneer and claim the Ark could not have survived in turbulent water, it could not have held all the necessary creatures and their food, its interior would have become foul and toxic, it could not have been built with the available technology and tools and by Noah and three sons – how their mental reasonings are stunted by Satan! Thor Heyerdahl succeeded in easily crossing the Pacific Ocean in 1947 on a raft of 9 large balsa logs with gaps between the logs from which they could pluck a fish to eat from the surprisingly numerous fish that chose to accompany them on their voyage. But the Ark was not a simple raft and its specification was that it be waterproof to preserve the dried food for an unknown duration. Heyerdahl later crossed the Atlantic on a boat made of reeds but again while reeds can be bundled and bound together to make big rafts it would be impossible to make a triple decked floating box with reeds. Yet again, the Marsh Arabs of Iraq live on floating islands made of great layers of reeds but they cannot build three deckers of the size GOD specified.

Traditional boat building has always needed an inner frame and outer planking and undoubtedly many impressive vessels were constructed and sailed successfully in various parts of the world before reaching their peak in the great nineteenth century naval and cargo ships that helped Britain establish its empire. Most of these vessels ultimately deteriorated as the action of waves caused 'hogging' which is the waves lifting first one end of the ship then the other so that its centre portion became unsupported and then loaded with the result that the timbers worked loose until eventually too many leaks made the ship founder or too costly to repair. Turner's 'Fighting Temeraire' illustrates such a huge battleship being towed to a scrapyard while a few similar sized wooden ships linger as rotting hulks in secluded lonely anchorages or where they were cast ashore in storms and hurricanes. The Ark had to withstand what Noah understood would be a flood or deluge but perhaps he didn't know the magnitude of the waters that would fall on the Ark and push it about.

Taking the Ark being simply constructed from squared tree trunks in the manner of log cabins or today's wooden houses, and giving the bottom and walls a good thickness of say one foot with inner floors, walls and roof of thinner section planks then by simple calculation it seems the vessel would weigh perhaps two thousand tons empty. We are not told if Noah had enlisted local workers to help with the building but if not he and his sons performed a miracle of logging and construction to fell, transport and build with that mass of timber.

It s claimed we do not know what the 'gopher' wood specified by GOD was. The 'experts' say maybe it was a tree that was eradicated during the flood or perhaps some timber treated in a special way. One suggestion based on a wreck claimed to be the Ark on Mount Ararat is that 'gopher' means laminated as laminated woods like plywood can be made immensely strong and shatterproof. We just do not know precisely but a few accept the translation of 'gopher' to be 'pitched' as in wood that contains pitch or resin. Interesting facts about the pine tree and especially pitch pine are that the trees grow tall, straight and comparatively slim with each one yielding a single great length as was exploited during the centuries of sailing ships when tall strong masts were needed. Cutting and trimming such trees would be quick and simple while cutting and trimming a chestnut tree such as the huge protected specimen that grows outside my home would be a major task needing considerable time and effort to yield just one short very wide baulk of wood. The inner heartwood of pitch pine is waterproof while the sapwood outer is permeable and if used for a vessel intended to float a long time would become wet and possible rotten leading to various problems if not catastrophic collapse and sinking. But by utilising pitch pine's easily felled and trimmed long straight timbers and then applying waterproofing pitch all round as GOD ordered a good solid vessel could be created as required. Steaming the trimmings would provide the resin or sap based waterproofing GOD specified. It would need six hundred gallons of a thin modern waterproofer but if Noah had to use thicker rudimentary tarry pitch then he would need a good deal more.

An apparent problem for Christians is the origin of the waterproofing 'slime' as if GOD had made the Earth 'Very good!' then surely it must have been imperfect to have slime pits into which creatures could fall and from which emissions of fumes or exudations of toxic matter could spread across the land and down the streams? The movies with monsters from the tar pits use tar pits close to Hollywood but other tar pits are found around the world these days and usually close to where oil seeps out of the ground or can be drilled for. Some of the seeps have been known since the Flood and have been exploited as long, others have been found and exploited as recently as in 1942 when during World War Two the Germans accused the British of pouring oil into Libyan wells and the British accused the Germans! In London the Romans found seeps and made use of it for many purposes. The seepings can be all viscosities of oil but when it thickens naturally as the vapours evaporate off it is known as pitch or tar. Really thick pitch rises daily in the seeping known as Trinidad Lake from which vast tonnages have been extracted and exported round the world to be mixed with gravel to make the tarmac we are all familiar with. I remember seeing road gangs melting Trinidad Lake asphalt slabs over 60 years ago.

Pitch pine trees are so called because when heated their wood drips pitch which can be thickened to become tar. Birch bark also gives a pitch when heated. Many other plants and trees give sap that is similar to pitch. In our gardens and countryside we can see bright drips of resin on many trees and bushes – all of this material can give some sort of pitch.

Evoquacks sneer that these seepages of oils and pitch proves Earth is ancient and the stuff is formed deep in the Earth by chemical reaction among the dry dust that accreted to form Earth billions of years ago...and it is the copulating tectonic plates like as what compresses it all...

As usual this is EVONONSENSE and the truth can be found in the Bible. In Genesis we read that on Day Three GOD ordered the Earth to bring forth all kinds of vegetation already bearing leaves, flowers and fruit. The Evoquacks really deride that! How, they say, can grasses, flowers and fruit trees grow without sunlight or produce fruit from flowers when there are no bees to pollinate the flowers – you Creationists will believe anything!

No, there is a very simple explanation: GOD being the designer of the entire universe and all its life evidently knew how to have fruiting trees emerge from the soil simply because He had designed them and as He had designed their reproductive and fruiting systems He must surely be able to create ready pollinated twigs on the branches of the emerging trees? We know bees gather a speck of pollen on their bodies as they land on a flower and transfer that speck to the next flower they visit? Surely GOD could design the unpollinated flower in the bud and then will into existence a speck of pollen to pollinate the flower before it opened? If His son can instantly regenerate the dead body of Lazarus we can be certain GOD would know how to pollinate an unopened bud or have the undeveloped bud be already pollinated by the time it emerged from the ground and opened. It might just have needed a little tinkering with flower DNA like as what the Evoquacks claim has been happening imperceptibly for hundreds of millions of years through all those plasti cenes? It is sad how Evoquacks all display the mental faculties of the monkeys and fish they claim they is evolved from?

Thus we can know that from the end of Day Three vegetation of all kinds would be seen on Earth and we can know much would have been opening leaves and buds and in many cases discarding leaf and bud coverings. Some of that detritus would fall onto the ground to begin the process of making it fully fertile by building up a good humus or rich topsoil. The detritus would initially just lie there unmoved by wind or creatures but moistened by the soil still wet from rising above the waters. Dead vegetation starts to rot to slime or dry and shrivel and crumble to dust to provide what Darwin called 'mould' but which I prefer to call 'humus'. On Day Three Earth had a very moist climate and no creeping things to speed breakdown of the detritus but a few hours later all changed and the process of recycling the detritus into lush new growth began.

On Day Four GOD put the sun and moon in the sky – the sun perfectly placed to provide carefully calculated warmth by day and the moon to provide warmth by reflecting sunlight but also by gentle underfloor heating thanks to piezo-electricity in the newly hardened crust. The climate and environment were now ideal for airbreathing creatures of all sorts. EUREKA!

On Day Six all the air breathing land creatures were created before Adam was formed from the dust of the Earth. Now the accumulating of pitch could begin in earnest – not by the Evoquacks' ridiculous fantasy of oil emerging from dry rocks but simply by the breakdown of fallen, unwanted vegetation detritus! EUREKA!

Perhaps most Evoquacks cannot understand the simple process that made the 'slime' Noah used because they are not gardeners or they live in hot dry climates where the soil lacks worms, moisture and vegetation and the land is bare, barren, hot sand? I do feel sorry when I watch the news report of farmers in impoverished areas starving because their soils are nothing but scorching sand and dust without a worm to aerate it for the scanty rains to soak down to the roots. At the same time some equally hot areas can have gnarled ancient olive and fig trees rooted into rocky stony ground producing fruit annually with nothing but a pruning by humans or animals. UK farmers still spread all their livestocks manures on their fields just as they have done for over a thousand years and their soil is still productive and healthy. As I've travelled and studied during my long life I've seen and felt many types of 'soil' – soil being the surface of the ground in which farmers attempt to grow crops – and I know that some soil is rich in humus and some has none. I tell my friends that when weeding their gardens they should hoe off or pull up weeds or waste foliage, tear it small and just tuck it around the plants and let the worms come up during the night to enjoy eating it and turning it into manure just as GOD designed them to do! My words fall on deaf ears and their soils are light dead stuff that hardens when dry and crumbles to dust. Such soil gives low germination of seeds, seedlings do not thrive, the growing plants can be susceptible to pest and diseases or be sparse and poor quality. My hairdresser friend has an apple tree by her gate that is so starved of nutrients by having to share its dry light grey soil with a large hosta that the few apples produced are very sickly bitter pitted things not fit to eat. I told her to get a sack of horse manure from her riding friends but I doubt she will.

Some soil with heavy amounts of humus can be very black stuff indeed and is often encountered in my moist climate in thick undergrowth that drops a huge amount of leaves each autumn. Working with this black stuff leads to black stains on clothing that takes detergent and scrubbing to remove. That staining is a sign of the first stage of creation of pitch, bitumen or as Noah knew it, 'slime'. Researchers have discovered that in Amazonia some isolated tribes throw all their human and animal wastes and waste vegetation on the soil and it is noticeably 'oily' to the touch and leaves black stains on hands and clothes and oil slicks appear on water the soil is mixed into.

The very blackest humus is actually the sediments that fill ponds, lakes and dams as surrounding vegetation drops masses of leaves and other parts in over the years. Sometimes an oily film can be seen on these clogged waters to show that the creation of petroleum is taking place on a small scale. During The Flood such deposits were compressed under one, two or more miles of sediments and water which speeded up the process just as today scientists can make petroleum in a couple of days in the laboratory using heat and pressure and organic material.

Noah merely had to go to the nearest place where years of rotting vegetation had turned into pitch.   EUREKA!

Chapter Forty Three

The Creatures enter the Ark

I'm sure most people have seen disparaging magazine articles and television shows with titles such as 'Problems with a Global Flood' with loads of woozle about the Ark being ridiculously implausible and TRexes could not have been kept alive along with lambs or pigs, etcetera, and I'd like to help you overcome some of the mental blocks you have that are stopping you accepting the reality of the Flood - your only alternative being to accept your grandfather was a monkey and Hawk'n'Dawk'n'Attie are wise?

Gathering a selection of creatures before the Flood? Noah took many years to build the ark so that allows plenty of time for even the slowest sloth or tortoise to arrive once they had received the divine command to make tracks to Noah's shipyard. No problem to anyone with a human brain and not a monkey's. The total number of creatures required for the Ark is actually quite low if Genesis 7:2 is read and understood.

There are many essential facts to know and understand when trying to visualise The Flood and the Ark.

One major fact being that Earth at that time was one large land mass surrounding one central sea in the proportions of perhaps 75-90% land to water. Your Bible will tell you that it was not until the days of Peleg that the Earth became divided. Peleg was the fourth generation born after the Flood in Noah's son Shem's descendant's line: Arphaxad, Salah, Eber and Peleg. Conceiving and bringing those generations to maturity would take a minimum of one hundred years and perhaps much longer. During those years the Earth flushed green as vegetation thrived in the rich sediments before the Ice Age started on each of the new poles and began to spread out like a white cancer...

This undivided land is how so many kinds of creatures were able to walk or crawl to the boat. It explains the survival of snakes, lizards, small and large rodents, spiders, beetles and similar that crept along the ground. Cattle, horses, pigs and sheep easily travelled long distance as did the many varieties of deer, goats, bears and big cats. The big heavy creatures like hippos and rhinos happily plodded along as fast as the cattle. Pairs of TRexes with stegosaurs and brontosaurs no doubt travelled as fast as the cattle and the other huge creatures that were labelled 'dragons' through the ages. Maybe some like the sloths and all the monkeys travelled there through trees without touching ground! Bats, bees and butterflies, moths and dragonflies simply flew there resting in trees or rocky crags as necessary. The vast varieties of flightless birds walked along with the other creatures of their own locations. The flighted birds could fly all day as now we know birds don't breathe in and out as animals and humans do but instead have a sort of circular breathing system that ensures they don't get tired and can remain aloft for hours, days or even months. Angels guided many of the creatures towards the Ark with other creatures of all kinds being rounded up or joining voluntarily along the way. Birds of all kinds forgot their migrations and normal routines to fly towards the streaming creatures and then circle overhead to see where the route lead. GOD ordered flocks of quails to fall on the Israelite's camp didn't He?

Boarding so many creatures? They would arrive in good order and walk or fly calmly aboard.

Many would be small as chickens or dogs, chosen to be young, small and best quality. All the flighted birds and bats would happily roost on beams or simple shelves and platforms as would the leopards and panthers, sloths, monkeys, gorillas, snakes and reptiles.

Special diet? I'm sure Christians and anyone who believes in GOD has to accept that Noah obeyed the order to harvest and store all the necessary food. GOD created all the land animals to be herbivores so basically only dry grasses and seeds needed to be gathered and stored. Dried fruits would provide the sugary juices many others need.

Predation and competition among the animals and birds and lizards? There would be none as everything was nice and friendly remember? The lion no doubt lay down with the stegosaurus - sculptures of which adorn standing buildings built before anyone suspected there were such things - so there would be no predation among the gathering Ark animals.

We don't know the internal arrangement of the Arks ramps or stairs for movement between the three decks but last night I watched a video of 8,000 cars being loaded on a ship in an English port for export and it was done overnight and involved lots of wasted time getting drivers on and off ship to get the next car so if Noah and family merely had to wave the animals aboard then the whole lot could have been directed into a secure place in a day or two. Once there they would settle down either naturally or with divine help and soon settle down to hibernation mode and thus need minimum food and water and producing little manure to dirty the floors and the air.

Seaworthyness of the Ark? Well the reason large wooden ships leak is because waves lift first one end then the other and open the joints unless the ship is built very solidly with maximum rigidity or with superb caulking. But who said there were heavy waves during the Flood? Didn't Jesus calm the frightening waves about 2,300 years later? Maybe the Ark was built on flat land - better for handling the masses of wood? - and when the rains continued it just floated up. Then again Noah would have had divine help or guidance to ensure the vessel was sound.

When the last creatures had been taken aboard GOD's angel shut the door and stood guard. Seven days later GOD gave Earth a might shake that caused great earthquakes to shatter the crust and The Flood began!

Over a year later when Noah let all the creatures off the Ark they could once again spread out and travel to the extremities of the single piece of land with each creature seeking its own congenial place. This is the explanation of why kangaroos and koalas are only found in Australia and why Madagascar has its own collection of unique creatures. Sadly the Satanist who compiled the Madagascar website I just checked wrote that: 'Madagascar has been isolated **for several millions of years**, therefore giving the animal and plant species to evolve and diversify in isolation. About 170 million years ago Madagascar was a landlocked country within the Gondwana super continent. As a result of the Earth's crust movement, Madagascar and India split from the South American and African plates and then from Antarctica and Australia. India finally broke off leaving Madagascar behind, and the island has been on its own for the past **88 million years.**' Several million now means 88 million? And he woozled about tectonic plates? Evoquacks will write any crap no matter how nonsensical!

Obviously the mass of creatures converging on the Ark was unimaginably and unknowably greater than the few creatures we have today. Vast numbers of creatures were refused selection for reasons best known to GOD and had to die during The Flood. This is why the fossils records is so extensive and varied with new and unique creature's bones being found regularly. So many fossils seem to be of reptilian or saurean design as to give the impression that once upon a time Earth had nothing but reptiles and dinosaurs giving Evoquacks the illusion they named the Age of the Dinosaurs. Many of the dinosaurs and reptiles were no bigger that today's dogs and sheep with no obvious reason for their fossil remains being common compared to sheep. Could it be that most of the big dinosaurs and grotesque creatures known only as fossils were all the result of the Satanic meddling and admixture of DNA by the fallen angels and seen by GOD as abominations to be exterminated? Or was the ante-diluvian Earth so destroyed - 'ruined! - that the furry creatures and pretty feathered birds were being hunted or starved to extinction by the scaly thick skinned dinosaurs and crawling or flying reptiles forcing GOD to bring about The Flood to preserve a selection of the unpolluted furry and domestic creatures for the enjoyment of the humans He would resurrect during the Millennium or or on Judgement Day?

If the domestic and furry creatures had become seriously endangered by the depredation of the Nephilim and their carnivorous hybrids then logically very few complete skeletons of them would be found in Flood sediments leading Evoquacks to believe they had appeared by evolving millions of years after the scaly dinosaurs. Nothing remains of the millions of American bison skeletons from the great herds shot to extinction about 150 years ago? The only reason Mary Schweitzer or Alice Roberts can find TRex today is because they were quickly buried in deep fine sediments during The Flood 4,350 years ago.

No doubt the real reason so few small furry creature's fossils are found is because their light delicate bodies were easily swept away and pounded to mush in the first rains while the bigger creatures were able to withstand the rising waters a while longer or struggle to high ground – just as flashflood and tsunami videos reveal. EUREKA!

Chapter Forty Four

GOD shakes Earth to start The Flood!   GNEISS and Shocked Quartz formed.
Eugene Shoemaker.

At the end of Chapter ???? angels had just slammed the Ark door behind Noah then mounted guard to prevent any human or creature attempting to gain entrance. Billions of natural creatures, hybrids, Nephilim and humans had been left outside to die in the most awful catastrophe of judgement yet seen. However not one of the doomed nor Noah and family knew exactly what would happen next or when.

For seven days Noah and family were kept busy inside the dim Ark settling the creatures down making sure each had its own preferred type of resting place whether open floor, corner, roosting on or hanging from the internal beams, safely tucked into nooks and crannies, warmly wrapped in their fur, wings, webs or cocoons. Some of the creatures had immediately curled and slipped into hibernation, others remained awake but dozing and quietly patient. Eventually peace and quiet spread through the three decks of the vessel.

Outside, the many creatures rejected by Noah had begun to wander away or fight, kill and eat each other; the offspring of the Nephilim and the ordinary humans hunted the creatures for sport and food and laughed at how Noah's GOD had provided sport and food. As the night drew on the Nephilim, hybrids and humans drifted away to their religious ceremonies and worshipped their own dumb and deaf idols of stone, wood or metals, and joined in the drinking of strong drinks and potions concocted from roots and berries that made them insensitive and encouraged unspeakable sexual depravities of all types. Seven days followed each other with nothing to suggest the disaster Noah had prophesied would really happen.

Each day crowds came and stared at the closed door and jeered about the foolishness of wasting so much time on making a silly boat and filling it with creatures that would be very nice in the cooking pots. Intimidated by the guardian angels the people slunk away to spend their remaining times pursuing their sinful life. We are not told if any did have the sense to realise the ark and its cargo of humans and creatures confirmed the truth of what Noah had been preaching, we have to accept that not a single person outside was worth saving. The Ark sat silent, high above the firmament filtered the sun to its usual silveryness, the light breezes stirred leaves and dust, all seemed well – it really was a lull before the storm!

In heaven GOD sadly watched the Earth days ticking by, unwilling to destroy the Earth and all its life He had been so pleased with less than two millennia previously. He saw the Nephilim chase down and slaughter the lovely creatures and the grotesque hybrids and gorge on their blood. He saw the humans offer more food and drink to the Nephilim and to strange idols and imaginary gods at smoking altars everywhere. He saw Satan smirking as he directed his army of fallen angels to mislead the humans and seduce the attractive ones – male and child as well as females they had first lusted after. He saw the Ark safely closed up and guarded by faithful angels and was gladdened by how solid it had been constructed and how thoroughly Noah had overseen the waterproofing with the thick black pitch – the stinking black residue that now polluted many places on what should have been a spotless Earth. He looked and grieved at how His creation had become such a rotten mess that He had to destroy it all except those in the big wooden box. His time ticked by as he delayed His hand and the moment when everything had to die.

Finally with a mixture of anger and regret He reached out and smacked Earth hard! The Flood had begun.

Earth spun out of control, moon and sun lost their tracks, sudden vicious winds screeched to life, people, Nephilim, beasts all flew about, tumbled head of over heels, smashed into each other or solid objects, howls of agony and fear sounded all round but were lost in the crunching grinding shatterings coming from the very soil. To add to the horror the rolling Earth began heaving and splitting emitting thundering bolts of electricity that seared and sizzled all it struck with its blinding lights. Rocks and dirt blasted out of flaming gouges in the Earth killing and maiming the terrified people. Then came water, masses of it, water in such quantities never thought possible. Hot water, bitter water, salty water, water white with lime and quartz, black with granite dust, ochre with sands – all erupted from the web of cracks racing across the wrenching, shaking booming Earth. Trees and buildings toppled bringing death and destruction. Idols on prominent altars or hidden in sacred groves and temples surrounded by offerings stared impassively at the unfolding disaster and ignored the worshippers begging for help until one after another the carved stones and wood and worked gold and silver toppled to the floor shattering in their impotence.

A worse terror came - cold water began slamming down from the sky, more water and more deadly than imaginable slammed down on all living creatures instantly forming new streams that joined to become mighty torrents sweeping the smallest and lightest creatures away before its inexorable power. The biggest fastest creatures ran for high ground or tall trees and buildings and found safety briefly before the water became aided by masses of debris and battered the refuges down with impunity to overcome the strength of the bigger creatures and drag them along the channels the water was finding in its headlong race over the heaving shuddering land. As the waters raced along they lifted debris from collapsed houses, temples and shattered trees and eroded the land to release soil and rocks, gravels and boulders that all converted all the floating matter into deadly mincing grindstones that ripped human and animal bodies apart and pulverised the softer parts to shreds. The waters from below and above made the hills dissolve and the valleys widen and soon black dirty waters were everywhere. Rivers burst their banks to wash crops and home and life away. Everywhere in it were the flashes and blurs of the dead and destroyed swirling along being ground ever finer.

Day turned into night but still the hot waters screamed skywards and the freezing waters pummelled down from the same sky. If the surviving humans had been able to look up they would see the moon and stars becoming brighter between the clouds of water and dirt racing up into the stratosphere and beyond. The first day merged into the second but the howling wind and beating rains and fountains didn't ease as Earth and sky emptied themselves in a destructive orgy unimaginable and not to be equalled until Jesus comes to cleanse the Earth in a few year's time.

In the Ark Noah and family heard the commotions outside, the beating of rain on the roof, the whistling of wind around the roof vent, the screams of the dying but remained calmly going about the tasks knowing GOD had designed the huge vessel and it would keep them safe. The creatures sensed the rains and winds and stirred uneasily but soothing words and a caring hand kept them calm. The days crept by without any cessation of water or wind.

Noah nor any other human saw myriads of majestic angels appear from heaven on GOD's orders to capture and bind Satan's gang of fallen angels and carry them off to a prison of darkness known only to GOD, Jesus and the other spirit beings. Satan was allowed to go loose but had to watch and mourn his own half human hybrid children's bodies drowning and being torn to shreds in the deepening filthy waters.

At some point in those first days the Ark lifted off its construction site. Noah and family or some of the creatures aboard may have sensed the lift-off, perhaps they suddenly felt themselves to be moving or falling about, perhaps creatures and items hanging from the beams began swaying gently, perhaps there was no sensation as GOD had remembered Noah and ensured an angel was stationed above the vessel to soothe and smooth the waves just as Jesus was to calm the waters and the disciples in the boat with him as recorded at Mark 4:39 'Then Jesus got up and rebuked the wind and the sea. "Silence! He commanded, "Be still!" the wind died down and it was perfectly calm.'

Shut inside the Ark Noah and family were spared the awful sights stretching to the horizons all around. The waters were still erupting from below but their dirtfilled geysers now had to force up through the growing depth and the debris mats of great trees that had reached hundreds of feet into the air and now floated in huge tangled masses covered with branches and vines. Great roaring volcanoes still belched debris into the heavens. Every bit of space between larger debris seemed choked with shattered rubbish. Noah could almost have walked on a carpet of floating timber.   He heard the falling waters continue pelting down, and the new wind whistling round the Ark. Maybe he heard rocks returning from space clatter against the timbers of the Ark.

Noah could have no conception that at the moment GOD smacked Earth - as later recorded by Haggai 6 "This is what the Lord Almighty says: 'In a little while I will once more shake the heavens and the earth, the sea and the dry land.' - all manner of strange effects had occurred but would not nor could be understood for almost 4,300 years until in these End Times knowledgeable people began to understand obscure Bible passages and strange peculiarities of the Earth and skies, weather and animal behaviour and human nature and biology and question the prevalent world views, beliefs and sciences – just as I am doing in this book! EUREKA!

Noah could not know that the great earthquakes generated billion volt electrical surges to instantly kill vast shoals of fish that became preserved in exquisite detail just minutes afterwards under layers of sediments, nor could he know that Michael Faraday and Ukranian scientists would one day prove that the new radioactive elements in the Earth's crust had all been formed by the same super electrical surges. He could not know that around the world the shattered dark granite crust was now filling with quartz that as it cooled and solidified became the streaky rock we call 'Gneiss' and 'shocked quartz'. He could not know of the billions of jellyfish and trillions of trilobites that had been baked in the new layers of limestone discharged from the great cracks in the Earth's crust. He could not know that the masses of shellfish of the sea and rivers were now dead and compressed under great depths of sediments and waters and just about 4,000 years later would start to be collected and examined and labelled as fossils hundreds of millions of years old. He didn't know of the incalculable amount of tree bark that was settling in the waters and would one day be labelled 'coal' and be exploited as a filthy polluting fuel and chemicals to once again ruin great parts of Earth and kill millions of people directly and indirectly. He could not know that the once huge variety of creatures had been ruthlessly culled with only GOD remembering how He and Jesus had created them in amazing variety for the delight of we humans and will recreate them down to their last hair just as revealed in Jesus's words: Matthew 10:30 "But the very hairs of your head are all numbered." He may have understood that the awful Nephilim and the hybrid creatures would all perish but could not have known that as the giant bodies died the half angel spirits emerged to wander the flooded Earth until such time as his sons gave birth to new bodies to inhabit as Jesus revealed at Mark 5:9 when upon asking the demon that was possessing a wild uncontrollable man its name it replied it was "Legion" because it was one of many living in the man and giving him the strength to break iron shackles and chains.

Too busy caring for the creatures Noah probably had no time to wonder what the future held and certainly could not know that Satan was still loose and hell bent on claiming as many human souls as possible in his mad futile attempt to upstage GOD. While the rain beat on the roof Noah and family kept checking all life and provisions on all three decks were safe and left tomorrow to take care of itself. The interior of the Ark was warm, dim and perhaps pungent with the smell of all the creatures and their foodstuff.

But where could all the water be coming from! Where indeed. Read Genesis 1:2?

Chapter Forty Five

FLOOD DYNAMICS.

As if the Six Days of GOD's Creation wasn't impossible enough for the Evoquacks to get their heads round the dynamics of Noah's Flood certainly is! It is essential to understand Flood dynamics before trying to understand how worms prove Earth is young and The Flood was a real catastrophe every recently.

To cause the Flood was a simple matter for GOD to 'shake the Earth' to crack the shell to create the upheavals and inundations of the Flood and cause the mass scouring of topsoil that washed into the waters and settled out as the layers we see around the world today and which fool all Evoquacks .

The myths, fables, lies and sneering doubts about the Flood are breathtaking and depressing in the way they show a total paucity of clear thinking – bad enough among the couch potatoes who blindly worship StarTrek, Planet of the Apes and Jurassic Park – but unforgivable in the ranks of scientists – the Evoquacks – Evolutionists - who have had the benefits of a college education that should have exposed them to the need for critical thinking! But No! the great unwashed masses and the sycophantic academics all stridently deny the Flood and its easily observable effects. Oh sure, they angrily denounce me and others like me who dare challenge their silly theories and misreadings of the evidence under their noses but do we ever hear one admit that they were wrong but daren't speak out for fear of ridicule from their peers? Never!

Before starting explaining the actual dynamics of just how a flood could have completely stripped all the topsoil and vegetation off the Earth along with every airbreathing creature and the human except Noah's family I will briefly touch on the various Flood myths and legends and collective memories from around the world.

From the north of Alaska to the south of Australia to the east of Africa to the west of South America many tribes retain and recount histories that include a great flood that swept away every living thing except a small family that escaped on a boat along with some animals. It seems indisputable that humanity retains a collective memory of a great flood in the not too distance past. All humans who have lived since Noah stepped off the Ark will have some sort of Flood legend though it may be very distorted and discredited.

These memories vary slightly depending on whether the tribe has a strong religious ethic or it is a more prosaic survivalist attitude and ethos. Thus some tribal elders when recounting the memory of the flood may attribute it to natural climate events while others will blame the catastrophe on the actions of mankind battling with a god or gods; and yet others will have the event as the outcome of the humans angering a great god in the sky.

The accounts rarely vary in the number of humans who survived and the size of their boat and whether or not they had saved a large or small selection of animals and birds.

Examples of these flood myths and legends are:

Sumerians,

Egyptians,

South Americans,

South Sea Islanders, etc.

There are so many that is pointless to list them all or their specific claims.

Thus as we look at our Earth today and note that it is about 70% water covered and actually may be 75% depending on whose calculations are accepted, we have to wonder how any planets can have such a water area with truly vast oceans and small seas with great depths while at the same time having huge barren deserts too hot for all but minimal life, and great ice caps mostly devoid of all surface life. How could such a planet have come about if it had its origins in Hawking'n'Dawkins' big bang of hot radioactive – hot what? Not air obviously, not vapour, not dust, not any form of water? Just hotness, hot Dark Matter? Hawk'n'Dawk's scenario is plainly for fools!

But, say the Evoquacks, but what about the strata exposed along valleys and on high mountains – surely all those layers of different rock must have taken billions of years and repeated accumulations of cosmic dust or constant movements of tectonic plates rising and depressing other plates before being overcome and depressed themselves? Just the Grand Canyon is a mile deep and Everest is over 5 miles high and both clearly show many layers of different rocks. Surely that is proof enough for a long history following Hawking's big bang? And what about the many buttes or chimneys of hard granites and basalt that stand in flat deserts around the world? Surely they are eloquent memorials to a turbulent past of burning eruptions and contortions of a new born earth being racked by growing pains as the scorching cosmic dust clumped together and formed hot spots that had to find release by blasting masses of melted rock up in a cone of loose earth?

Well, as the big bang is cuckooland fantasy and Earth's core is dry rock we can immediately ignore it and simply state that if actually a big bang had blasted cosmic dust everywhere - to make a universe! – and if it had decided to clump into lumps we need to have an explanation of how our Sun clump became a perfectly tuned ball of gas that ignores all the laws of gas physics dreamed up by the Evoquacks.

No-one can deny there are immensely tall peaks and columns of very hard rock sticking up all over earth and therefore a scenario is needed in which these peaks can be originally formed and then exposed and revealed. Cosmic dust isn't responsible but The Flood is!

The buttes are simply ejections of liquid matter extruded up through a vent in the crust some time after the first phases of the Flood when vast layers of sediments were all over the land with a great thickness of water over them. The butte material extruded up through the lowest sediments that were already thickening and hardening. This is easily demonstrated with a container filled with 50% water and 50% soil or sand and gravel. Shaking the container will keep the solids suspended in the water but immediately the shaking is stopped the solids will begin to settle and after many hours the bottom layer with be very dry and compacted. During the flood such settled sediments would have settled over vents and cracks in the crust and as the blocks of crust shifted vertically of laterally jets of new material would bubble up through the firm lower sediments and up through the soft sediments up to the water surface. The rising material lifts a slab of the hardened lower material which if thrust above the water level will dry and harden into rock. For many weeks the buttes would then harden surrounded by the hard and soft sediments until the drain down phase began. The draining away of very soft, firm and hard sediments would leave the harder dry deposits on the buttes exposed. The height of the butte is just less than the maximum depth of Flood water. Look at the Condit Dam videos to see how sediments slump and flow away.

Far too many Evoquacks lack the ability to visualise the Flood! And that is despite them having their heads filled with disaster movies showing every kind of flood, tempest, mountain slide, earthquake and volcano that the wildest Hollywood imaginations can dream up!

Practically everyone in the cold or temperate parts of earth have experienced sudden icy cloudbursts that drown, freeze and destroy people, animals and property so why they cannot imagine a huge continuous rain and its effects is baffling and only understood as Satanic brainwashing!

Until the advent of worldwide instant news bringing us disasters practically as they are happening many people could be forgiven for thinking that natural disasters were small local events. Thanks to the emergence of Youtube about 2005 we can research flashfloods dating back many decades. Here in Britain we watched the flashflood that hit Boscastle a few years ago as it progressed from scenes of the little stream through the town growing and rising to make crossing it risky and then it rising to cut the town in half before becoming a torrent that swept up and over its channel and began carrying cars out of the carpark and used them as battering rams to smash down buildings. Luckily no-one was hurt but the amount of water that came along and its power was and impressive demonstration. In June of 2015 a similar flash flood swept down Tblisi in Georgia and destroyed lots of property and left a great depth of mud on the streets – again it is obvious that water has enormous power to reshape land in a very short time!

The Boscastle (and Tblisi) flashflood was caused by heavy clouds sitting over the hills beyond the town and then dropping lots of water for a few hours. For a while the land soaked up the water but eventually it became waterlogged and the new rain had to start running down the slopes. At first much of it filled the many little natural streams and hollows until they could soak up no more and the water had to move on downhill. As the rain continued the flows became greater until eventually all met above the village and then filled the town's river until it became the raging torrent we all watched with fascination.

Naturally such a vast downpour over a hilly area could and did have enormous destructive power. The flooding river was filled with dirt washed off the land and it casually washed away a big tree with all the soil around it. The debris and vehicles smashed their way downstream until eventually meeting the main rivers.

Experts estimated that about 3inches or 75mm of rain fell from 1-3pm and then a continuous but lesser amount until about 7pm. Total rain has been estimated at about 2 billion litres or 400 million gallons! Water weighs about 10 pounds per gallon so the mass was about 2 million tons – an amazing quantity and weight to try put in perspective!

Anyone who saw the Boscastle flood expressed awe at the amount of water but accepted the expert's view that the disaster was due to Boscastle sitting in a valley surrounded by the hills that had caused the clouds to stop, pile up and dump their water. That is a perfectly reasonable and accurate explanation but trying to get people to believe that such a vast amount of water could fall all over a flat low lying country is impossible! Yet if it is accepted that soaked grass and top soil is heavy and will try to slide downhill then logically it will happen on any slope even if the slope is hardly noticeable. If Boscastle's slope was say 7% then the same amount of water on a much wider 1% grade would behave the same way and be an irresistible force – as any wide lowland river will prove!

This is the stumbling block of The Flood. Yes, everyone says, a cloudburst over hilly areas can cause small local flooding but it impossible for the weather to drop vast quantities on wide expanses of flatter land to cause the same raging torrents to strip away trees, people, houses and topsoil. Those people and learned Evoquacks who should know better all say it is cuckooland thinking to believe some GOD could order the weather to do that all round the world in one simultaneous worldwide downpour!

It is bad enough that Satan's followers believe and promote the flood as being merely some mythical retelling of a small Boscastle-type flashflood with a bit of pagan religion thrown in. Didn't Thor or Vulcan or some other Norse GOD cause huge storms to kills his enemies they say? But to have the ministers and clergy of the world's religions all smirkingly dismissing Creation and The Flood as fairytales just shows how completely Satan has fooled them all.

It therefore behoves me to explain to everyone just how the Flood came about and had such unimaginable enormous power that it destroyed all humans except Noah and family and all land creatures except those taken on the Ark by Noah.

First some water facts:

Water weighs about 62 pounds per cubic foot or almost exactly 1 tonne/metric ton per cubic meter.

A column of water 2000 feet high weighs about 56 tons per square foot or about 58tonne per 30cm square.

This equates to 2inches of water on one acre weighs about 200 tons or about 125,000 tons per square mile or about 50,000tonne per square kilometer.

If it is accepted that at the height of the flood Everest was fully covered and is today 29,000 feet high then the weight of water per square foot of land at the old sea level would be about 800 tons or about 75 tonnes per square meter – plenty enough to compress vegetation into coal. Mists occur when the air is wet and we know that the Superdomes have rain from all the moisture exhaled and sweated out by spectators at big sports events. Experts have calculated that in some areas about 5inches or 125mm of dew falls per year – if the Earth's original antediluvian climate was mild and moist then in fact all the land would be well saturated and unable to absorb much of the downpour and so the Flood effects would happen faster and as the moist soils were already in a good wet and loose states they would start flowing very quickly.

As the Earth's original climate and environment aided the growth of vast forests of vast trees ferns it is logical to think that the first rains of the Flood washed the edges of each forest into the nearest or newly made depression then a layer of general foliage and forest floor humus arrived followed by the ferns from the next band of forest and then the humus and so on until some valleys could have many layers of trees and vegetation all buried under the subsoils, sub-subsoils, rocks and boulders from collapsed cliffs.

The loose heavy boulders will settle first followed by the smaller rocks and then the gravels while the green vegetation and roots and any creatures will float for quite a while and be pushed out to sea. But during The Flood there could have been enormous and unknown currents surging around the planet so what we today see as oceanic currents and the Gulf Stream may not have been in existences or could have been modified completely and the boulders, rocks and general topsoil could have been pushed enormous distances over the sea beds

In these End Times the mechanics and progress of flashfloods are very well known.

A flow of warm moist air from a sea or ocean moves over a land area and rises up a slope or hill condensing into light cloud and then darker clouds. Clouds are water vapour and the darker the cloud the more vapour is up there. Eventually the top of the cloud reaches so high it gets cool and the water vapour condenses and starts to fall.

As it falls it gathers other droplets until it falls as the huge drops we are all familiar with during downpours. As the cloud was formed by wind pushing water vapour along so it continues to form as more air follows in due to the fallen water leaving space for more vapour. More vapour rises, condenses and falls as heavy rain. Continue this for many minutes and we have the 'cloudburst' that drenches everything quickly. Let it continue for hours over a concentrated area and the situation is ripe for a flashflood.

Flashflood damage can be shown by a simple experiment with some turves, some soil, a sloping board and a water hose. Fix the board at a shallow angle of perhaps 20-30 degrees, sprinkle a couple of inches of soil on the board, add the layer of turves. Start spraying with the water hose set to a fine spray. The turves will absorb the water initially and slowly release it into the soil until the soil is saturated to become mud and excess water flows off the board. Adding more water will eventually make the saturated turf so heavy that the wet soil begins to act as a good lubricant. As the the wet turves weight increases it eventually passes the critical point and the bond between the soil and the turf will break and the turf will try to slide down using the wet soil as lubricant. As the first piece of topsoil and grass tears loose and moves it adds it weight and bulk to that below it and that too eventually tears loose and so on until the grass sod is moving. Eventually the mass of turf and water will become greater than the 'stiction' of the slope and inertia and all will slide down the board.

Do this on a natural hillside many yards wide and a major landslip will occur. Over hillsides many miles wide the result will be catastrophic.

It may happen much faster if the water has made rivulets that concentrate water onto patches of grass.

The sudden slippage of the top layer of turf will make a disastrous landslide as videos show, but if the rain continues falling or is falling over a wider area it soaks and loosens the exposed subsoil which then liquifies it to mud that then starts to flow downhill. As that flows away it exposes lower layers to the rain which then also liquifies and runs.

Deep-rooted trees or vegetation on the slope will remain in place at first but as the rainwater adds weight to the canopy putting strain on the roots, and soil continues to be washed off the rootball the trees of all sizes will begin to loosen and collapse into the mud flow.

Continue this constant collapse, exposure, liquification, erosion on a wide area and soon a vast heavy mass of vegetation, mud and water is moving downhill.

If the rain continues over one or more days the total landmass under the rain will lose its topsoil and vegetation, then the subsoils and then the sub-subsoils until all the loose material has been washed away and only bedrock or very large boulders remain.

At first the mass flows round each obstruction in its path but as the flow continues larger pieces of flotsam get caught on the obstruction and increase its size – and thereby increase the total pressure of the mass on it until it collapses – which is why houses and bridges get pushed over.

(From Xmas to New Year's Day 2016 the UK had very bad flooding with several road bridges washed away and at lunchtime on New Year's Day I looked over my apartment's carpark wall to see what was happening with the little stream and saw it was very high and a small tree possibly 15 feet tall had been washed off land higher up and had been swept along until it jammed against the large sewer pipe across the stream. Not knowing who was responsible for either the stream or the pipe I could do nothing but hope the pipe held up under the force of water pressure. Luckily it did so but the vast amount of water running that day continued to run for several weeks as the rains continued and the soaked lands drained out.)

The Flood caused these enormous landslides all over the world but as the rains continued for 40 days and nights all the world's loose soils would have been turned to liquid mud loaded with humans, creatures, vegetation, trees, debris of all kinds and all would flow down to the nearest river and on into the new great depths under the Earth's crust. Debris from near areas would naturally arrive first while that from distant areas would take much longer to arrive.

Here is how to understand the sequence of events that occurred during the Flood.

Divide a hillside into five horizontal sections A,B,C,D,E with A at bottom of slope. Each section will have its own variety of living and dead vegetation, detritus, top, subsoil and sub-subsoil thicknesses with each having natural variations of soil and rock particles. Let rain fall equally on all the hillside. For some time the water will soak in where the soil is exposed or vegetation slows the water flow. After many minutes the topsoils will be wet enough for water to soak down into the subsoil. This will happen in random, unpredictable areas due to natural variations. The weight of the topsoil will now be very heavy and exerting a downward force restrained only by the grip of the roots in the subsoil. As the subsoil becomes wet and loses strength the roots will pull loose and slabs of topsoil will begin to slide downhill. If the slabs are big enough and the subsoil is very slippery mud the slab will slide downhill and continue until being slowed by obstruction or reaching the lowest point. As each topsoil slab slides downhill the subsoils are exposed to the full rain and will become saturated, turn to mud and slide- exposing the sub-subsoils. Next the sub-subsoils turn to mud and slide to expose lower sub-subsoils repeatedly right down to bedrock – but all this will happen in a random sequence.

Loose vegetation and detritus will start sliding first, followed by topsoil A, followed by topsoil C, followed by subsoil A, topsoil D, subsoil C, topsoil B, sub-subsoil A, topsoil E, subsoil B&D and sub-subsoil C, subsoil E& sub-subsoil D – and so on with topsoils, subsoils and sub-subsoils all arriving to the base of the hillside in quick succession. The layers at the base would be TA, TC, SA, TD, SC, TB, SSA, TE, SBDSSC, SESSD. The entire mass would be a total mixture of topsoils, subsoils and sub-subsoils interlaced with loose vegetation and detritus.

Have two opposing hillsides and the resulting mass of material would be even more mixed – exactly as happened during The Flood and confounds the Evoquacks today!

As the mass of material reached the base of the hill it would then seek to flow along the lowest possible route- down a river or valley and onto the sea or lake.

As the first mixture arrived at a low place it would settle in the layering seen all over the world such as in the Grand Canyon. **THEN WHEN MATERIAL ARRIVED FROM MORE DISTANT PLACES IT WULD FLOW OVER THE TOP OF THE FRST SEDIMENTS AND ALSO WASH AWAY SOME OF THE FIRST SEDIEMNTS TO GIVE MORE LAYERS WITH VERY DIFFERENT MATERIALS TO GIVE THE CONFUSING LAYERS SEEN IN SEDIMENTS IN MANY AREAS.**

However what all Evoquacks fail to understand is that The Flood was not just one heavy storm and then 40 days afloat in a boat but was 40 days and nights of rain accompanied by vast earthquakes and great geysers of water and minerals from under the crust as detailed in an earlier chapter. After the water stopped falling and emerging from inside the crust it was then driven about by wind to stir up the sediments again and again until they seem as mixed up as we see today. During this time the top layer of sediments under the waters would form a smooth level plateau over most of the land and smooth level floors in the great new valleys of the crust. Connecting the plateaus and the valleys would be slopes heavily cut with rills and gullies just as we see in the barren lands and the Grand Canyon today. The Condit Dam video illustrates the formation of plateau, valley, rills, gullies and scree perfectly.

Coal is usually found in many thick and thin layers separated by various sands, shales and stone which clearly is the result of the continuous stirring and re-stirring of the 300 days of active water action during the Flood. This strata can be seen all over the world and it is likely that if samples could be collected and analysed it would be possible to recognise the different stages of the Flood in them.

Those Evoquacks who claim the multi-layering is due to plate tecxtonics needs to go have their heads examined for signs of inability to understand the fizziks involved in climbing plates.

Not only that but when GOD decided to divide the Earth up after the Tower of Babel incident the crust slabs or tectonic plates – complete with all those live shellfish - were thrust up and down to create the mountains and the deep valleys and this would also cause further slumping, stirring and mixing of the sediments.

We know radiation is bad for humans and so it is logical to accept that Noah's atmosphere was much thicker and a canopy held so much water the sun was just a bright light rather than the blazing ball we see today. In England in spring we do have days when the sun is a pale silvery disk that we can look directly at and this is due to the amount of moisture in the air.

It was this moist thick atmosphere that made earth super productive so that all the animals had plenty of vegetation to eat. The main hook on which all evolutionary theory hangs is simply the fossils found buried deep in rock around the world. These, claim all the Evoquacks, are sure proof that there has been many stages of life on earth and each stage evolved from the previous one that is why there are only seabed creatures in the lowest rock and then a progression through fish, lizards and reptiles to birds and warm-blooded furry animals before finally along came human beings – all over about 260, 350, 450, 550 million years. And dinosaurs all died out about 65million years ago!

From these silly claims it is obvious that the Evoquacks have no ideas or consensus about evolution times and just pull numbers out of a hat as they poke about with old bones in the bottom of holes in the ground. JURASSIC PARK AND ALL THAT NONSENSE!

I designed my Fast Sedimentation Experiment to allow anyone to quickly see how nonsensical are all claims about the fossil chain being hundreds of millions of years old.

To perform your very own FSE all you need are a large glass jar that held fruit or preserves or else get a storage jar – any glass or clear plastic jar with a screw on lid will be fine. Go to the nearest field – preferably a field that has never been levelled or ploughed or bulldozed about – and get a shovelful of soil from about 8 – 12 inches, 200-300mm down. Break this into fine particles and remove all small pebbles. Get two raw chicken wings from the store and tear them up. From a pet shop get a couple of frozen mice as used for feeding hawks or other birds and roughly chop them into about six pieces. Go clip some flowers, leaves and twigs from a bush and tear it all up into small pieces. Look for some dead insects in the window bottoms or go kill some as they fly around the garbage bin. Turn over some stones to catch some beetles. Now put soil, chicken and mouse pieces, insects, leaves and twigs into the jar and fill to 70% with plain fresh water. Then vigorously shakes the jar as long as you are able while watching a clock with a second hand or a digital clock will do. As the clock shows the start of a new minute stop shaking the jar and set it down and watch to see how long it takes for anything to settle on the bottom. It will take less than a second!

That is right – in just one second some gravel will drop stationary while sand grains will settle out and then over the next minutes and hours all the other mineral, animal and vegetable matter will settle out.

After a few days carefully syphon off any water left on top and then study the contents of the jar.

At the bottom will be large gritty sand, next will be sands of various sizes and mineral content, then will come soft muds of large particles, then finer mud of very fine powders. But mixed in all these layers will be found bits of the plants, animals and insects.

Now do the experiment again but this time before shaking the jar put all the items and water in and then let it stand for a few days to thoroughly soak everything before doing the shaking and sudden stop and watch the slurry settle out and compare it with what happened to the first jar full.

The two jars will show identical sedimentation of the soils and the minerals in it. But the animal and vegetable matter will show different patterns of sedimentation.

The first jar will show lower layers of sediment almost completely of soil, sand and other mineral matter. Above these will be the first life material. A complete mix of animal pieces, fur and feather, with leaves and bark, and insect bodies. Above these will be very fine sediment and on top will be bits of woody twig and insect wings.

The second jar that was given time to thoroughly soak all the contents will have more vegetable and animal matter low in the bottom layers of sediment. Other less solid vegetable, insect and animal matter will be mixed in a thick layer of fine soft sediments or mud if you like. On top will be the hollow parts of animals, insects and vegetation that has internal air spaces or fermentation of natural juices to gas to provide buoyancy.

Dry and bake the jar's contents carefully and cut vertical slices from them and take them to an expert on sedimentation and biology and ask the question: How old are these two layers?

Both samples will show identical sediments of the soils – sand is sand and clay fines are clay fines and have to settle at their own natural speeds.

The estimate that each acre of land would have 2100 creatures does allow each one 2 square yards and we know that densities of small rodents can be much higher than that if there is plenty of food. Many farmers have hundreds of rats or mice in small feed stores? Make 1050 of the creatures small rodents at a density of ten per square yard and automatically the larger creatures have over 4.5 square yards each. Make another 500 medium size animals like squirrels, rabbits, cats, dogs and give them a square yard each and that gives the bigger ones get about 8 square yards each. We know that the US prairies had very tall grass and the weight and food load on an acre of such grassland would easily feed vast numbers of small animals especially as the vegetation would be growing as fast as some of the tropical plants do today. However the estimate of 2100 creatures per acre is a bit simplistic and does not allow for the fossils being those of animals washed across the land from now submerged land as that could reduce the 2100 down quite appreciably.

At this stage I have to introduce Charles Darwin's book as proof of origins of the great coal measures. While we Christians denounce Darwin as a tool of Satan because of his more famous EVONONSENSE book on Origins of Species it is his last book on Worms that explains the vastness of the Earth's pre-flood vegetation that became the coal measures.

This explains how the Evoquacks who try to estimate the potential topsoil wash off of the flood are totally wrong in their calculations.

The topsoil would not only turn the rivers into liquid mud but as much of it would be muddy humus and small green stuff it would settle out quickly to make the great coal beds and not be light and washed clean with rain to allow it to float for a long time.

The Time Team Phenomenon shows topsoil arriving at a rate of say 2 feet per 1400 years so in the 1600 years from creation to flood the whole earth would have grown a vast tonnage of loose easily soluble topsoil covered with dense vegetation.

Add to this tonnage the subsoils and ground down rocks washed off exposed hills and it is no wonder there are great depths of sediments with many layers of fossils to be found around the world.

Not only that but we know many fish fossils show the fish to have been buried alive which points to the water being more mud than water.

The 3 or more vast layers of flints in the chalk/limestone are the cavities left by all the jellyfish – they died and their bodies being 95% water quickly evaporated into the lime and then silica trickled into the impression. Interesting many flints contain bits of coral or shellfish which was the creatures last meal while many alos have a core of 'flint meal' more limestone which would no doubt have been ingested by the fish at the throes of death.

To try grasp how the Flood's Fast Sedimentation and Diagenesis fools the Evoquacks it is necessary to look at the nonsense they constructed around the various layers of rock, soil, clays, shales, sands under our feet.

These – the strata shall I say - seem to be a good means of dating Earth's history as the same strata can be found over large areas and even either side of seas and oceans.

As we walk on the topmost of these layers - and in England the top layer is almost always rich soil with plenty organic material, small creatures and the remains of generations of small creatures - it seems logical that if a different layer of material is just below the topsoil then it must be older; and if another layer is different material again then that must be older and so on until the geologists arrive down at 'the bedrock.' They gave each of these layers an impressive sounding name, then over the years they bandied these names about and publish their thoughts about them in the academic press or in lectures at universities until the names have accreted a whole mythology with supporting fairytales that make StarWars seem quite sober.

The list of the names they assign to the layers include the Cambrian and others listed previously.

Many of the layers seem quite distinct and separate from lower layers and seem to have their own range of fossils. In some areas the range of fossils in a layer are different from the same layer elsewhere which really baffles the Evoquacks though they generally make up some EVONONSENSE to cover up. The excuse will usually be to date the layer differently as according to their thinking a few millions years is sufficient for one creature to evolve from something else so the new fossils must obviously be from something that evolved a bit over time.

To add to the confusion many of the layers of strata are uplifted or sunk as though earthquakes had been a regular feature of prehistory.

Further confusion comes from some layers having loops and swirls as though it had all been moved while still wet. I know exactly such a tight loop in Scotland where a new road has been cut through a small hill. The wet layers looped up about ten feet over the lower level layer. The lower layer must have been set and dried when the top layers flowed over it then the flowing sediment was blocked and piled up in several layers to make the loop. The looped material then dried and hardened before being covered with further material.

All over the world are vast expanses of barren desert and bare rock as well as deep frozen tundra and icecaps that show no signs of growing a covering of hardy plants let alone and trees to provide homes for birds and shade for animals and humans; but all over the British Isles there are quiet graveyards that are lush green oasises in the middle of busy bustling overcrowded towns.

In the far corners of any graveyard can be seen the outline of untended graves complete with stone slabs and edgings. The grass and weeds that covers the graves is thick and on a base of rich humus that provides excellent rooting and water control. This vegetation has developed in as little as 40-50 years – easily dated by the lack of respect so many British feel to their dead these days. Slow motion filming over those 40-50 years would show that the bare stone slabs seeming to offer no fertile ground for grasses the grass nevertheless managed to slowly creep over the stone and build itself a fine soil as it did so. A grass plant sent out roots to explore, they found support in the roughness and joins of the stones and settled in and pushed out shoots that reached to the sky to take the sun's energy and they grew strong ad fed the roots to encourage them to send out other shoots and other roots and soon the grass had captured a bit of the gravestone territory all for itself. And slugs and snails and creeping things smelled the new grass and scaled the stone to browse the tender shoots and in season the birds and mice came seeking seeds of the grass and all these slithering , sliding, walking and flying creatures left their own fertiliser and the grass accepted it all and converted it with sunlight and water to more grass and that brought more creatures and one days the last speck of gravestone disappeared under a probing root and the grave became a green grass hump among scores of others.

Hitler's V3 gun site at Mimoyeques and many other gun bunkers give more proof of how Earth is young and GOD's designing it to be evermore fertile. Video shows the solid concrete has grown a nice mat of vegetation in 73 years. How thick will it be in another 1,000 years?

Another item of proof also dating back to World War 2 is the abandoned airfield featured in the video 'In Search of Bomber Command' made in 1978 and quite clearly showing the concrete slab roadways about 33% covered with grass as no vehicles had been through the gates since 1946. If that airfield remained locked shut until today it is certain the roadways will be entirely covered with vegetation and topsoil just as so many of the Roman and

Norman ruins are when shown on Time Team.

Here in Driffield, East Yorkshire an abandoned concrete products factory yard is now like a jungle after just 8 years. I contacted a few people about having it made a project by local school to catalogue the total variety of flora and fauna but there is no interest in it.

When I was little 'trainspotting' was all the rage and any spare time in holidays and weekends found scores of kids sitting alongside the railway lines watching the big dirty steam engine chuffing past as they had done for about 120 years. It as a scene we British grew up with and in about 1956/57 we could no more imagine that in just a few years the last steam engine would be dragged off to the scrapyard to close an era than we could imagine what it would be like to live in one of those luxurious white mansions by a sparkling blue sea. I tagged along with my brother and his friends and drifted down to the railway but while he and they tried to sneak into the engine sheds to clamber onto the engines I liked to go sit on the abandoned railway banking behind the working line's banking.

By banking I mean the long mound – like a 'prehistoric burial mound' that was the waste soils moved aside to make a railway line through my areas hills. Back n 18?? it had been decided that the town could benefit from having the railway come through and get people, raw materials and finished goods moving and an optimistic promoter had organised surveyors and contractors ad soon gangs of navvies had been hard at work cutting the line through the hills and as aesthetics was not a consideration the best place to put all the waste material was alongside the new trackway. There was so much material that it was piled up and took the form of a tall barrow perhaps 40 feet high with a top about 2 feet wide and stretching hundreds of yards – just like a long barrow in fact. And then for some reason or other the management decided the new line wasn't direct enough and it was decided to survey and excavate another line almost parallel to the first and so they did and that line also gained itself a long barrow of waste material before the whole scheme suddenly seemed superfluous and was abandoned. All these years later the two barrows stretch along just as the long-dead navvies left them and almost as I remember them. What fascinated me over 50 years ago was that the twin barrows – the 'banking' to us kids – was practically devoid of cover. It was all loose grey shale crumbly and slippery to walk down. Only in one spot of the second barrow was there any vegetation. A slight depression had managed to hold moisture, a seed had settled, sprouted and prospered and soon made itself a clump. Bird or wind had brought a alder seed and soon the spreading clump of grass had its own sapling – alder being one of the few trees that thrive in really poor locations. The grass was not the soft green stuff to delight a beast's mouth – it was dark thin shooted spiky stuff but it was grass. Rabbits obviously came to take it and birds came for the seeds, and all fertilised it and so over I don't know how many years the little green oasis had become established and extended to perhaps one square yard/metre on the thousands of yards of bare grey shale. In 2014 I went to look at the bankings again and found the original clump had trebled in size and the sapling was now a typical scrub alder ten feet tall and low spreading. The grass was still the dull spiky stuff. Rabbit droppings showed it had some attraction for them.

Walking about both the bankings I found other small clumps of the same rough grass had managed to get a start and built their own clumps but most were less that a yard/metre square. Somehow all that grey shale was supplying their needs enough to think that in perhaps another 2-300 years the whole top of the bankings might be grassed over with small stunted trees ? The sides of both bankings were as devoid of vegetation as ever. It will take a miracle to get grass to root and sprout on those barren slopes but given several hundred years the top grasses might spread out and start down those sloping sides?

Ivy can grow up walls and over marble and granite grave stones by its roots being able to lock into the tiny interstices in the stones but also by secreting an acid that can dissolve the stones so that when removed the ivy will have left an image of itself. Curiously enough this is another World War 2 link as the penetrative power of roots is why after 70years the great thick concrete roofs of Hitler's underground bunkers are falling to pieces with curtains of roots hanging down from the cracks!

If bare rock and loose shale cannot regenerate but old graveslabs can be grassed over in fifty years then in say 10,000 years since last ice age the topoil should be 10000 divide by 50 = 200 = 200cm = 2metres or about 6.6 feet.

Whither all the bare rock and sand?

As I type this I find a post from LiveScience entitled 'Missing Mega Poop Starves The Earth,' The article is basically saying that lack of large sea fish, overfishing of smaller fish, lack of large animals and lack of fish eating seabirds have all combined to make the oceans and land much less fertile than 'at the end of the last ice age' due to lack of whales, elephants and fish to move nutrients from the deep oceans to the shallows and from land onto the seas and vice versa.

While all this may be basically true we have to remember all the other reports of nutrient rich rivers carrying excess fertiliser into the seas as well as much of today's mass production food facilities pumping vast quantities of nutrients into the water chain either directly via waste into sewers or indirectly as human sewage.

It is a fact that a century or more of inshore trawling has removed much of the seaweeds that would have converted sunlight into food and captured various nutrients coming down the rivers of from birds and fish but equally the increased amount of seasonal and unseasonal flooding will be washing vast quantities of nutrients off flooded land with deep snows doing the same.

The report author fails to understand the problem has been building up since The Flood not the ice age as the flood sediments would have effectively blanketed much shallow sea vegetation as well as stripping all the land vegetation and creatures and buying the vegetation and trees as well as the bodies in deep sediments where their nutrients are locked in and could not be recycled. Perfect examples of this are in the 'muck' above Alaska and Siberia and in the Gulf of Louisiana where rich sediments lie up to 2miles or more deep and must be holding incalculable nutrients captive.

While overfishing has emptied the oceans we have to remember that the vast shoals of fish don't actually release their nutrients for the benefit of the land to any great extent as smaller fish are eaten by bigger and bigger ones when dead are eaten by small scavengers. Seals and seabirds that deposits vast acreages of guano get that guano by eating masses of fish in a fairly closed cycle – humans breaking that cycle and harvesting the guano to put on land may seem to be decreasing the overall cycle load but then we have to remember that putting guano on plants does cause the plants to grow more and collect more sunlight to produce their mass and their roots to gather minerals from the subsoil and rocks. Feeding the plants to humans, animals, fish or birds puts the guano back into circulation as these creatures all deposit their wastes. Burning or burying wastes does take nutrients out of the chain.

The Sargasso Sea is often spoken of as extending for over 1.2 million square miles and is seen as a placid area where top and lower water do not mix much. It is usually thought the weed originates in the Gulf of Mexico and slowly drifts north and then circles into much denser mats in which small boats can get stuck and big ships can be slowed down.

It would seem that the topwater drift of sargassum weed ought to be mirrored by lower stratified layers of richer water to feed the sealife in the are of the Sargasso but this is not seen – almost as though the water underneath is fairly sterile.

My screensaver is closeup view of Grand Canyon with isolated bushes establish. Lack of worms means each bush cannot spread far or fast but each one confirms that GOD created seeds to be able to germinate and grow in very hostile situations.

In my area we have lots of maidenhair fern and a small purple flower that grows in cracks of stone walls despite logically having neither nutrition of reliable moisture as we an have 2-3 weeks without rain.

Chapter Forty Six

Princeton Evoquacks fooled by The Flood! EUREKA!

UniverseToday.com has big woozle headline: **BACTERIA FOUND DEEP UNDERGROUND.** The article continues: A Princeton-led research group has discovered an isolated community of bacteria nearly two miles underground that derives all of its energy from the decay of radioactive rocks rather than from sunlight...self-sustaining bacteria found in deep gold mine...seems to live on 'geologically produced hydrogen and sulphur compounds for nourishment'...the bacteria are truly unique...analyses of the water they live in showed that it's very old and hasn't been diluted by surface water...in addition the hydrocarbons in the environment did not come from living organisms, as is usual...the source of the hydrogen needed for their respiration comes from the decomposition of water by radioactive decay of uranium, thorium and potassium...because the groundwater the team sampled to find the bacteria comes from several different sources it remains difficult to determine specifically how long the bacteria have been isolated...the team estimates three to 25 million years... implying that living things are even more adaptable than once thought.'

To strip all the EVONONSENSE and woozle out of this report: it boils down to researchers went down one of the very deep South African gold mines and in a sample of water in the cracks in the quartz found some bacteria that like anaerobic conditions.

As the team toss in terms like 3 to 25 million years it is obvious they are all addicts of Planet of Apes and that colours everything they believe.

Knowing that the gold miners are drilling for gold in quartz that is very cracked, unstable, experiences many small earthquakes each day, and is full of hot water – exactly as the Germans found when they drilled deep into the Earth – and knowing that to start the Flood GOD gave Earth a great shaking to shatter the granite and quartz to release all the water locked under the crust since Creation Day 3 – it ought to be obvious that the bacteria got down to where they were found by the floodwaters draining back into the fractures in the quartz. The same fractures are filled with the fine gold that was dissolved out of the Earth by the hot water at the start of the Flood.

The Flood sequence was: Crust shattered – high pressure hot water dissolved gold and blasts up out of Earth - gold is deposited in the shattered quartz - the quartz sets into gold bearing reefs as found all over the world – as hot water is exhausted the space it emptied from fills with cool water containing anaerobic bacteria draining down through cracked quartz –

German and South Africans drilling deep find hot water running through shattered quartz - University team finds the bacteria and go into full Evoquack mode - Rose White debunks Princeton Evoquacks and their silly ancient bacterias. EUREKA!

    Footnote: Is these bacterias related to those that live around black smokers?

Chapter Forty Seven

SATAN, FALLEN ANGELS, NEPHILIM and DEMONS

    The confusion and disbelief about the reality or existence of Satan, Fallen Angels, Nephilim and Demons is such that anyone attempting to clarify the matter in these End Times is treated as a fool or mental defective and their explanations are met with condescending smiles, or more likely, Schopenhauer Stage One and Two resistance. However being long used to such attacks and having the Spirit of Knowledge and Truth and the armour of GOD to protect me I'll clarify in the simplest terms the origins and reality of Satan, his gang of Fallen angels, their offspring the Nephilim and their inhuman spirits – the demons!

    To verify the truth about Satan, fallen angels, Nephilim and demons it is essential to read the Book of Enoch and understand the role of Enoch in the Bible chronology. We have to thank Satan for the Book of Enoch not being included in what we now call the Bible although Jude 1:14 refers to it. Satan made sure his slaves who assembled at Nicea in 325AD dismissed Enoch as being nonscriptural when really it details his sin in plotting against GOD and corrupting heaven and many other angels. Unfortunately for Satan and all his friends, slaves and promoters we can be sure Enoch is true as fragments of it were found in the Dead Sea Scrolls.

    SATAN. GOD created Satan as a wonderful heavenly spirit named Lucifer before Creation of the Universe Day One. Being spirit needed neither food, water or sleep he was free to instantly fly through space and time to roam the entire new universe and be able to walk through fire, water or solids as easily as the new humans walked through the oxygen of Earth.

    Satan began to feel slighted the moment GOD and Jesus created Adam and Eve and expressed themselves well pleased with the humans. He had no reason to be resentful as he could seek the direct company of GOD face to face while Adam and Eve had to remain on Earth and needed oxygen, food, water and sleep. GOD ordered him to be a guardian cherub to watch over the new humans which stung him to his core. He allowed his imagined slight to fester until he arrived at a comforting solution: he would make himself a throne higher than GOD's! This is in Isaiah14:13- "You said in your heart, 'I will ascend to heaven; above the stars of God I will set my throne on high; I will sit on the mount of assembly in the far reaches of the north".

While invisible and watching Eve wandering innocently about the Garden when Adam was off working elsewhere Satan spotted his chance to ruin Creation. We don't know how 'old' Eve was by then and maybe she might even have been just a day or two 'old' if she didn't realise that serpents could not speak? Satan stood invisibly behind the harmless serpent and by synchronising his words to the serpent's constantly flicking tongue he insinuated to Eve that GOD had denied her the forbidden fruit because He didn't want her to be like Him in knowing right and wrong – but what could she know what 'wrong' could be as she and Adam had been told to enjoy themselves in every aspect of their lives? Nevertheless Satan's words seemed reasonable so she plucked and ate some fruit and gave some to Adam. Satan then stood back to watch what would happen when GOD came to spend time with the human pair.

GOD realised they had eaten of the tree and cursed both of them and all the Earth. He reserved His worse punishment for the real criminal,Satan, by telling him that Eve's offspring would one day crush the head of Satan's offspring. Satan as a spirit has no crushable head but will be tossed in a lake of fire after the final battle of Gog and Magog. (Read Revelation 20 for the timetable of Satan's end.) Neither Satan nor Eve had offspring at that time which shows GOD had determined to allow Eve to procreate just as Satan was left loose to wander between Earth and heaven and produce offspring of his own. These Satanic offspring were to be the Nephilim which are now practically extinct but survived by vast numbers of their spirits – the Illuminati and the demons.

Jesus can trace his ancestry right back to Adam and Eve and so is directly her seed and destined to crush Satan completely on his return to Earth. Genesis 3:15.

We are not told why GOD allowed Satan to survive thousands of years before being destroyed or why he was allowed to roam freely over Earth and heaven itself. It is tempting to think that GOD granted Satan six thousand years to try subvert GOD's rule of Earth by equating one day of Creation with one thousand years as in the oft-quoted 'One day with GOD is as a thousand years.' But no matter how the date is calculated it seems it may be well over 6,000 years since Eve fell. Maybe GOD is being merciful and extending His deadline in order to have as many humans hear the Gospel as possible as this aim has been obstructed by the many communist tyrants of the last century. Now as Muslims rampage across the world and the pope demands Christians join Jews and Muslims and himself in establishing a 'New World Order' to rule Earth it seems that Satan's days of destruction are about to reach their awful finale.

For some hundreds of years after Adam's creation Satan seems to have bided his time while constantly plotting how to harm the humans and have them turn away from GOD. Genesis 5 gives the genealogy of Adam and his descendants during this quiet time and how they had multiplied and spread across the land. Adam and Eve's children and the next few generations were all still quite perfect in their physical shape and naturally as the highest of GOD's creations would all have been fine attractive physical specimens no matter what actual colour their skin, eyes and hair.

Though all of the first gang of fallen angels were bound in chains we can know from Daniel 10:13 that there was a powerful angel who detained the angel sent to Daniel until Michael the archangel came to help. 10:13 "But the prince of the Persian kingdom resisted me twenty-one days. Then Michael, one of the chief princes, came to help me, because I was detained there with the king of Persia".

Daniel wrote this about 165 BC. Revelation 2:13 tells us that Satan was the king of Persia – and controlled the king of Pergamos! Therefore Satan had not been bound along with the fallen angels.

Satan's Earthly kingdom was based at Pergamos in those years as the priests of the Baal/Sun had left Babylon after the Persians overran it. This is an excellent example of how Satan rules both sides and plays both sides against one another! Satan had raised up Nimrod and induced the Babylonians to worship the sun and especially in a great ziggurat we call the Tower of Babel. After GOD destroyed the tower and confused the language the Babylonians and their worship of the sun as Baal moved to Pergamos and remained there until about 133BC when they moved across the sea to Rome and there infiltrated the Roman church to make it the devilworshipping demonic institution it is today with the sun worshipped as Mary queen of heaven.

Writing in about 570BC Ezekiel the prophet 8:16 'And he brought me into the inner court of the Lord's house, and, behold, at the door of the temple of the Lord, between the porch and the altar, were about five and twenty men, with their backs toward the temple of the Lord, and their faces toward the east; and **they worshipped the sun toward the east**. 17Then he said unto me, Hast thou seen this, O son of man? Is it a light thing to the house of Judah that they commit the abominations which they commit here? for they have filled the land with violence, and have returned to provoke me to anger: and, lo, they put the branch to their nose.'

The branch was some sort of aromatic plant which the priests burned and inhaled just as so many druggies inhale various herbs and weeds today – and no doubt they had the same hallucinations and euphoria - and their sun worship with churches aligned to the east as Muslims mosques are is the true **'abomination of desolation'** before the **'rapture of the faithful true Christians'** Jesus spoke of at Matthew 24:14:

14 And this gospel of the kingdom shall be preached in all the world for a witness unto all nations; and then shall the end come.

15 When ye therefore shall see the abomination of desolation, spoken of by Daniel the prophet, stand in the holy place, (whoso readeth, let him understand:)

16 Then let them which be in Judea flee into the mountains:

21 For then shall be great tribulation, such as was not since the beginning of the world to this time, no, nor ever shall be.

22 And except those days should be shortened, there should no flesh be saved: but for the elect's sake those days shall be shortened.

31 And he shall send his angels with a great sound of a trumpet, and they shall gather together his elect from the four winds, from one end of heaven to the other.

Satan is leading the world rulers into joining the pope to set up the One World Government or New World Order which will institute the building of a Third Temple in Jerusalem - the moment it is finished and the opening ceremony is the worship of the sun then the end will come.

FALLEN ANGELS. Using the word that is so much used in the news these days we can say Satan had been trying to 'radicalise' other angels with promises of a wonderful life if they followed him down to Earth until eventually two hundred of them swore an oath to him and followed him down to Earth to land on Mount Hermon. They looked down and saw how the humans were farming the land and building houses and towns but also saw how the humans took every opportunity to indulge in delightful exhilarating sex that seemed the most wonderful experience and produced perfect replicas of the humans. Envious and believing Satan's fantasy of establish a kingdom on Earth the angels determined to try this sex activity to raise their own nation on Earth.

They materialised gorgeous bodies to impress the women and walked among them flirting and seducing all they wished. These angels told the women how eating and smelling the mandrake root would make them feel amazingly aroused – Adam and Eve surely had no need of aphrodisiacs – the angels bodies would have continuous energy for unlimited sex. Many sex addicts today worship men and women who have great sexual stamina and endurance so perhaps the angel's physical attraction was backed up with offering the women better sex than the men? Also they instructed the women in what roots would cause abortions and also prevent fertility – so the women would remain constantly available for sex! This is written in Enoch but of course Satan ensured Enoch was ridiculed at Nicea. The angels 'married' all the women they wanted and of course they could have multiple wives around the whole area by simply using their angelic powers to instantly travel from one place to another wherever they had wives waiting or new conquests available. Several Bible verses compares the fallen angels to trees having many branches and each branch being a wife.

Lamech was the seventh in line of descent from Adam and it was he who first took two wives contrary to GOD's decree of one man-one woman. Adam was still alive at 874 when Lamech was born and his taking a second wife by say Year 900 would neatly fit in with the Earth by then being filled with attractive women to tempt the watching angels to sin. Could Lamech have been susceptible to 'radicalisation' by a fallen angel in a human body telling him how delightful it was to have two or more wives especially as the first wife may be unavailable for many weeks of late pregnancy and while nursing? Or did he take two wives due to using mandrake or other aphrodisiacs? (Interestingly Lamech was aged 777 when he died five years before the Flood came on the Earth.) Lamech came down Cain's line and was cursed and the offspring of the first murderer, and if genetics are believable then like-father like-son could allow Lamech to think it a perfect good idea to take a second wife. The daughter of a fallen angel would perhaps be well versed in beautifying herself with antimony eye makeup and the use of herbs and roots and just as today's gangster's molls flaunt their beauty and dress provocatively it is likely Lamech's choice did so too. If his wife was tainted with the fallen angel DNA and Ham chose one of her daughters as a wife to live out the Flood it would explain why Ham's line became the Canaanites and had so many giants associated with them and why GOD so contemptuously told the Israelites to slaughter them all. Noah and Ham came from Seth's line and was righteous during the terrible pre-Flood times.

A few weeks ago I called at the antiques and decorator centre about two hundred yards from my home and find they have a modern marble sculpture of a winged angel with prominent genitals seducing a lovely young woman! This replicates the procreation of the Nephilim! But who is buying these Satanic pieces? Are they being bought naively as simple artworks or specifically by people who actively worship or promote Satan? Go on the internet and surf for Mozart's Lacrimoza to see a similar statue.

Were fallen angels the first to experiment with homosexual, lesbian, pedophilic and bestial sex and passed these immoral urges on to their offspring?

The materialised bodies were complete in every degree to fertilise the women's eggs but the DNA of Satan and the angels was an imperfect match and the resultant babies were apparently all male and oversize and quickly grew to enormous size but worst of all they had evil temperaments and motivation. It is logical to think that trying to birth these babies would cause injury or death to the mothers but the lure of the angels was so strong that they could find new conquests until Earth had enormous numbers of giants – the Nephilim.

The fallen angels used their superior intellect and their secret knowledge to teach the humans how to discover ores and minerals in the soil to make metal for tools and weapons.

They began crossbreeding animals with each other to produce odd hybrids for amusement, sport or warfare just as today Americans breeders fierce fighting dogs, Mexicans breed fighting cocks and the Spanish breed fighting bulls. The angels then moved on to mating with animals to produce the half human creatures such as the horses that had the top half of a human body or the goat with human top half that is today worshipped as Baphomet by the Freemasons and Jehovah's Witnesses. Their hybrid lion body with a woman's torso and head is epitomised by the sphinxes of Sodom and Giza. Once again the art centre close to my previous home had such a marble statue – was it just a cute novelty decorative item or for serious occult use?

GOD watched all this sinning until deciding to intervene and He caused The Flood and had all the fallen angels bound until the end time. 2 Peter 2:4: 'For if GOD spared not the angels that sinned, but cast them down to hell, and delivered them into chains of darkness, to be reserved unto judgement."

As Paul at 1 Corinthians 6 says some humans who survive to live and rule on Earth with Jesus will judge angels it may be that the Prince of the Persian kingdom is one such angel who forgot his place and duties.

NEPHILIM. These are Satan's evil seed that Eve's seed will eventually crush. They were half angel-half human babies with defective DNA that matured from spoiled and demanding children to greedy gourmands eating all the usual food until only flesh and blood of the domestic and wild creatures would satisfy their cravings. They grew into fierce and warlike adults banding together to build cities complete with giant structures for defence or idolatry. The monstrous stones they built with still exist as testimony to their size and strength. Many Bible and non-Biblical books speak of their size with many described as tall as cedar trees and the tallest given being about 500 feet. This sounds like complete fantasy but is confirmed by a brief study of the many stones in the Baalbek temple and unfinished ones in the quarry ranging in weight from 500 to 1500 tons with the heaviest being 2000 tons proving the Nephilim must have been huge enough to lift and place stones as easily as modern human can work with stones weighing perhaps one-twentieth of a ton.

Their hunting and destruction of the creatures was distressing to the humans but the Nephilim were uncontrollable and eventually began hunting and eating humans and drink the blood.

We are not told of the sexual habits of the Nephilim but hopefully these great men did not try to mate with the women.

Enoch wrote eventually GOD realised that unless He acted the whole human race would perish and so with a mix of sadness and regret He determined the time had come to act and after Noah had finished his work and was safely closed up in the Ark He smacked Earth.

The end of the Nephilim also came during The Flood as their half-human bodies drowned. Victory over Satan's angels was simultaneously accomplished by Jesus and the faithful angels rounding up all the fallen angels except Satan and binding them 'in chains of darkness' awaiting Judgement Day. Their destruction as recorded at 2 Peter 2:4: 'For if GOD spared not the angels that sinned, but cast them down to hell, and delivered them into chains of darkness, to be reserved unto judgement."   Just this afternoon I was looking at the website of www.GODandscience.org and saw the demons of the Bible times and today described as being Satan's gang of fallen angels!   This is a lie and woozle as in fact as Peter wrote above the fallen angels are still chained! How easily most Christians prefer lies rather than accept the truth given in GOD's scripture!

When Noah finally stepped back onto land 371 days later the world was transformed, gone were the people, their homes, shrines and idols; gone were the hybrid animals and the giant Nephilim; gone were the sacred fires and sacred trees; gone were the familiar fields and rivers, valleys and hills, trees and bushes; now Noah saw huge mountains with volcanoes and steaming geysers, raging rivers roared between jagged cliffs; at his feet green grasses spread a green sheen over most land with seedlings and saplings trying to raise spread their shoots, above his head were billowing clouds driven by strong winds; the sun seemed brighter and felt hot on his face, but best of all high across the sky under the clouds was a diaphanous arc of red, orange, yellow, green, blue, indigo and violet colour – the rainbow sign of GOD's covenant to never drown the Earth again.   He was unaware that now the apparently pristine clean Earth carried an invisible army of deadly demons intent on causing whatever mischief or evil they saw fit – perhaps even as directly ordered by Satan!

The DNA of the Nephilim appears to have passed through The Flood as stated at Genesis 6:4- 'The Nephilim were on the earth in those days--and also afterward--when the sons of God went to the daughters of humans and had children by them'.

We have to accept that either Nephilim DNA was present in Noah's son Ham's wife leading to more Nephilim being born to cause their trouble in the centuries after The Flood ,as recorded at Numbers 13:33 and Deuteronomy 3:11 and at Joshua 3:12 which also tells of the Nephilim living in Ashtaroth, or else accept that Satan was able to fertilise more human women after the Flood. We are not told if Satan was somehow unable to mate with more women after the Flood but there are many accounts of women submitting to an angel during occult ceremonies which generally have a high sex content. The constant reports of incubus and succubus activity and the physical reality of these beings does suggest they are fallen angels rather than demons. The Witch of Endor was truly frightened by the reality of what claimed to be the reincarnated Samuel. We are told Satan was free to walk about the Earth so obviously he had the chance to seek mates and it is conceivable that he did so to continue producing the many hybrid and alien creatures on and around Earth. Can the aliens in UFOs seen by astronauts actually be either more fallen angels of Satan's hybrids as it seems the demons are restricted to Earth as they need either a human or creature's body to inhabit? EUREKA!

As GOD described Noah as righteous and blameless but many of Ham's descendants were described as having giants in their ranks as the Israelites of the Old Testament found when they were attempting to possess Canaan, somehow the Nephilim DNA had polluted Ham's bloodlines known as Raphaim, Emims, Canaanites, Amorites, Zuzims. Numerous times Ham's descendants were described as great men, of great size, fearsome and terrifying. This shows they had once again resorted to tricking the humans into building altars to Baal and erecting Ashtorath stakes beside them. Today's Jehovah's Witnesses Freemason leaders may carry Nephilim DNA or are possessed by demons as they insist Jesus died on an Ashtorath stake as a sacrifice to their god Baal whose image they constantly subliminally insert into the publications!

History is full of the names and crimes of evil rulers and generals who displayed gross venal appetites during their rise to their peak and while they stay in power, and their tortures and murders of innocent victims as well as all who cross them is as sickening as today's news images coming out of ISIS controlled territories. Just yesterday I watched a program examining a castle built in England about 900 years ago and one of its lords was described as so barbaric that his favourite pastime was having men hung on hooks and left to die in agony as the hook dug deeper into their bodies; he also liked gouging out the eyes of people who had angered him! He must have possessed by one or more demons but the psychoquacks would just say he needed help controlling his appetite! Do such evil people actually carry the DNA of the fallen angels?

DEMONS. Unfortunately while the Nephilim and hybrid creature's airbreathing bodies drowned the angelic spirits survived and were set free to roam upon the Earth. Reports of hauntings, visitations and apparitions tell us demons can materialise bodies of every degree of attractive, handsome, homely or hideous appearance. They can easily enter into the body of any willing or susceptible human or creature as shown in videos of the charismatic and Pentecostal church leaders such as Kenneth Hagin or pastors of Holy Laughter churches .

They can induce the unclean sexuality which is so common in so many religious communities and organisations. It may be the demons were the first to experiment with homosexual, lesbian and pedophile sex or did they learn this immorality from their fallen angel fathers? They are still lusting after human woman and men sex by being the incubus and succubus that can create solid bodies to pester either males or females in their sleep or when demons are deliberately conjured up in black magic, spiritualist, clairvoyance or similar situations. Just a few weeks ago I read of a widow being wakened by the embrace of one of these incubuses pretending to be her dead husband coming to comfort her. She said the parting kiss was wonderful!

Demons can somehow give their human host amazing strength and powers in ways we cannot understand. Some demon possessed people can be uncontrollable by groups of today's police officers or medics just as the seven sons of Scleva were no match for the demon they tried to cast out of a man. Jesus also cast demons out of man who had the strength to break iron shackles. How can demons make puny human muscles do these amazing feats? Read Acts 19:11 and Mark 5.

 Demons can also somehow make their invisible bodies have weight and substance enough to move heavy furniture or cause loud knocking noises as well as speech and laughter but how they do it we cannot know any more than we can know how Jesus could materialise in the upper room in front of Doubting Thomas.

The demons are enjoying themselves doing Satan's tricks while they can but can easily be sent away by using Jesus's name to make them shudder with the horrific knowledge that soon they face total destruction. Read James 2:19. This is, once again, something the church leaders do not wish to get involved with and any mention of the reality of demons will be swiftly deflected. All church leaders claim that special permission is needed to cast out demons – that's if they don't dismiss possession as merely being a person having some bad dreams or reaction to drugs and rock'n'roll.

I and many other Christians who tell the truth of Fallen Angels, Nephilim and Demons get sneered or cursed for it by all manner of religious leaders, scientists and evolutionists who cannot do the research, just as Schopenhauer said: All truth passes through three stages. First, it is ridiculed. Second, it is violently opposed. Third, it is accepted as being self-evident.

I'll restate my truth: Satan induced many angels to follow him in sinning with human women and produce the Nephilim whose bodies were then drowned during The Flood to release the demons who live in human and animals bodies until the body dies then they seek another body.

The evil spirit world hierarchy is Satan, fallen angels, Nephilim, demons.

Jesus too can operate as a spirit creature beyond the reach of humans. 1 Peter 3:18 tells us:18 'For Christ also hath once suffered for sins, the just for the unjust, that he might bring us to God, being put to death in the flesh, but quickened by the Spirit:
19 By which also he went and preached unto the spirits in prison

At the moment Jesus's body died on the cross his spirit was set free and able to go preach to the fallen angels!

After all, just like the Nephilim, Jesus had a human mother and a 'spirit' father – GOD! EUREKA!

Chapter Forty Eight

Life on The Ark.

Life on board, food, water, sanitation. We aren't told much other than Noah was told to take on board food for all the creatures but the vast majority including the humans would be happy eating dried food. As all creatures had been created to eat grass and herbs their digestive systems would be well satisfied with dried vegetation, nuts, cereals and dried fruits even if today many creatures seem to have very specific and restricted dietary requirements.

The manure situation could have had some simple discharge chutes or perhaps the animals were put into hibernation for the duration. Many animals can hibernate without emptying their bowels for many months so could GOD not make their bowels stop as long as needed? How many times does the Bible tell of GOD altering time and physical states such as with Lazarus or Jesus Christ himself?

The days and nights would just pass with only the noise of water and debris against the timbers and the stirrings of those creatures that didn't sleep the time away. The humans would take turns patrolling the vessel, feeding the creatures that wanted food and water before going to eat and sleep themselves. Maybe the pterodactyls flew about or the humans unfastened the dinosaurs and exercised them back and forth.

We are not told and don't need to know about life on the Ark but once again, all concerns and questions about Ark life can be satisfied by just remembering Noah was acting under divine instruction and direction – Evoquacks, devilworshippers and monkeys just can't understand that?

While researching the latest fairy tales about volcanoes and the gasses emitted by them I came across the EVONONSENSE woozle that 'During the Cambrian period fantasy of 500 MYA the concentration of carbon dioxide is claimed by researchers to have been 7,000 parts per million compared to today's 407 ppm...the present concentration is the highest in at least the past 800,000 years and likely the highest in the past 20 million years'.

Obviously this is woozle on several points as the Cambrian is the bedrock of Creation Day 3 about 6,000 years ago, and there has not been any past 800,000 years or twenty million years.

We know the Flood was a very real cataclysmic event about 4,350 years ago and so there should be a link between the 7,000 ppm level and either Creation or The Flood – if the 7,000 ppm is genuine and not just another flight of fancy by some Evoquack in an ivory ivy clad tower. One explanation can be the CO2 built up over the whole 371 days of the flood with the massive volcanic activities in the first days contributing massive amounts

Take the 7,000 ppm Cambrian as being a true and verifiable fact – but is it? NO! The silly Evoquacks who published this figure based their research on ice cores from Antarctica...the same ice under which is buried modern grass and flowers that had been growing very nicely in the rich sediments left as the flood waters drained off the land! Those plants in Antarctica along with the mammoths of Siberia grew and prospered for many decades after The Flood until the Ice Age built great weights of snow and ice on the unbalanced Earth until it flipped over at high speed burying everything in superchilled air and fresh ice. Thus we know that the 7,000 ppm figure might be totally wrong.

On the other hand GOD told Noah to construct The Ark with a roof and Noah did not look out for over 300 days. Was that to ensure they were not poisoned by the atmosphere or injured by falling debris returning to Earth as meteorites, asteroids and the same size of giant hailstones that we are seeing today and will be used to slaughter many unbelievers in the tribulation? Did direct sunlight and all the volcanic activity cause a massive short term rise in carbon dioxide or or was all the oxygen really depleted by the rotting of unimaginable masses of vegetation and bodies washed off Earth during the flood? EUREKA!

Chapter Forty Nine

Landing thee Ark, Evacuation, Altar sacrifice.

One constant criticism of the whole Ark narrative is the apparent inconsistencies of the claimed duration of the event but a careful tally of the given times allows the total duration as being 370-371 days that can be broken down into several phases.

Naturally, one of Satan's outlets - Rationalwiki - casts doubts on the whole event by muddying the waters and listing the duration as 40 or 150 days! Accuracy is not Rational's strong point.

Phase One was the seven days after the door was shut. This occurred 4,350 years ago. *1

Phase Two was the initial 40 days and nights of intense rains as the Earth's crust shattered to release vast hot water geysers and the waters above the Earth began to fall as great rains. *2

Phase Three was the next 110 days during which Earth and sky continued to add waters until the mountain tops were under fifteen cubits. *3

Phase Four was the Ark drifting about over the debris filled water for 150 days until running aground on a mountain top that was invisible beneath the water. *4

Phase Five was the water beginning to subside under the wind until on day 224 the mountains were visible under it. *5

Phase Six was the forty days to day 264 of waiting for the water to drain enough to make land appear. Noah sent out the raven but it could not find a place to land and flew continuously about. He also sent out a dove but that bird could find no tree to perch on and returned to the Ark. *6

Phase Seven was day 271 when Noah sent the dove out and it returned in the evening with a fresh plucked olive leaf showing an olive had either rooted and sprouted or had perhaps survived on a high mountain. *7

Phase Eight was Noah waiting seven days to day 278 before sending out the dove again but this time it did not return as presumably it had found a resting place and food as doves do love eating soft new shoots of plants. *8

Phase Nine was day 313 when Noah removed the covering of the Ark and looked out to see the land was now free of water. *9

Phase Ten was day 370 when the land had dried enough to walk upon and GOD told Noah to disembark all the creatures and free them to go off in their herds, flocks and groups to breed and fill the earth again. *10

Phase Eleven was GOD telling Noah He had arranged to have rainbows in the sky as a covenant with humans to never again flood Earth. *11

My 370 total may be a day out depending on how the start and finish days are taken. However the total of 370/371 makes a nonsense of all the teachings and preachings of a 40 Day Flood! And it makes fools of Hawk'n'Dawk and all the ranting Evoquacks!

Noah wasn't to know that during the following 350 years until his death he would feel another great cataclysm as the Earth went into its Ice Age as the new climate allowed extremes of heat, rain and cold. These factors combined to move water from the new lowlands and have it condense as snow over the mountains and poles. Eventually the ice had built to such great depths and weights that Earth suddenly overbalanced to cause super tsunamis and hurricanes that swept civilisations and herds of mammoths away and buried them in deep rich muck or pure loess that froze into what we now call the permafrost of Siberia and Alaska and Antarctica.

Clarification of the Phases:

*1 Phase one: This waiting period seems to go unnoticed by most especially the Evoquacks. We aren't told the reason but can surmise it as to allow the creatures to settle among themselves and perhaps drift off into hibernation.

*2 Phase two: To start the Flood GOD smacked or shook the Earth causing the crust to shatter in many places and release the hot water inside as great fountains or geysers. GOD's design of Earth was so ingenious and inspired as to include undersoil piezo-electrical heating powered by the sun and moon and the weight of tides. The shaking was so hard that the waters were so explosively heated and compressed they shot high and either escaped Earth's gravitational pull to create the comets and pummel the moon and nearby planets or else they drifted around and began to fall back with the rain of the condensing canopy that had previously diluted the power of the sun and its harmful radiation. During the shocks of electricity new elements were created to give the dangerous radioactivity we find today. This phase needs an understanding of piezo-electrical generation, Newtonian physics, hydrodynamics, high speed filming of water impacts and tides. Evoquack palegeologists find the shattered crust and quartz shocked into tiny fragments and fantasise it was all shattered by innumerable meteorites over billions of fantasy years. EUREKA!

*3 Phase three: During this phase the hot waters continued to pour out of Earth and returning waters poured down from the sky until the mountain tops were below fifteen cubits of very dirty water. Even so because the Earth was emptying itself and the crust was shrinking and falling the hills could be classed as mountains by virtue of the height from the sunken land to their tops. The hot water leached and dissolved minerals and forced it into cracks in the crust to make the quartz veined rock we call Gneiss and liquid gold filtered into cracks in the quartz just as miners find it. The temperature and humidity aboard the Ark may have become a bit uncomfortable but the roof vent GOD had specified would have allowed warm air to escape.

*4 Phase four: the Ark would be pushed by the strong winds and drift among the debris over what may have been a great part of the Earth but so dense would the debris be that it may not have been able to drift far and may only have drifted from the starting location of Iraq up to Ararat. We know icebergs drift far but when they start their journey they are wedged among other bergs and have to wait for the foremost bergs to drift off before they can start moving.

The Ark eventually grounded on land but so dirty was the water the land could not be seen. Pour grass or vegetable matter into a tank of water and after a few days the water will become thick and green and remain so for a long time and it it reasonable to assume the waters on

Earth were laden with rotting vegetation and flesh. It is likely by Phase 4 there were worldwide plankton blooms thanks to the superabundance of decomposing matter in the warm waters. During Phase six much of this matter and the plankton itself would become fertiliser to make the dried land very fertile and productive. As the Ark stayed grounded the humans would have been apprehensive but did not open the covering to check what the situation was either from fear of more rain or of angering GOD.

*5 Phase five: In this period the water began to drain back under the crust while the floating debris was sinking quickly and the fine particles of sediments were falling to allow the waters to be transparent. But so immense was the amount of debris and sediments that it took 74 days until the mountain tops could be seen under the clearing waters. The Sargasso Sea and the North Atlantic Gyres have long been though of as immense floating mats of seaweeds and rubbish but they were trivial compared to the flood debris mats. Some seas can be clear but some cloudy with plankton. The canal close to my old home never did clear as the sediments in it were so fine they never settled out. As the water receded from the higher lands it carved out canyons and river systems.

During those 74 days Earth would have to cast off or absorb about 166,000 cubic miles of water. Much of it would percolate back under the crust to refill the voids made as the hot water flashed into steam and escaped into inner and outer space.

*6 Phase six: This was the waiting time for the waters to be blown away after GOD ordered the wind to begin blowing to evaporate the waters . Noah sent out the raven but ravens are birds that prefer to live and nest on high peaks and mountains it could not find anything to land on but continually flew about. He sent out a dove but these birds need to rest in trees and having found none the bird returned. To try arrive at the amount of water to be blown away needs the calculation 4/3 pi-radius-cubed or 1.33 x 3.142 x 4000 x 4000 x4000 minus 1.33 x 3.142 x 3.996 x 3.996 x 3.996. This formula takes the water as being about 4 miles deep and the Earth's diameter as exactly 8,000 miles. Subtract the lower sum from the higher and that gives the cubic miles of water that had to be removed to make Earth dry again as we are used to nowadays. My calculation is about 800 million cubic miles of water had to disappear but where could it have all gone? Take today's Earth diameter of 7.912 and water only two mile high gives 390 million cubic miles. All this water would continue to percolate under the crust but much would make the new clouds and rains and the snows on the new peaks such as Ararat.

*7 Phase seven: This seems to indicate the dove found a newly sprouted olive seedling that had germinated in exposed soil but in fact olives quite happily grow over many centuries in rocky soil and the bird may have found such a tree's twigs poking above the water that had survived the flood by being securely rooted in rock and had survived immersion in what may have been quite neutral water. This is plausible as the bird did bring the leaf back to the ark because if it had found a secure perch it may have remained there rather than return to the Ark. The ancient bristlecone pine tree could also have survived in this manner.

*8 Phase eight: Noah had another seven days wait until again sending out the dove which this time found a safe perch and food and did not return signifying the waters were draining noticeably and some parts of the land was able to support the weight of light birds or else numerous seeds and trees had survived and floated to root in various places.

*9 Phase nine: Noah opened the covering and looked out with apprehension not knowing what the view would be. He must have been stunned to see nothing familiar with soft mud rutted with draining rivulets stretching on all sides with new green shoots of plants sprouting everywhere.

*10 Phase ten: Noah waited another 57 days before GOD told him to leave the Ark. This seems an extraordinary length of time but anyone who has ever played around rivers and bogs will know that a thin carpet of grass can conceal soft mud very dangerous to walk upon. GOD knew that the sediments would be very deep though appearing dry enough to support grass and would be too dangerous for heavy humans and creatures with small hooves and feet to walk upon until it had dried deeper and really firmed up. The horses, bison, dinosaurs and giraffes would have grown very large during the interim and their small hooves would need the ground to be quite dry and firm. Also the grass needed to get well established and tall enough to withstand the grazing of many creatures concentrated in the vicinity of the Ark. Study the Condit Dam video to see how deadly great depths of soft sediment can be and try to estimate how long it would take for such sediment to dry firmly.

*11 Phase eleven: Noah offered sacrifices to GOD before being told that from thence forwards a rainbow would appear in the clouds to signify GOD had covenanted to never flood Earth again.

Noah may have selected the offerings based on knowing some of the remaining creatures would be pregnant and the chosen one would not be essential to spread their DNA to further offspring.

Rainbows need direct sunlight, water droplets in the air and a backdrop of dark clouds – all these were new after The Flood. See the chapter on Genesis 2:6 and 9:12 to see if you can understand the pre- and post-flood atmospheres.

Mount Ararat is stated as the place the Ark grounded on as is confirmed by Genesis 11:2 which says the humans journeyed from the east to the land of Shinar where they decided to build a city and great tower to serve as a marker for a central civilisation contrary to GOD's order to fill the entire Earth.

The Flood is so simple to understand yet how few people actually believe and are prepared to defend it!

I like watching old movies to see the long dead actors and actresses and many of these are cowboy or war films and quite often outside location shooting will have a backdrop of mountains with quite visible layers of sediments with rills, gullies and screeslopes exactly as left by the Flood 4,350 years ago. Cowboys' horses always raise clouds of dust as they gallop across the bare sands that cannot regenerate due to shortage of worms, water and food for the worms. EUREKA!

Chapter Fifty

Timber and Coal. OOPARTS.

Poor Kevin Boyce and his Pangea!

Origins of coal is same as baked jellyfish?

Did the Flood really turn masses of timber into coal just about 4,350 years ago or is coal the result of plate tectonics pushing continent sized plates over Earth's peat bogs hundreds of millions years ago?

Satan's internet service – Wiki – publishes 'Coal is formed from compressed peat and is known from most geological periods, except there is little in the Permian-Triassic extinction event but it is found in the pre-Cambrian strata and that coal is presumed to have formed from the algae that thrived in the seas that miraculously appeared in the dry dust thanks to the gravity of Hawking's fantasy BigBang'.

As Cambria is Wales and Wales had an enormous coal industry until the Illuminati closed it all down to enrich themselves it may seem accurate to think the coal came from the algae in primordial soup seas about 540 MYA. But as we now know Cambria is an area in Wales showing the barren granite bedrock that appeared on Creation Day 3 any rock labelled Cambrian is merely a part of the Earth's crust that was raised during The Flood and any coal on it must be made from the finely ground up vegetation, dinosaurs and humans washed off Earth in Phase two of The Flood 4,350 years ago. Another fact beyond the understanding of the Wikiist is that many fish love eating algae and the supposed great masses of it would be unlikely to exist.

The Wiki piece states there is a lack of coal in 'The Permian-Triassic' but this is due to the strata labelled such is actually more bare bedrock scoured during Phase Two and Three of The Flood.

Wiki continues woozling: 'At various times in the geologic past, the Earth had dense forests in wetland areas that were buried underneath more and more soil. The plant matter turned to carbon in immense peat bog that were eventually covered and deeply buried by sediments'. WOW! What a load of EVONONSENSE! Trees turned to peat and kept being covered with soil? From whence cameth all this soil? Did it drift slowly in from outer space attracted by Hawk's gravity? And coal is compressed peat? Not really as now we can see coal is mostly finely ground bark, twigs and leaves that settled out in the last phases of the Flood. We know this thanks to the new lake that formed after Mt St Helens blew up and blew millions of trees down. Those trees jammed against each other in the water causing all their bark and small growth to abrade off, sink and be buried by the sediments being pushed into the lake by the continuing eruptions. If that lake hadn't drained itself one night when no-one was videoing we would see a huge black mass of rotting bark stream out as the waters raced away.

An online encyclopedia prattles this woozle about coal: 'If peat is allowed to remain in the ground for long periods of time, it eventually becomes compacted. Layers of sediment, known as over-burden, collect above it. The additional pressure and heat of the overburden gradually converts peat into another form of coal known as lignite or brown coal. Continued compaction by overburden then converts lignite into soft coal and finally, into hard coal. This has occurred many times in the past, but most abundantly during the Carboniferous and Super Cretinous Ages between about 300 and 100 million years ago'. The Evoquack that composed that website gives no origin for the overburden and how it can build up very slowly but quick enough to heat up the peat -presumably readers are to imagine tectonic plates climbing over each other like turtles. Wozzle, woozle, all is woozle!

In an article from January 22, 2016 titled 'Stanford scientists discover how Pangea helped make coal', Evoquack Kevin Boyce at Stanford University woozled this EVONONSENSE about the origins of coal complete with a picture of a '360 MYO' sample of coal: 'Coal deposits during the Carboniferous period are closely tied to a unique combination of tectonics and climate conditions that existed during the assembly of Pangea'. Or in simple terms Kevin fantasises lotsa plants grewed up and got smothered as giant slabs of Earth climbed on each other like mating turtles. Pangea is the old sci-fi movie land where the Archinons fought over Raquel Welch in her rabbit skin skimpies?

Maybe Kevin should try figure out the Newtonian physics involved in getting a few billion tons of Earth to climb on another without having a few billion tons of push? A 'tectonic plate' of Earth about 100 miles by 100 miles and just 5 yards thick would weight about two hundred billion tons dry. Add lots of water raining down from that fantasy primordial sky and the slab's weight rises to perhaps 300 billion tons as water is like heavy stuff and known to fill the interstices betwixt grains of soil? And can childish Kev explain how did the ascending plate get pushed up over the descending plate without some Newtonian physics to supply the ascending plate's push and some fixed block to resist the pushing? Is it any wonder the universities turn out nothing but Evoquacks like Kevin when the students are taught nothing but woozle? EUREKA!

Scientists were able make coal in the laboratory in just a few days way back in 1926! They used high pressure and heat! But as the developed world was already deep in the stuff there was little impetus for further large scale trials. Sadly all websites dealing with the origins of coal all pretend it was formed many millions of years ago and by masses of sediments putting plant material under high pressure and heated by the liquid iron core of Earth and with the aid of natural catalysts!

Surely this is exactly what happened in Phases 2,3 and 4 of The Flood just about 4,350 years ago? But no, say all the scientists in their desperation to promote EVONONSENSE: turning vegetation into coal needs catalysts and a great depth of vegetation to live and die over hundreds of millions of years and be buried under masses of hot sediments...and they shut their eyes to the Bible account of The Flood and all that pesky deep hot water! Then I read that one of the catalysts for coal can be iron...and how many trillion tons of that stuff is there in all Earth's topsoils washed off with the vegetation during The Flood? Many other sources also state turning coal into vegetation needs catalysts but give no further details as presumably the secret is locked in an old book of alchemy magic like in Disney's Saucier's Apprentice cartoon.

Related to coal is oil shale and digging into the overburden of woozle piled up by Evoquacks in their studies of oil shale reveals some fascinating facts about the vast sediments in the water during The Flood!

One report on the Haynesville shale near Carthage, Texas, claims it to be from Upper Jurassic Park 147MYA and lies at drilling depths ranging from 10,352 ft to 10,702 ft with possibility an additional 7,500 ft or sediment washed away over the years. The seam has an average thickness of 475 ft and is poor stuff of only 2-4% organic matter with a temperature of a temperature of 271F. The report continues on to claim that the low temperature is due to the shale being buried during the Cretinous Epok with a temperature of 350F.

This silly report can be shredded like this:

The great depth of existing and eroded sediments were obviously laid down during the late part of Flood Phase two when most of the soft vegetations had been washed off land and into the depths and we know this because of the low organic content. In other words the whole Haynesville sediment is secondary eroded matter from after the first rains had stripped off the primary lush topsoil from the lands upstream of Haynesville and were eroding the subsoil from which worms had already removed all the organic matter that makes the Kimmeridge shale so rich. During Phase four the whole subsoil settled around Haynesville long enough for the lower layers to firm up until Phases 7, 8 and 9 made the top 7,500 feet of still wet sediments flow away exactly as Condit Dam shows. The present day temperature of 271F is unrelated to the organic content but is a function of the piezo-electrical underground heating so thoughtfully provided by GOD to prevent us all freezing from the day He stretched out the heavens. It would be really interesting to dig a deep shaft and then tunnel out under the Haynesville shale sediments to see what dinosaur and human remains are down there!

The origins of the vast quantities of oil, gas and coal around the world baffle the geologists who like to dig and drill for it. Blinded by Satan's EVONONSENSE the exploiters flounder about for an acceptable answer to the question of where such masses of liquids, gas and solid materials could possible have come from.

Some of the most rabid Evoquacks claim the stuff has seeped all the way to the surface from its origins deep inside Earth, some claim that like themselves the stuff is miraculously formed from rocks interacting with each other- Evoquacks frequently invoke miracles to explain their illogicalities! - others say it is all produced from the bodies of the dinosaurs that were wiped out by the mythical meteorite 65 millions years ago.

Geologists, oil, gas and coal producing and exploration companies cannot agree on the total amount of these three commodities in known fields or how much has been extracted since industrial production began, or how much is lost each day through seepages or uncontrolled wild fires, or how much may still to be discovered, or how and why some oil and gas fields still produce commercial quantities while others can be very productive for a short time and then dry up.

The 'Door to Hell' gas fire in the Karakurum Desert in Turkmenistan has been burning off masses of gas since being set aflame to burn troublesome methane off in 1971. The whole area has gas fires burning constantly although their size reduces as natural gas wells reduce the gas volume and pressure. Photos of all these fires show the typical sedimentary layers from the Flood. The burning cliffs of Azerbaijan have been known from antiquity.

An hour spent looking at Youtube will show many old wells still producing oil or gas in sufficient quantities to make it worthwhile to keep the well open for some hours on a regular basis. Many people own whole or part of an old well first drilled 50, 75 or 100years ago that as regularly as daily or weekly can be pumped for an hour or two to give a very nice return in cash – for practically no outlay. On the other hand some major oil or gas wells just keep on pushing out immense quantities that bring a bonanza for the owners.

Here in Yorkshire were once many coal pits ranging from little scrapings in the side of hills to vast industrial deep mines, now they are all closed with one kept as an intriguing museum and others being totally eradicated.

I grew up one hundred meters from a brickworks quarry and was used to seeing the layers of sediments from age 5 or so as I used to be free to toddle down the lane and into the unfenced quarry to look at the rocks and watch the big old shovel at work. The quarry's excavator driver told me he did not want the yellow clays or reddish sandstone rocks or the loose soils but only the grey shale that had the desirable organic matter incorporated in it. The quarry walls were twenty to thirty feet high – 6-9 meters and the sediment layers seemed to be a complete mix of colours, thicknesses and textures. On one side of the lane a temporary 'opencast pit' had been opened in the grim years after World War 2 when the crazy socialist government had nationalised, bought and closed many pits only to find there was soon a shortage of coal! The opencast coal was poor stuff as it had not been compressed under deep soils. The former pit owners laughed all the way to their offshore tax havens.

The brickworks only wanted the 'carboniferous shales' which was grey crumbling stuff desired because it had a high proportion of organic matter – shredded wood, leaves, creatures – in it which allowed the stuff to be easily formed into bricks that when air dried and stacked in the circular continuous Hoffman kiln would actually burn and bake by virtue of the organic matter being good fuel. The Hoffman kiln worked on the principle of having many separate chambers arranged in a circle with small connecting vents so that one chamber would be fully burning and hot and the preceding one dying down and the proceeding one being heated by the heat of the burning one until it too caught fire and so on all round on a continuous process. As each chamber started burning the kiln operator poured small quantities of coal into an inspection hole in the top to encourage a good hot burn. More vents were opened and closed as needed to allow the cooled and damp exhaust gasses to divert off to the chimney which was a prominent feature of the works and always had a thin plume of smoke lazily climbing from it. The coolest chamber would be emptied of fired bricks and then refilled with damp ones that several days later would once again be sufficiently dry to set alight. Making bricks with other types of material would need large quantities of natural fuel such as coal, gas or wood.

So just how much timber was on Earth at start of Flood? Evoquack researchers have tried to estimate the potential mass of trees available on the Earth to give the masses of gas, oil and coal but coming from a EVONONSENSE point of view limits their calculation totals. By comparison a Creationist who uses published studies of tree densities on today's temperate and tropical lands and the evidence of the various stump forests exposed in mining or from recent erosions can arrive at some reasonable figures of the timber load of the antediluvian world. Here is my estimates:

First essential is to choose a figure of the total land area then and now. Lots of figures are banded about by many people who claim inside knowledge of ancient sunk civilisations such as Atlantis, and while Atlantis may be nothing but myth, it likely has some actual origins. Indeed the many deep megalithic ruins and roads indicate that the sea covers many former cities.

Today's Earth is about 29% land and 71% seas with suggestions that based on the sunken ruins the original land mass was directly opposite and 71% is perhaps near the mark.

Taking the 71% figure for lack of any more precise figure gives an area of about 360, 000,000 square kilometers or about 140,000,000 square miles.

What spacing of trees pre-flood? In today's rainforest tree spacing is about 5yards, northern pine and fir is the same, temperate woodlands can average 10 yards, overgrazed savannah can be 50 yards. However these are all with today's climate not the ideal moist one Adam and Eve knew which from evidence of giant tree ferns may have allowed a much greater load of trees. I pass one piece of neglected waste land on which saplings are spaced one foot apart.

Assume a mature tree to be about 30 yards tall and 3 feet diameter and ignore the branch mass to allow for shorter thinner trees = about 25 cubic yards wood per tree.

Spacing at 5 yds = 123,000 trees sq. mile = 18 trillion trees x 25 = 450 trillion cubic yards.
Spacing at 10 yards = 30,000 = 4 trillion = 100 trillion.
Spacing at 50 yards = 1200 = 168 billion = 4 trillion.

Assume 45% solid cellulose fiber, lignin, resin and convertible liquids after water removal = 5yds = 200 trillion cubic yards of solids,
10yds = 34 trillion.
50yds = 2 trillion.

Calculate conversion into methane, oil and coal at say 15% and arrive at:
5yds spacings = 30 trillion cubic yards.
10Yds = 1.5 trillion.
50Yds = 150 billion.

It is impossible to know the actual percentage of methane, oil or coal in each underground deposit around the world.

The figures show that there should be vastly more gas, oil and coal stored under land and sea than we could ever use.

If the original land mass was only 50% the totals would still be vast.

At 33% land mass the figures would be half that above but still mind boggling!

But it has to be borne in mind that the figures are just for standing timber – not for the mass of grasses, bushes and leaves on the ground between trees. As it is impossible to know what mix of trees was on the Earth pre-Flood no figure can be arrived at so I have not included any figure but it must add a significant amount. We have no real way of knowing.

I did think of asking some agricultural researchers if it would be possible to wrap some various large trees in fine net towards autumn to catch every shed leaf or needle to weigh and calculate solid mass and potential oil, gas and coal producing matter.

One very real extra source of methane, oil and coal deposits is that locked in the rich humus that had built up on Earth in the 1600 years between Creation and Flood! This is a spectacular quantity no-one has ever been able to estimate – mainly due to lack of knowledge of how worms build up humus and because very few people believe in Creation and Flood.

However in the interests of science take 14,000,000 square miles of land surface and assume 1,600 years of humus rich top soil complete with masses of worms and other creeping things = 140,000,000 divided by 1760 = 80,000 cubic miles of topsoil. Wash all this topsoil off Earth into deep craters or cracks in the Earth's crust, add the pressure of masses of fine sediments and the weight of water to arrive at ??? how many million cubic miles of gas, oil and coal?

But then another unacknowledged source of organic sediments has to be considered and it looks like I'm the first researcher to describe it. This is the amount of gas, oil and coal bearing sediments and strata laid down in the time when the Earth's crust was raised, lowered and split open to create the map of the world we have now! This happened 'during the days of Peleg.'

Peleg and the dividing of Earth is recorded at Genesis 10:25. The Earth reshaped itself either under GOD's command or direct intervention or more likely due to the crustal resettlements as the build up of ice and snow at the poles squashed the central equatorial areas causing the crust slabs to lift up and slide down on the water that had soaked down in the many years after the flood.

Darwin's meticulous scientific research and my own observations are that worms build topsoil in immense quantities and that topsoil holds a mass of leaf matter, roots and the worms and other creatures. All this mass of vegetable matter and creatures was washed off Earth during the flood and swept along into the depths to make more gas, oil and coal!

Research into Amazonian Black Earth shows it includes a good percentage of oily matter.

Another unknown mass would be produced by the fermentation of all the billions of fish, animals, birds and humans that drowned and were buried in the sediments! Who could possible estimate that total! There is vast quantities of fuel still to be discovered but when Jesus returns and sets Earth on fire it really will burn ferociously! Read 2 Peter 3:10 'But the day of the Lord will come as a thief in the night, in which the heavens will pass away with a great noise, and the elements will melt with fervent heat; both the earth and the works that are in it will be burned up'. EUREKA

Noah's Slimy Pitch?

The slime pits? I've personally burnt shrub trimmings that were high in volatiles as are today's eucalyptus, rhododendron and similar bushes and trees. Many trees exude resins and natural rubbers. Paper mills have to ensure this build-up is removed from the pipes, tanks and fittings that raw wood chip and pulp passes through. Grains, nuts and seeds have oil in them. Add the undigested fractions of the vegetation in TRexxies droppings, cover with dust from deforestation and other ruinations of the Earth and the mixture would make the pitch that GOD knew would waterproof the Ark outside against the seawaters and inside against the animals' droppings.

It logically follows that far from needing a source of pitch to seep up from some deep ancient underground region as the Evoquacks claim all that is required to make the slime pits is the natural annual leaf, fruit and bark fall of the various trees over the 1600 between Creation and Flood.

The La Brea Tar Pits are not ancient as thought and the sabre-tooth tigers found in them were obviously descended from the pair on the Ark. EUREKA!

Chapter Fifty One

The Silence of the Clams.

Shellfish fossils on Mount Everest and all other mountains.

During Flood Phases 9 and 10 the waters drained off the land leaving a good depth of sediments through which innumerable streams trickled down to the rivers, lakes and the main central sea. So vast was the quantity of water in the land that it remained as thin mud that could and did flow as easy as the waters. To see this drain down in perfect detail go to look at websites of 'Rill and Gully Erosion' and see photos from all over the world of gullies that are perfect miniatures of The Grand Canyon and Bryce Canyon that were cut during one rainstorm. All a canyon needs is a sudden heavy flow of water, one hour for a gully across a field, a night to scour out the Mt St Helens Canyon, a month to cut the Grand Canyon. Or look at the videos on Youtube of the draining of the Condit Dam.

The Condit was a dam built to provide water power for a mill in a valley in East USA on what was a river noted for salmon until the dam blocked their migration to spawning beds. Once the mill closed and the dam became redundant the local wildlife conservationists petitioned that it be removed and so one day its base was dynamited to allow all the water to flow out. The dam was quite deep and the escaping water had tremendous power at first but as it slowed a little it became really black with the sediment of a hundred years. Time lapse film shows the water level dropping to the top of the main areas of sediments and then carving a valley through them. This valley widened dramatically as the soft waterlogged sediments slumped under their own weight into the central valley where water dissolved them and washed them away. As the hours passed the valley became wider and deeper with flat topped sides and base with rills and gullies between the two – in fact a perfect example and replica of the Grand Canyon, Mt St Helens Canyon and vast numbers of similar water cut courses – including the dry moat around the Sphinx beside the pyramids! And exactly as with Mt St Helens Canyon the Condit Canyon was carved out in just a few hours! Next day the Condit showed the flat topped sides where water had swirled and eddied about, the steep sides made as water dropped taking sediments with it, the wide valley made by the soft sediments levelling under its own weight just as any thin mixture will, and from side to base were the unmistakable gullies and scree slopes. Condit is a textbook illustration of Phases 4 to 10 of The Flood – and best of all it is available for study in slow motion!

This type of plateau, table, scree and valley landscape can be seen all over the world where waters have drained through soft sediments but curiously they are not seen quite the same in soft sand, perhaps because sand is porous and water can soak through it rather than run over as it does on sediments composed of fine organic matter and fine silts. My computer's opening screen is a panorama of the Grand Canyon and shows this Drain Down topography perfectly yet the tiny Colorado River of about 300 feet wide is supposed to have wandered side to side and carved the canyon to maximum width of 18 miles over millions of years! Curiously today's rivers prefer to cut a groove and stay in it; even when they flood they always subside back into their channel. The fact is the Grand Canyon was cut in just a few weeks when a great Flood lake that covered the lowered central plateau of the USA suddenly breeched the lake edge and spilled down the nearest sloping land.

As usual Wiki panders to Satan by woozling this: 'The geology of the Grand Canyon area includes...nearly 40 major exposed sedimentary layers...ranging in age from about 200 million to nearly 2 billion years old...deposited in warm, shallow seas with long-gone sea shores...marine and terrestrial sediments..including fossilized sand dunes...from an extinct desert...at least 14 known unconformities...in the geology.' This is a perfect example of EVONONSENSE as what it basically says is that no-one has a clue to the real age of the canyon but Evoquacks date its rocks by the fossils found in them and date the fossils by the rocks they are found in. No Evoquack dare put forward a theory explaining how most of the canyon walls show many layers of Flood sediment but some layers refuse to conform and prefer to be actually upside down!

Psalm 104:7 in your Bible describes how Phase ten of The Flood made the unconformities: 7 'At Your rebuke they fled, At the sound of Your thunder they hurried away. 8 The mountains rose; the valleys sank down To the place which you established for them'. EUREKA for GOD!

Today we can still see this Drain Down topography all over world and it is indisputable evidence of a worldwide flood. Feeling lazy this lunchtime I watched a Spaghetti Western - actually a war film – that seems to have been shot on location in the deserts of Italian Libya and the landscape was perfectly and impressively Phase 9 and 10. There were flat topped mesas either side of wide flat valleys with rills, gullies and scree leading from the higher to the lower surfaces. Everything was powdered sand with no rocks to be seen except in one scene shot in a deeper gully. This exact landscape can be seen in innumerable places around the world such as in videos of US marines fighting in Afghanistan and surrounding areas, or cowboy films shot in Monument Valley, train lines across Canada. The Drain Down landscape can also be seen in numerous places around my part of UK and I get a delight in pointing out that it can only mean the entire world was once under a great depth of moving water for along time and the water all drained off quite quickly in exactly the same manner.

Evoquacks sneer at this.

As it is indisputable that fossil shells and marine creatures are found on all the high mountains I just wasted a few minutes on the latest thinkings of these shells and came away saddened by both the childishness of the articles and gullibility and ignorance of those who wrote them. I just read this woozle: The 'great' nature writer John McPhee wrote 'In 1953 the climbers planted their flags on Everest in snow over the skeletons of creatures that had lived in the warm clear ocean that India, moving north, covered and the remains had turned into rock. The summit of Mt. Everest is marine limestone'. McPhee is another in the long line of fools unable to see the Young Earth evidence of worms for the lack of worms in their own backyard. Hilary and Tensing did find fossil shells on Everest and like McPhee all Evoquacks claim the shells got up there thanks to those mysterious 'plate tecksonixt'.

Another promoter of the tecponix woozle in Stewart Green who had his head packed with tecitonix woozle spouted by charlatans professing woozle at the University of Colorado and was such a devoted devotee of their tectoniical woozle that they awarded him a 'BA with Greatest Distinction' - kind of the Medal of Honour of Woozle? Now he feels it incumbent to indoctrinate impressionable initiates into tectonix woozle. He does this by woozling that once upon a time there was a land called Eurasia where India now is and that a tectonic plate called India, weighing, what - 50 trillion tons? - decided to wander north and hit the 'Eurasian' plate which refusing to move -what! - forced the Indian plate to sink under it while the Eurasian Plate buckled up to make the Tibetan mountain areas. Green's fairy tale includes a lake he calls the Tertis Sea full of shellfish and Little Mermaids that was trapped between those darned plates and having nowhere to go was pushed up higher and higher to make Mt Everest at the rate of 4 inches per year! I kid you not – he really does believe that! This is fantasised as starting about 65 MYA – just when the dinosaurs are fantasised as being wiped out when a fantasy meteorite hit Earth – and by today the mountains is five miles high. Do the calculation of 5 miles divided by 4 inches per year and you get: 79,200. Now this figure does not equate with 65 MYA so Green's chimera has to be filed with Piltdown Man and Bigfeet to raise those shells up high so quick. Poor Old Indo-Magnon spent millions of years lugging great sacks of shellfish to the ever rising peak of Everest just so one day Stewart Green would get the Evoquacks' BA of Greatest Distinction – presumably writ on dull parchment to reflect Green's IQ?

That scenario makes perfect sense to all Evoqacks like Green brainwashed by being force-fed Planet of Apes along with baby formula. Unfortunately it is so childish that Green, the writer of it ,should be ashamed of himself but I feel sure the rewards Satan showers on all those who promote his lies will ensure Green feels no shame like all other Evoquacks rewarded for their EVONONSENSE. He actually received a BA with greatest distinction! Amazing!

Perhaps Green and his ilk should look at Niagara Falls or Hardraw Scar or any of a hundred other impressive waterfalls and note how the waters cut them all back but if the water had been falling for millions of years each of those falls would now be far receded into the back of a deep gorge?

The website of the 1953 Everest expedition also promotes Everest as being pushed up from under a Tehyts Sea but occurring about 30 to 50 MYA or half of Green's fantasy. This reveals that when Evoquacks rough out their fairy tales they can be 50% out in their dates and no-one will question it - well I do! The website says the Everest peak is made of limestone and dolomite from the Ordovician era but Wiki states the Ordovician Period lasted almost 45 million years, beginning 488.3 million years ago and ending 443.7 so during those 45 MY lots of planktons swam around above the countless trilobites and then they all got squashed under limestone and dolomite for about 380 to 420 million years before those darn textonics plates pushed the whole lot up in the air and made Everest! In the land of the texting plates the woozler is king!

Now to explain to all the Evoquacks and Planet of Apes fans who cannot understand how seashells seem to be on top of all high mountains and even whale fossils have been found high in mountains: the key to the riddle of the seashells is worms.

The mystery is: whence came the push that pushed the Indian plate into the Eurasian plate as Newton's physics say to push a textonix plate needs a pusher that can be pushed by something that can resist the push it is pushing the pushed one with. Or: Plate #1 must be pushed onto or under Plate #2 by Plate #3 which in turn must be pushed by Plate #4 which must butt up against something in order to be able to push Plate #3. This is why turtles have to walk onto the beach to mate as the daddy turtle #2 cannot push itself up on mummy turtle #1 in the sea as the sea will not offer resistance to daddy's back legs and attempts to mate in the sea will merely result in daddy turtle pushing mummy turtle about like a tug boat #2 pushes a giant ship #1 with the push coming via the thrust#3 of the propellors of the tug. In all the textonnix fantasies the narrator – perhaps Old Guiseppe – has to decide whether Plate #1 is to slide below or climb upon Plate #2 as sometimes in tekztonics yarns the plates does one and sometimes the other. Now do you understand the sillyness of Stewart Green's Plate Tectornix? Does the University of Colorado really hand out 'BA with Greatest Distinctions' to fools like Green for childish idiocies like texictonic plates!

Obviously as Psalm 104:7 says, during the Flood of Genesis some SLABS OF CRUST - falsely labelled textoniks plates – could be expected to be pushed into and onto others to give the curious layers of sediments seen in mountains and cliffs.

Another Evoquack who pushes tekctonix plates about to fit his silly fairy tales is Simon Morris of Cambridge University who practically echoes Green and Boyce in prattling on about imaginary tecconicz plates lifting up and squashing imaginary seas full of lower life forms. Having noticed that when The Flood water spewed across the Earth it buried all the creatures on the sea bed and preserved them in fine sediments that often harden into quite hard rocks he fantasised that the process must have taken hundreds of millions of years in keeping with the central message of the the Planet of Apes movie that was all the rage when he was an impressionable midteen. He seems to have become so possessed with the Satanic message of the movie that he has made it his life's work to prove himself evolved from an ape over hundreds of millions of cloned movies. Sadly, just as Lucifer inflated his own importance so too Morris puffs himself up and declares: '**My work is central to the palaeobiology of the Cambrian Explosion**'. Wow – such foolish modesty! His pix shows his head is about average size. His fantasy is based on kindergarten woozles that the Cambrian is the first spark of life on Earth in which primitive looking creatures are found when all Christians know that 'the Cambrian explosion' was Day Four of Creation during which GOD created all manner of sea creatures – and naturally as I've pointed out elsewhere - GOD being GOD and supremely intelligent He designed creatures to crawl about the sea floor recycling all the detritus He knew would soon trickle down there. The most common and well known of these are the Trilobites. Morris pokes about the Maotianshan Shales of China and finds it full of mini recyclers labelled 'Naraoids' similar to trilobites. Wiki says of these creepies: 'All Naraoids were benthic marine bottom dwellers...probably lived as scavengers...deduced from having spiny basal segments of the legs...likely used for chewing...with a relatively small digestive system, that indicates high nutrition value food'. So Morris finds countless well preserved creatures similar to trilobites and decided they mooched about the seabed chewing nutritious detritus with their legs! I agree they ate nutritious detritus to recycle it for all the water weeds. But I have to say GOD designed everything above amoeba with mouths to chew with rather than legs ...but on Morris's Infantile Planet of Apes anything goes.

As we know shale is sediment a glance at details of these Maotsetung Shales of China claims the shale stretches over thousands of square miles and is believed to be ancient sea bed messed about by texxonix plates. The creatures Morris finds are, like Mary Schweitzer's TRex, still fresh and not fossilised, and are found in thin layers less than an inch thick that Morris dreams were laid down in the fabled sea in successive times 500 MYA. To anyone with a brain who can read the Bible and knows that before The Flood there was just one central sea for about 1,600 years, the thin layers of the Maotsetung Shales were obviously laid down in Phase three of The Flood about 4,350 years ago. Simon Morris is borderline lunatic.

The monkeys of Planet of Apes live on a barren dry landscape so why so many of today's academic monkeys prattle on about ancient seas and tetkonic plates is a mystery.

Did any Planet of Apes movie ever have the monkeys rooting about for edible fossilised shellfish on high hills? And what did the monkeys horses eat as Planet of Apes looks to be noting but dust and rocks?

As it indisputable that fossilised closed seashells and various fish are found on all high peaks it is essential to work out how they got up there. Every cookbook and recipe for clam chowder or mooles marinayre says to ensure all shellfish are closed - any that are open are dead and ought to be discarded as potentially poisonous in some degree. Oregon State Environmental Health says: 'Clams, mussels and oysters in the shell are alive and the shells close tightly when tapped...gaping shells indicate that the shellfish are dead and not edible'. Many other sources say shellfish such as clams will stay alive AND CLOSED for for a couple of days or possibly even two weeks but only when kept cool and wet and in the shade – certainly not in warm, dry, sun! THEREFORE: CLOSED SHELLFISH IS ALIVE: OPENED IS DEAD? Got that? How can we get Simon BigHead Morris to get it?

Knowing that a live shellfish can keep its shell closed while a dead one loses its strength and the shell falls open, and having mountaintops covered in vast numbers of fossil shellfish all closed tightly can only mean the creatures reached the high peaks alive and died very quickly? It is ridiculous to even consider silly Evoquack Green's theory that Tectonix Plate India pushed the shellfish up at the rate of 1.7 inches a year or 1,000 miles in the next 10 million years! This is such childish woozle it ought to have been kept for Planet of Apes remakes. Green apparently thinks a bit of seabed carrying plenty of shellfish was pushed out of the sea at the rate of 1.7 inches per year - but assuming the shellfish was even able to survive until being pushed above highwater mark they only had about 7 days before dying off en masse as they dried out. This is ignoring the possibility of them dying off in the thinning waters due to the land being either warmer or colder than the water they had been happily living in? If the water had been too many degrees either side of their tolerance range they would have all died off. In the world of the Evoquack the most idiotic is king. Hail King Green!

The top third of Everest is known to be composed of limestone and dolomite – dolomite being limestone with added magnesium. Magnesium is also the central compound in the chlorophyll that enables plants to absorb energy from light. Seawater is full of magnesium. Dark green plants also have lots of calcium carbonate and iron in them. The moment before The Flood the Earth had masses of green plants, the seas were full of plankton and all those stromatolites carried masses of cyanobacteria. Can you now sense a link between seashells embedded in a limestone-magnesium mix on Everest and the Flood dynamics I described earlier? Lots of trilobites and closed clams are found up there proving the peak was definitely the bottom of a sea. The mountains were lifted up at some phase of The Flood. And how did the clams die en masse?

Shellfish die when dropped in hot water... Pour water on lime and it heats up by chemical reaction... My trilobite sample has the poor creatures set in limestone... Lime precipitates out of hot water... Flood a bed of clams with warm limey water and before they can run for the stewpot they will be set and locked shut in limestone just as solidly as that schoolgirl's fingers were set in that Plaster of Paris... If the clam's bed was on the corner of a crust slab that was lifted up after The Flood the closed clams would find themselves on top of Mt Everest! About 4,300 years later '*The great nature writer* John McPhee' would claim the clams is 400 million years old! QED.

Curiously, despite several hours trawling the internet for every aspect of Everest, flints, dolomite and limestone I just cannot find any direct reference to flints being found on Everest in the same limestone as the trilobites. Yet, both creatures lived in the same sea and died en masse when the seas were heated during the first days of The Flood 4,350 years ago.

Why are there no flints embedding in Everest limestone?

How did the seas heat up? Lime gets hot when wet. Wait until you get to my final chapter THE FLOOD!

Chapter Fifty Two
Volcanoes and Black Smokers?

Some Evoquacks claim volcanoes is vents releasing magma and gasses from magma chambers thought to be one to seven miles below the Earth's surface, the chambers being filled from the magma hundreds of miles below the crust; most subscribe to the fairy tale that magma circulates up all the way from what they think is Earth's molten core. As GOD quite clearly created Earth as a water covered ball of rock it is hard to understand where the molten core idea came from unless it was in some medieval Planet of Apes comedy. GOD also created the other planets in the same manner but my explaining that is for another book.

A quick read of Genesis and some basic knowledge of how to calculate and extrapolate volumes of spheres and the pressures compressing masses of rock together would quickly rubbish the 'rising magma hypothesis'.

Wiki does its usual woozling with EVONONSENSE like: 'Volcanoes occur where Earth's 17 tectzonic plates is diverging or converging or where the plates is thinning out...which is called the plate hypothesis'. 'Hypothesis' being a clever word for 'a supposition, theory, fantasy or proposed explanation made on the basis of limited evidence' which really means a fairy tale.

Wiki continues woozling with: 'Hotspots is the name given to volcanic areas believed to be formed by mantle plumes which are hypothesized to be columns of hot material with tectonic plates sliding across them...the volcano gets blocked as the tekxtronic plate advances over the postulated plume '. Wiki's authors think eventually the plume will burst through the plate and make a fireworks display. Those pesky tex plates get blamed for everything even though they are only a figment of Evoquack's imaginings! Stepping out of the movie cinemas where they loop Planet of Apes 24/7 for hordes of excited Evoquacks the reality of volcanoes eruptions is not some imaginary magma plume pushing through a teskoincx plate that fell asleep on its eternal travaille but each eruption is due to the power of the steam heat finally overcoming the weight of the plug of old material on the vent – a volcano is sort of like a steam boiler with a faulty safety gauge. Each eruption has to be bigger than the preceding one!

Thankfully Evoquacks have moved on a little further than the primitives who thought volcanoes was angry underground gods who needed regular offerings of children, chicken, goats or vegetables as still practised by Indonesians who have live volcanoes everywhere.

In actual fact no Evoquack or' gist' – geologist, paleogeologist, volcanologist or similar, can know what happens deeper than the 40,230 feet drilling achieved by the Russians in 1989 before they had to abandon it as the weight of drill string and the drilling mud mixture was drying out by the temperature in the region of 350F and the drill was in danger of locking fast and breaking off.

In January 2017 an Icelandic geothermal experimental project drilling into the edge of a post-flood crust crack reached only 15,000 feet and encountered water heated to the supercritical temperature of 800F. How much higher the temperatures and pressure would be lower down is hard to imagine.

Why does 40,230 feet = 350F but 15,000 = 800F? Obviously the Icelandic drilling was close to one of the underground fires left by The Flood while the cooler Russian drilling was into solid rock away from any fire.

Removing all words like 'hypothesis, believed, posited and supposition' out of the description of volcanoes and black smokers all that is left is: 'volcanoes and black smokers is jets of either steam, smoke, lava and ash billowing out of the ground or sea bed from holes and cracks in the Earth's crust'. Some volcanoes blow out great quantities of methane – obviously from organic matter aka dinosaurs and humans from the Flood. Geysers blow out hot water with dissolved minerals. SCW can dissolve solids and is responsible for the gold 'speci' nuggets.

The whole of Earth has been warmed since Creation by the piezo-electrical underground heating GOD thoughtfully built into Earth. Read Genesis 2:6.

The truth of super hot and destructive volcanoes is that **'THEY ARE GREAT FIRES UNDERGROUND WHERE VAST QUANTITIES OF DINOSAURS, HUMANS, ANIMALS, FISH, SEAWEED AND VEGETATION ARE BURNING IN SUPER PRESSURISED HOT WATER AND PERIODICALLY THE PRESSURE BUILDS UP AND ERUPTS UNTIL THE PRESSURE SUBSIDES FOR AN UNKNOWABLE TIME.'**

Naturally everyone sneers at this simple truth as they have heads stuffed full of Planet of Apes EVONONSENSE little better than that of Indonesians.

The fact numerous volcanoes are found along 'fault lines' is a good pointer to them being fuelled by dinosaurs as of course the fault lines are really the edges of the great chunks of Earth's crust that were all shattered by GOD to start The Flood 4,350 years ago.

During the initial phase of the flood GOD smacked or shook Earth so hard that the crust shattered into innumerable pieces like an egg or apple hit by a bullet. The point of impact seems to have been on the Atlantic side as there are comparatively few volcanoes in that side of Earth while on the opposite side there are many thousands of active and dormant volcanoes, black smokers and geysers. GOD knew that He had to make the impact at ninety degrees to the place where Noah waited expectantly in the Ark. Obviously the impact could not be close to the Ark as if we assume that the Ark rested somewhere in the Biblical Middle East an impact on the Atlantic side of Earth would cause maximum damage 180 degrees away while Noah at the halfway 90 degrees point would feel little impact. High speed film of tennis balls hitting racquets shows this perfectly.

As the impact compressed the Atlantic side of the crust it naturally transmitted a shock wave into the under crust waters but being incompressible waters they passed the shock on until eventually reaching points of least resistance under the Pacific area then it began lifting chunks of crusts along with vast quantities of rock, soil, water and all the living creatures and vegetation above each releasing. Fast moving and compressed waters do funny things. Study videos of bullets hitting eggs.

Due to the water already being quite warm thanks to the piezo-electrical heating systems the shock added so much heat that the waters became heated to supercritical state. This is an interesting state for water to be in. It becomes like a powerful explosive only needing to open a crack to display its latent energy. In their 2009 paper in 'Nature' Christine Wu and team recognised that supercritical water may display exotic chemical reactivity and possibly 'Detonations of high explosives containing oxygen and hydrogen produce water at thousands of kelvin and tens of gigapascals, similar to conditions in the interiors of giant planets'. Wu and team are just one of numerous researchers trying to work with this SCW and find uses for it but so far none seem to have caught onto the mass of SCW that since the Flood has been burning dinosaur and human bodies. Sadly WU has bought into BigBangEVONONSENSE and doesn't know how GOD created Earth and the other planets to make them also have SCW under their crusts.

Study boiler explosions to see just a few hundred gallons of water can have the destructive energy of a similar weight of dynamite. Depending on how high its temperature reaches water can become a solvent for many minerals – this explains why gold, silver and copper veins are found in quartz – as well as many other minerals. Yellowstone's geysers have lots of minerals.

At only 40,000 feet down the Russian drilling the water temperature was 350F and the Icelanders found it to be 800F but what it is further down under more overburden pressure and the twice daily input from the moon can only be guessed at. Luckily the 15,000 feet of rock would take an enormous amount of pressure build up to become a volcano but as each day the moon's gravity gently flexes the crust and all the quartz gives off piezo-electrical energy the pressure will be building.

GOD designed Earth to be a super boiler easily able to safely contain the temperatures and pressure He determined would be best for we humans and all the other living things.

The bedrock was strong enough to contain the intended water pressures and the sun and moon were so finely placed to add just enough direct radiation and piezo-electrical warming that Earth had an optimum temperature everywhere. Satan ruined Earth's purity until finally in order to wipe out all Satan's Nephilim, hybrids and followers GOD shook Earth and the resulting agitations made a spike of piezo-electricity that raised the water temperature and pressures until inevitably Earth exploded just like an overheated old boiler. Bursts occurred in many places on land and underwater and also where the giant 'meteorite' craters are seen today as revealed by the fact the crater centres do not indicate a mass of nickel rich 'meteorite'.'

Adding to the direct bursts in the crust was the water hammer that has been shown to cause much more damage that expected due to sudden spikes of energy. Water hammer accompanies a quick release of pressure as can be heard in house plumbing when a tap is suddenly snapped shut and the slug of water in the pipe has to stop suddenly. Plumbing Supply Company's website article says that water hammer in domestic pipes can cause a short pressure peak ten times higher than the working pressure which is easily able to burst copper pipe. Earth's granite and quartz crust could not withstand such pressures and shattered into many pieces – now laughingly described as shocked quartz and textonikx plates. Between the slabs of plate emerged the first volcanoes as superheated water and minerals burned. The pressures heated water to form the first geysers.

'After 150 days the fountains of the deep were stopped' says the Bible. While we cannot scorn the idea that GOD could actually put His finger on the holes the waters were blasting out of, the fact that the waters precipitated vast quantities of lime indicates it was super hot water under great pressure from deep under the crust. The pressure was initially so high as to prevent water flowing back down under the crust but eventually the reservoirs were emptied and water was able to surge down into these many great chambers – and suck along vast quantities of sediments full of dinosaurs, humans, fish, terrestrial and marine vegetation.

Here is my plan for a simple experimental rig to display this action. Take a glass fish tank full of water with a good amount of suspended small floating stuff 'floaters' through all the water. Make a cylinder with fittings for temperature and pressure gauges and a screw cap and put a little water in it. Screw on temperature and pressure gauges to monitor conditions in the cylinder. Around part of the cylinder put a metal jacket in which a gas flame can be lit to heat the water in the cylinder, the jacket to have a tall chimney higher than the water in the fish tank. Insulate the rest of the cylinder with soft polystyrene or similar. Connect the burner to a gas hose and light the flame. Put the cylinder in the tank and weight it down to the bottom. Watch the temperature and pressure in the cylinder rise. Video the experiment in slow motion with close ups of the gauges. Depending on how strong and thick the cylinder is the longer will be the wait. But at the critical pressure point the inner steam will burst the cylinder wall, burst through the foam insulation and a strong jet will burst up through the water and above the tank. The slow motion video will show as the cylinder empties itself water and floaters will be sucked in to fill the vacuum. Now repair the cylinder wall so the floaters are trapped inside and turn on the heat until eventually the cylinder bursts again. This is exactly what happened during The Flood.

The actual rupturing of Earth's crust would resemble the way in which boilers can do enormous damage when they burst from one or two separate causes: explosive water hammer or explosion of hydrogen gases reacting with minerals and iron in supercritical water as Christine Wu and team researched.

Explosive water hammer can be demonstrated easily by shaking a can of fizzy drink and using a sharp knife try to cut the can open down its length. As the knife is applied and dragged along it first scores the metal until force is added to pierce the can at which point the pressurised drink begins to squirt out the hole but that immediately causes surrounding liquid to flash into bubbles and so on until all the potential energy in the drink is released in a split second with sufficient power to rip the can open! When this happens with a boiler the boiler can be blasted high and wide.

Exactly the same sequence occurred during The Flood but at various dates and places depending on the size and depth of the crust and its water cover.

Some of the Flood sediments around the world can have a very high percentage of organic matter with the shale in Kimmeridge Bay in the UK being up to 70%. Sadly the researchers into the Kimmeridge swallowed Planet of Apes woozle hook, line and sinker and prattle on about it being deposited and starting to move textonnixally in Late Jurassic Times aka about the first week of The Flood 4,350 years ago. They deny the Flood really occurred and have coined a new phrase of EVONONSENSE to describe the starting cataclysms of The Flood and now refer to those days as a time of 'Rapid Global Geological Events'. At Kimmeridge was found the skull of a giant creature from 'sometime between Jurassic Park and the Cretinous Epoqy but of course it died during The Flood'.

After the flood the reservoirs would still be very hot and the Earth's continued flexing under the sun and moon's gravitational pulls would raise the temperatures and pressures again to supercritical level but this time huge quantities of combustible material was available. It burned. It is still burning. As it burns it causes steam to feed itself to burn hotter. Each day the moon pulls on the crust and the quartz crystals compress and release their piezo-electrical energy. The Earth warms. The fires heat up. The steam pressure climbs higher. The steam tries to lift the crust and volcano peaks and the plugs of old lava and ash and causes seismic senses to dance. Time passes. The tremors grow worse. The pressures reach maximum. The volcano erupts to release the pressure. Out pours vast quantities of methane, steam, rocks and ash from the burned rocks. The eruption runs its course until once again the pressure is unable to maintain an opening and the cone closes up and the fires settle down to slowly burn Flood material from outside the cone's circumference. The fires slowly expand, the heat and pressure builds up away until once again reaching some point at which it must once again erupt but in a much bigger eruption than all previous ones. All volcanoes have the potential to become Mt St Helens or Krakatoas thanks to repeated eruptions building bigger and bigger cones over wider and wider areas and thanks to the square rule formula.

However having described the workings of volcanoes there is the matter of their distribution and frequency of eruption to consider. If a detailed historical map of eruptions could be produced for the entire Earth including the increase and decreases of black smokers it might be found that an eruption in one place while being detectable by seismographs is linked with other potential eruptions.

The Ring of Fire has the most active volcanoes and seems to operate like a giant plate with bevelled lower edge being slowly forced down into opening – just like a tapered bung fits in a barrel? As researchers claim the crust chunks aka teskoninic plates is about 62 miles thick but a minute with a calculator would show that the Ring of Fire chunk is bound to have lots of volcanic actions – but not from the slipping and sliding mating-turtles action fantasised by such woozlers as Nicholas van der Elst, a seismologist at Columbia University's Lamont-Doherty Earth Observatory as he describes the mating turtles theory as being: 'due to convection in the mantle. Hot material near the Earth's core rises, and colder mantle rock sinks. "It's kind of like a pot boiling on a stove". Hmmm, has he calculated the loss of pressure of the fantasy rising magma? My pots boil over and mess up the stove - though 58 years ago I did see potassium permanganate crystals send plumes circulating through warming water.

As the Ring of Fire is definitely NOT due to van der Elst's mating-turtles plates the cause must be more logical and needs an understanding of the weight of water and the dynamics of car crashes as revealed by slow motion videos.

Water is heavy stuff. As the ring of fire has been burning under the crust slab for about 4,350 years and there have been many eruptions and at least 40,000 noticeable subterranean volcanic cones with related lava flows at this moment it logically follows that there will be a general lack of material under the entire ring of fire or Pacific slab. However most of this missing material was blasted out into space from under the slab the day GOD shook Earth as when a bullet strikes an apple the opposite side bulges before bursting. After the initial impact the continued egress of high pressure water caused a great loss of solid material by entrainment – as in shot blasting. The entrainment would be tremendous as the velocity of the escaping steam would be constantly rising and ripping material off the crack it was flowing up. The loosed material would be blasted out into outer space carrying terrestrial life which is now returning as meteorites carrying terrestrial DNA. As the released water escaped it left superheated steam behind and the tremendous hat in the rocks.

Lava expands in volume as it rises and pressure reduces while smoke and fumes from the burning material and water will also rise, turn to steam to join with the lava in blocking cracks around the vent pipes and ensuring further water cannot slip past to refill the chambers. This is rudimentary lava, water and vapour action that anyone should know. Once the volcano fire has burned for some time and is running out of fuel the volcano will subside into a smoking, steaming shadow of itself. Eventually water and possibly gas and oil from decomposition of Flood debris in surrounding strata will slowly seep back towards the vent, aided by the continued pressure of the slab and water above it. This new fuel will begin to increase in temperature from piezo-electrical energy. During eruptions the piezo-electricity will ignite the oil and gas vapours as well as making some impressive lightning in the eruption plume.

As a big volcano begins building to eruption pressure and especially when supervolcanoes like Vesuvius, Krakatoa, Mt St Helens are building up they might be having an unknown effect on crust slabs a good distance away. Each eruption will cause a degree of altered pressure and size that will travel around the surrounding slabs and depending on the severity may trigger tremor, medium or large earthquakes or increased or reduced pressure in other volcanoes.

Each big earthquake from crust slabs settling must have a compressive shockwave effect to cause other earthquakes or make a volcano more likely to erupt.

There is no way to escape these unnatural natural events and so we just have to keep our fingers crossed. There have been minor tremors in my part of England but whether due to a big slab moving or beginning to crack cannot be known.

Many old volcanologists of 100 or more years ago did not have the benefit of today's worldwide news dissemination and perhaps could not understand the sequence of event that triggered volcanoes.

It has been known for centuries that sparks can jump out of Earth during earthquakes but the facts of immense piezo-electrical discharges were unknown although many researchers have claimed lightning sometimes flies from Earth into the clouds.

The San Andreas Fault is about 750 miles long and Wiki says tekonix plates are about 62 miles thick. The Fault is known to move one to one and half inches per year. 750 X 62 = 46,500 square miles of rubbing contact every time the plates – actually slabs of crust – move. Now anyone who knows the piezo-electrical voltage generated by quartz can calculate the possible energy under the Fault.

It was also known for centuries that coal mines and coal spoil heaps can burst into flame and burn under the surface without an apparent supply of oxygen and now this can be carefully studied at Centralia, Pennsylvania – just a couple of weeks ago I saw an old miner from northern UK telling of how the burning local pit heaps were like a beacon for German bombers during World War 2.

Now that supercritical water can be demonstrated and the evidence of worms is known it is easy to knock down the entire EVONONSENSE edifice that has been built up over so many years.

Read my chapter discussing the amount of vegetation and living creatures that the Flood buried and sucked into the Earth.

Chapter Fifty Three

Post Flood Ice Ages and Mammoths buried in rich frozen sediments.. Worms.

For many hundreds of years the natives of far Siberia and perhaps Alaska have been finding and trading ivory salvaged from frozen mammoths found exposed on seashores or river banks in what was a vast area of deeply frozen soil – the tundra. This soil is rich in organic matter showing it had once been highly fertile and ideal grazing for animals. It is now melting as the climate warms up. The number of mammoths found could only mean that huge herds had once roamed Siberia or wherever the soil had originated. To further confuse the issue the tundra lands are dotted with hills in which the soils are alternated with fine sharp sand deposits, great ices slabs and lots of mammoths but also the bodies of animals still living today but not in Siberia. Often the remains are so complete a complete dissection reveals the last meal and cause of death but most of the bodies are found torn to pieces as if by massive trauma. This mutilation mimics that of the great herds of dinosaurs found in the boneyards of USA and South Africa but their ruin was great surging Flood waters while the mammoth's is due to one great tsunami at the height of the Ice Age.

After storms the Arctic sea tosses up masses of fresh but frozen trees from underwater forests the locals name Noah's wood to confirm the tsunami.

The Evoquacks all snottily dismiss the mammoths as having died during teccstonic plates activities in some dim distant past with Wiki's Evoquacks laughable claiming mammoths split from a similar creature about 400,000 years ago in the Plastocine Era, and despite needing as much water and grass as a modern elephant they all lived in the cold barren area and plodded around the snowy wastes trying to find enough food to stave off starvation which is total EVONONSENSE unless ape-ladies knitted them woolly muffs for their trunks?

The mammoths lived in a mild fertile habitat as is deduced by the stomachs of some mammoths containing C3 Plant matter. C3 plants tend to thrive in areas where sunlight intensity is moderate, temperatures are moderate, carbon dioxide concentrations are around 200 ppm or higher,and groundwater is plentiful as these plants lose 97% of of the water they take up by transpiration. Wiki claims: 'C3 plants originated during the Mesozoic and Paleozic eras' – of course this is more rubbish as GOD created all plants on Day 4 and and C3 vegetation still represent approximately 95% of Earth's plant biomass. Practically all the common grains, leafy plants, vegetables and common trees are C3 kinds while the common weeds and nuisance grasses tend to be C4 types. It might be that C4 plants result from GOD cursing the Earth after Eve obeyed Satan and ate that fruit. Therefore we know that mammoths lived in a lush fertile temperate zone not the chilly tundra.

That mammoths lived in an environment rich in C3 plants is proof they lived after The Flood when the new climate was moderated by the new winds, the air was moist and the ground had plentiful springs to promote a heavy growth of grasses and sedges. One estimate for one site is it contained about 1,000 skeletons indicating a giant herd living closely together without need to space out to find sufficient food. The sheer number of mammoths confirms that they were breeding well and if today's elephant fertility rates applied then the herds must have had at least two centuries from The Flood to build to high numbers until the day came they all died in some great catastrophe that literally killed them as they stood eating with many being torn apart and all engulfed in thick black muck interspersed with bands of sharp sand loess that froze solidly within a very short time – perhaps only two or three hours.

Knowing that the mammoths and the deer, bison, lions and similar 'modern' creatures buried with them, had an abundant supply of C3 plant food the question arises of how they come to be frozen in enormous quantities of 'muck' – muck being the name as used by scientists and keen gardeners for good rich soil containing lots of organic matter.

Evoquacks naturally invoke plate tekskotonics and dates of 20-65,000 years for the catastrophic burial and freezing while some Creationists are just as ignorant and claim the mammoths died during The Flood 4,350 year ago.

Using logic, evidence, knowledge and the gift of wisdom freely given to all who believe Jesus is the son of GOD I can state that mammoths were buried when the Earth rolled perhaps two or three hundred years after The Flood – or about 4,000 years ago!

The key to the mammoth mystery is the same key to the correct age of Earth and Creation: worms! Or specifically: worms, muck, and the sharp sand loess.

Practically no-one can understand how worms clarify the age of Earth and the date of The Flood, while no-one except me can explain how worms date the burial of the frozen mammoths.

Muck is dark soil filled with plant matter as can be found in any land that is manured and carries good crops – like a cow field, a good vegetable or flower garden. For the mammoths to be buried in the muck something remarkable must have happened to send gigantic amounts of the stuff sweeping over them suffocating, squashing and burying them.

Mary Schweitzer's and Alice Robert's imaginarily fossilised dinosaurs were buried during vast flows of stratified sediments in The Flood about 4,350 years ago; mammoths were buried and fast frozen in just one quick tsunami of mixed muck about 4,000 years ago at the height of the Ice Age when the Earth rolled.   EUREKA!

This cannot have happened during The Flood as the muck is not stratified into gravels, sand, and organic matter topped with fine silt but is literally a messy mixture of gravel, sand, silt and plant matter just as can be seen at the head of flashfloods or tsunamis – ergo it was never sifted back and forth in deep water as are the sediments that all fossil dinosaurs are found in.

After GOD finished giving His promise of the rainbow he left Noah and family to stay close to the Ark with some domestic creatures and watch as the other creatures they had spent so much time with were now spreading out with barely a backward glance. The domestic creatures remained close around the Ark as they had plenty grass and water and no urge to move away from the family's protection. The cows and goats duly responded by supplying rich milk; the chicken and ducks gave eggs for food or breeding.

As darkness fell Noah and family looked at the stars and tried to determine their new location and timing and wondered why the moon now had conspicuous dark patches and craters over much of the surface that had once been a brilliant white reflector through the nights.

Freed from the burden of heavy work tending a mass of creatures in the claustrophobic Ark the four couples maybe found they had energy to spare and as the day ended they all retired to bed and started the task of refilling the Earth.   In the darkness the multitude of creatures did exactly the same thing. GOD's angels glided over them checking all this activity and reported favourably to GOD.

In the following weeks, months and years Noah and all the creatures became accustomed to the new extremes of heat and cold, the wind and the rain, the labour of making new homes and planting crops, but they had their health and felt little stress. All the rain may have been as much a nuisance as we find it today. They all would have been unaware that the heavy rains were eroding the Earth at the same time as it fell as snow on the new high ground. They had enough food and the wild animals would not become predators for some years and we can surmise that life would not have been as worrisome as in the last years of the Nephilim Age.

Despite the new extremes of the climate the populations of creatures increased dramatically until vast uncountable herds of big and small creatures covered all the lands and above them flew flocks of birds and clouds of insects. Rivers and lakes swarmed with great shoals of fish. Fruit vines, bushes and trees provided their crops but only during the relevant seasons and not continuously as before. Everywhere grew great forests but now they were different trees to those before the flood.

Noah and the humans were not to know that the soil was initially only superficially fertile and productive from lack of worms with the result that the new grasses made a thick carpet but it was not deeply anchored due to lack of worm tunneling giving roots deep anchorages.

Nor were the vast quantities of new sand – quartz – grains being rounded off from passing through worms. EUREKA!

The humans too increased in number. Japheth remembered GOD's order to fill the Earth and he and his descendants spread west and became the Gentile races whom Jesus would later tell the disciples to preach the Gospel to - Matthew 28:19. Shem and his line headed east to become the eastern peoples. Only Ham disobeyed GOD and stubbornly established cities in the small area of Canaan which the Old Testament constantly records was inhabited by the last Nephilim and people who worshipped Baal and Ashtorah. Today Ham's area is still inhabited by people who worship the renamed idol Baal as 'Allah' and are the deadly enemies of Christians, Jesus and GOD Himself!

Perhaps as the humans spread across the land they saw the distant peaks of the new mountains were turning ever more whiter as snow and ice built up and didn't melt away in the warmer months.

Knowing nothing of Earth's dynamics and physics the humans could not know that the Flood had left great masses of new mountain ranges in many locations and most of these were gaining great weights of snow and ice each year. Because the mountains weight acted on a bigger radius than the low lands they exerted more of a turning effect that increased as their snow and ice load increased just as Archimedes would one day prove when he said 'Give me a long lever and the moon for a fulcrum and I'll move Earth!' 'EUREKA!' Earth had not been designed to support such a mass of weight and gradually and imperceptibly to the humans became unbalanced.

Earth is not a perfect sphere these days and maybe never was, or perhaps GOD had created it as a perfect sphere that slightly bulged at the equator as He set the Earth spinning. Then again perhaps the Earth retained a perfect sphericity until the shattering crust at the start of The Flood allowed it to sag into its present shape the way a fat woman sags when she steps out of her corsets? Scientists now make precise measurements showing Earth is slightly flattened to be wider at the equator. No-one can know if this flattening was less or more extreme in the centuries after the flood when Earth was entering the Ice Age as the snow and ice crept over the north and south and all the high mountains but slowly the climate cooled and the snow and ice grew deeper.

Maybe some animals sensed this increasing chill and migrated to warmer areas just as the humans felt the chill and had to wear more clothing, build warmer homes or also trek to warmer climes. Neither humans or creatures had any experience of any danger posed by the snowbound areas and though life in the cold areas may be less congenial all seemed tolerable as the Earth spun through its days and the sun and moon rose but weren't as useful for following the years as pre-Flood.

One day the snow and ice mass surpassed the critical limit and Earth's equilibrium was overcome. EUREKA! Without warning the heavenly bodies flew to strange orbits, the ground underfoot wobbled and groaned, animals and birds howled and screeched in fear, winds began blowing and lifting light objects and dust skywards. Earth had rolled many degrees.

We have no written record of this event. Just as with the initial impact by GOD that began The Flood the Bible writers may not have felt a great movement if they lived close to the swivel point while other humans living in the direction of tilt would have suddenly travelled thousands of miles. Hold a ball between finger and thumb and move its equator to the pole to see how the swivel point hardly moves while the outer point moves much faster? If a globe in a stand is available spin it and note the distance and velocity of points on it as it is spun.

As the Earth began to topple the huge winds blasted all loose soil into the air to choke mammoths and other creatures just as dissection of their bodies show. The trees tore out of the ground and, strangely, huge mats of grass lifted off to go sailing through the air. This debris rolled and bounced along the ground being shattered and mixed into a jumbled mess. (Is this scenario proof of geocentrism?)

Inertia made the waters initially build into giant heaps as the land advanced underneath until finally it all moved just as videos of tsunamis show. The tsunamis that buried the mammoths were vastly more terrible than the ones that desolate Japan and other areas in these Last Days.

The water was displaced for thousands of miles not just a few miles as today's tsunamis. It was many times worse than anything seen since the Flood. Roaring over what was north pole at that time the waters stripped more of the land and rolled the mammoths, other creatures and debris masses with great force to shred and pulverise until eventually as the Earth slowed and stopped its roll the debris settled in a thick black mess. Where the land had been nothing but sand left by The Flood the wind lifted great clouds began to settle in thick layers as the muck eddied back and forth. Finally all was still and began to get very cold.

The soil was rich and organic but the sharp sand grains that make up the bands and lenses of the loess reveal that it had buried the mammoths before worms had managed to travel to the mammoth grazings and reshape all the sand grains to have rounded corners! The sand was new sand made from the masses of quartz ejected from Earth during the upheavals of The Flood.

As worms cannot survive in cold soil and also drown under deep de-oxygenated water logged soil which the flood waters would be for perhaps the first 200 days we can know that much land would be practically worm free for many years after the flood so that when Earth toppled in the Ice Age the grass layer would instantly come loose and all the sand in it would be sharp and unprocessed by worms. EUREKA!

We know this from Darwin's meticulous research on worms over many years from which he concluded that worms can build topsoil at the rate of one inch per five years during which time they ingest all the sand in the soil and expel it with its sharp corners rounded off! He studied how worms undermine solid objects and bring soil up so that grasses can establish and then become food for various creatures whose droppings fertilise the plants to make them grow better ad infinitum!

Adding to Darwin's research was that of mid 20th century Dutch engineers who having reclaimed land from the sea noted worms reluctant to move into new polders as the polders were created by pumping sand and silt from offshore. The new polder land was practically the same as Noah found when he stepped out of the Ark. Only when the Dutch added vegetable matter to the sand did worms begin to travel across the new soil. Today the Earth's extensive deserts cannot regenerate for lack of both water and worms.

Chapter Fifty Four

Repopulation, Babel, Tribes, Baal-Allah, Rome.

The matter of the human race repopulating the Earth after the Flood and the building and destruction of a tower at Babel has generated immense quantities of woozle! Most Evoquacks deny the Tower of Babel incident just as they deny The Flood yet it merely takes an understanding of the Bible in Genesis 10-11 to follow the sequence and perceive the truth of all present humans being descended from Noah and family and how that family grew then fragmented out across the whole world while all the land was one big mass before it separated into islands and continents!

The Evoquack's silly testroniks plates theories has allowed all sorts of EVONONSENSE be imagined and made fact by constant woozle!

Here is what happened after the flood waters had drained and the land had become dry.

Immediately Noah, his family and all the creatures exited the Ark GOD told them to once again multiply and fill the Earth with its vast lands to explore and utilise. Fresh grass, food and utility plants and seedlings were popping up on all sides. The sun warmed them and the rain wet them but promised to make the ground fertile and offer a wonderful future. The humans and creatures eagerly proceeded to fulfil their GOD given task.

It had only been a little over a year previously Noah and family had looked in disgust, anger and sadness at the lawlessness, idolatries and perversion that surrounded them as fallen angels led humans, Nephilim and hybrids to ever more idolatry and evil sins. Now the Flood was over the little family could see all the former evil beings had been destroyed and washed away by the great waters. Noah's own brothers, sisters, nephews and nieces were among the perished! His wife, sons and son's wives may have looked around and been shocked to see that their relatives had vanished along with all trace they ever existed. They thought they had the Earth to themselves. They were wrong but perhaps did not know it at first. Just as the invisible Satan had watched Eve now countless invisible demons watched Noah and family to try spot means of deceiving them just as Satan had deceived Eve.

We cannot know if GOD had told Noah all the fallen angels would be captured by holy angels and bound until being released for a short while at the end of the future Millennium as the pope's new throne portrays, or if GOD had told him that though the Nephilim bodies would drown the fallen angel spirits in them would be freed to wander Earth again.

The dried out land had only one major sea and this was probably today's Mediterranean – which actually means 'in the middle of the land' - as it's geology and deposits certainly seem to be Flood-made despite Satan's Wiki doing its usual blasphemies by claiming the Mediterranean is millions of years old, was once dried up to make massive salt deposits, flooded again and again without dissolving all the salt, was formed by an African tekxtonix plate pushing north, woozle, woozle, etc... This woozle is further added to by the author of the Bible History.com website who states that the families migrated to the head of the Persian Gulf forgetting that there was no Persian Gulf until after GOD had caused Earth to split into separate lands and seas.

Repopulation and The Tower of Babel.

Noah's sons duly produced many children and soon they drifted off to find room to live away from Mt Ararat on the plains to the east and south of the great sea. There they farmed and prospered with Noah himself planting a vineyard and getting drunk on its wine. Genesis 9 tells of the trouble wine can cause.

After some years Noah reminded the three families that GOD had ordered them to spread over Earth but for some reason – clannishness, fear of what may exist in distant land, satisfaction with the food resources of the area, rebelliousness and refusal to obey GOD, influence of Satan - they gathered at a place where they decided to build themselves a mighty tower perhaps as a grandiose marker or rallying point visible from great distances. It may have been one leader's attempt to elevate himself to vastly higher status that all others.

The ringleader appears to have been Nimrod, son of Cush and grandson of Ham and great-grandson of Noah. This gives reason once again to believe that Ham's wife was carrying the DNA of a fallen angel, and naturally Nimrod being rebellious opposer of GOD he had no qualms about defiantly ordering the building of a tower as a central gathering place for his kingdom and the worship of his imaginary sun-god Baal.

Trying to arrive at an estimate of the population at the time they decided they needed a tower is difficult as no formula seems to give a logical result. Using population calculators with a starting population of just 3 breeding pairs and a birth rate of just one baby per year as is achievable today produces under one thousand souls after four hundred years. Doubling the birthrate only gives about 7,000 by 300 years after The Flood. By allowing for twins, triplets and 10 months spacing and double the birthrate would give about 16-30,000 people after 400 years. Is that enough to cause a need to spread out for living, farming and grazing land? And is such a number viable to spend a good deal of time and effort on building a tower as a beacon or rallying point? What if the available time had only been 250 or 300 years and the populations much less? We just do not know.

We know even less about why the population rise since the Babel incident has not followed any formula. Wars, plagues, famines, droughts and failing human health have all conspired to modify what should have been a steady rise in the population. We do not know what the population was on the day Jesus was born. We cannot guess what it will be on the day he returns but we know that after he is done slaughtering all the evil it will have plummeted to a tiny fraction of today's. We cannot guess what it will be at the end of the Millennium when all the Raptured have been busily procreating for one thousand years. We cannot know how many billion will come out of their graves to discover their names are written in the Book of Life and with new bodies live happily ever after.

Numerous government and research institutions release estimates of present and projected populations with and without overrides on the figures depending on the availability of food, water and resources to ensure healthy populations. A basic calculation is: 7.6 billion live on one-tenth of the Earth's surface today. If Earth is remade as one land mass with a small central sea the nine-tenths of land could support 9 times 7.6 or almost sixty billion happy people and uncountable creatures. EUREKA!

Wiki panders to Satan as usual and calls the Tower of Babel account a myth. Hordes of explorers, writers and occultists have written of the tower just as writers feel drawn to write woozle about the Great Pyramid. Regardless of Wiki the people began to build their tower. The site had clay or mud that could make brick and evidently was well wooded to provide fuel for burning the bricks. There were also seeps or lakes of asphalt of some kind to serve as mortar.

When the tower had risen many feet GOD came to see it. Realising the humans were now capable of combined actions and also intended to stay in the area and not spread over the entire Earth He instantly made them all suddenly speak 'in tongues!' We cannot know if the affliction was so severe that husbands and wives or brothers and sisters lost the ability to communicate or if it was done on a family by family or tribe by tribe basis. Nor can we know if they had a written language at that time and suddenly found themselves writing and reading gobblydegook! Today while we can all easily fly around the world, we know when we get there it will be impossible to communicate with the native inhabitants except by sign and pantomime and the results will be hilarious, annoying or deadly.

(This speaking in tongues is about the same as what happens in some pentecostal churches that have a strong emphasis on speaking in tongues being the sign a person has the gift of holy spirit. I have my doubts as I have often heard such people speak in tongues but when speaking English they utter lies and false doctrines...including their belief they all fly off to heaven at death!)

Naturally as no builder could understand what his fellow workers were saying it would have suddenly been immensely frustrating although perhaps quite comical in the way we have become accustomed to slapstick comedy in this last century. Unable to coordinate their labour they stopped building the tower and began to drift away in individual families or small groups that could understand each other. They began the great migrations that ultimately filled the Earth.

Japhath's family became the Gentiles and spread westward and northeast around the central sea. Eventually the sons built their own dynasties and named nations and regions after themselves. After GOD broke up the landmass they reached the British Isles and then over the ocean to Canada and USA.

Shem and family did not spread eastward very far but preferred to colonise the entire lands today known as Palestine, Iraq, Iran, Persia, Syria and on to China. They settled what became Sodom and Gomorrah and the other three towns that were to become the epitome of idolatrous sexual depravity and be eventually destroyed by fire from heaven. It is from Shem's line that the Israelites came and through whom Jesus's bloodline can be traced right back to Eve.

Ham may have had dark skin and his descendants migrated south to today's Ethiopia, Egypt, North Africa, Arabia where the people are mainly devilworshipping Muslims of many dark colours. It seems Ham being the son of Noah was genetically as perfect as could be at that stage of humanity but his wife must have been a descendant of one of the fallen angels: she could perhaps even be related to Satan which would explain why Ham's line became so constantly idolatrous and evil that GOD had no qualms about sentencing them to annihilation at the hands of His chosen people and Jesus will slaughter all of them when he returns unless they choose to learn the truth of the Gospel and become his followers?

The fallen angels had taught the women how to beautify themselves with dark eyeshadow and as Ham seems to have been a literal and figurative black sheep such a woman may have had the same allure for a man like him as do the painted ladies who haunt the criminal fleshpots of today. Ham's grandson was Nimrod who is described as a great hunter – but a hunter of what is not clear unless it means the usual hunter of beasts and possibly the dinosaurs which had survived aboard the Ark. Or was he the father of all the cannibals that have lived and may still live in Africa and around the world? There are videos of Muslims slaughtering Christians and other Muslims like cattle. Nimrod seems to have been the builder of Nineveh while Ham's second son Mizraim and his offspring colonised Canaan. The Old Testament details many Canaanites who were giants. Ham's sons really do seem to have had satanic angel genes! EUREKA!

By some unknown date the human race and all the creatures had successfully migrated to the ends of the Earth. A brief sample of the migrations of the creatures includes such as: kangaroos had hopped all the way to what would become Australia while duck billed platypuses plodded slowly behind but got there eventually, lemurs had scampered to Madagascar, komodo dragons to Indonesian islands, giraffes to Africa, bison and skunks to North America, pandas to China. Lists of these isolated species are easily found but as expected the Evoquacks like Attenborough stupidly use phrases like 'evolved in isolation over 150 million years...'

GOD watched the humans and creatures disperse to their present day locations, and deciding the time to intervene had come, caused the single huge land mass to separate out into small and great blocks of varying sizes with deep water seas, oceans and channels between them all. Just how He did this is a mystery but from the evidence of the mammoths being quickly buried and deep frozen in masses of rich muck it might be that the dividing occurred as the Earth rolled at the height of the Ice Age and GOD merely helped things along while ensuring the bloodline from Adam was preserved in Shem's extended family. All we can learn from the Bible is that 'In the 239 year lifetime of Peleg the Earth was divided.' His lifetime extended from 100 to 339 years after The Flood and by his death the Earth was deep in the Ice Age which neatly correlates with a post-Flood ice age. EUREKA!

Chapter Fifty Five

Sumerian worldwide flood Myths.

Lots of immature fools have swallowed the lies of the Evoquacks and Satanists who claim the Bible account of The Flood as being a rewriting and mishmash of old flood myths from many other places and cultures; they usually claim the Sumerian account as being the original one that Moses plagiarised. These are generally the same people who also claim they is evolved from monkeys and deny GOD created the universe so their calling GOD, Jesus and Moses liars is not unexpected. Obviously they lack understanding but are a delight to Satan – who they also deny exists! Most religious leaders also deny the Flood was real and subscribe to it as being a small local flooding such as the Nile was known to do each year – this is how to show allegiance to their real god Satan.

A star example of this halfwitted denial is shown by this quote from Robert R. Cargill of UCLA: 'In order to even entertain the possibility of a worldwide flood, one has to bypass all laws of physics, exit the realm of science, and enter into the realm of the miraculous, which many believers are willing to do.'

Wow! Although a halfwit he mentions physics, science and miracles! What a great shame the worms make an utter fool of him as it needs great faith in miracles of science, physics and biology to accept and teach EVONONSENSE!

I'm not sure why he sneers the Flood is outside 'physics' in his childish ignorance unless he is a bit dumb but the Bible explains how GOD used physics to start the Flood: read Psalm 60:2. Or my Flood chapter. His second sneer that the Flood was outside science shows himself to have missed the science class as scientists who have studied Earth say that if all the mountains and seas were levelled out and the ice caps melted there would be almost two miles depth of water everywhere. Obviously having evolved from a monkey that evolved from the dust of the bigbang and being brainwashed with the barren drynesses of Planet of Ape films he watched so often as a baby he is unable to use his GOD-given brain to visualise that if Earth had accreted from dust it would be fairly flat with no reason or means of making the teccstonical plates he and all other Evoquacks is so fascinated with...and if it had accreted from dust and was flat how did it get covered with water? Ignorance really is a wonderful thing! If Cargill wasn't an ardent supporter of Satan he could have watched a few videos of flashfloods and seen how just a few hours heavy rain can turn into raging floods that strip everything right back to bedrock. Cargill's problem really is his desire to promote Satan and show his own stupidity because if he had the power of thought he would be able to start compiling a list of the miracles inside science and physics that were needed to evolve everything from the big bang to his own body with all its amazing design – and all without a shred of proof or repeatability! No doubt he'll snarl viciously at these words of mine!

Cargill is thus one of those people Schopenhauer wrote of: All truth passes through three stages. First, it is ridiculed. Second, it is violently opposed. Sadly Cargill will never reach Stage Three.

The area of Sumeria does show evidence for many localised floodings over the years but not the total scouring of land back to bedrock as do lands that have never recovered from the real Flood of 4,350 years ago.

Any tale of a Sumerian flood is really a lie and traceable back to the linguistics problems following the Tower of Babel debacle.

EUREKA! The Tower of Babel incident is the root of all the Flood lies!

Sumeria! Place of myth, legend and mystery so beloved of Satanists, atheists and Evoquacks! History.com states: 'Sometime around 4000 B.C., ancient Sumerian culture emerged on a sun-scorched floodplain along the Tigris and Euphrates rivers in what is now Iraq.' That is nonsense as Earth was only created around 4,000 BC and Sumeria came into existence after The Flood! Lots of other researchers document the rise, prosperity, infighting and ultimate downfall of the Sumerians in the years after the Flood! Basically not many years after The Flood of 4,350 years ago some of Noah's descendants migrated to and settled the area between the rivers that would have been green farmland still abundantly fertile from residual mud from the Flood and the flush of lush vegetation that the rich soil and damp air promoted. They lived well, had no reason to form armies and battle each other for land or possessions. Everyone spoke the same language that Adam and Eve had spoken although their looks would range across the central spectrum of humanity before sectarian and ethnic inbreeding brought about the extremes of colour and physiognomy we see today.

Unfortunately the Babelites had forgotten GOD's edict to Noah as recorded in Genesis 9:7 "7 As for you, be fruitful and increase in number; multiply on the earth and increase upon it." Instead of spreading across the entire Earth that was still mostly one connected landmass they chose to gather in Sumeria and eventually settled on Babel as being an especially favourable place with abundant food and water.

As Satan and many demons and possibly some fallen angels had survived The Flood it didn't take them long to seduce the Sumerians into inventing and naming various imaginary gods along with inventing all the ceremonies, priests and acolytes all religions seem to spawn. As he did with Eve and that first edict concerning the forbidden fruit Satan stepped forwarded and insinuated that rather than spreading across the Earth as GOD ordered they should surely build up Babylon and in fact construct a great tower visible over great distances as a sign and symbol of their prosperity and influence. They were to put a temple on top of the tower and use it for abominable sun worshipping ceremonies. Satan contradicts everything GOD says! As Eve did with his suggestion about the fruit the Babylites thought Satan's tower idea was perfectly reasonable and set themselves to erecting a great tower.

As they all spoke the same language and were well fed the construction went quickly ahead. The rivers provided vast quantities of clay and evidently there was plenty of wood to fire the clay bricks. Flood-buried vegetation was degrading into tar or similar matter and seeping to the surface for easy collection for use to stick the bricks together. Perhaps they failed to realise that as the tower grew larger it would suck in more labourers demanding more bricks that all consumed great amounts of firing fuel – and demanding more food from a workforce that had to intensively crop the land, hunt edible creatures and fish the rivers. It wouldn't take long before their efforts were despoiling the land and preventing them expanding across the Earth. The tower rose to impressive heights – 300 feet according to one account, 700 feet in another and even 8150. The volumes of towers of such heights compare with the pyramids but the pyramid's design was different perhaps due to the natural difficulty in setting the limestone blocks and the bits of rubble used to level odd shaped blocks in the pyramids while baked bricks could be made a uniform shape and give a more dense load bearing structure.

GOD saw the tower rising and knew that the Babelites were ignoring His edict to spread over the empty Earth and confused the language. Unable to communicate and no doubt getting exceedingly frustrated trying the builders stopped their work and drifted away with their own little family who could understand them.

The moment GOD confused their tongues they would face immnse difficulty in trying to invent new words for everything in their heads! Suddenly a cake of bread is a ??? A fish is a ???? A mud brick is a ????? That man's name is now ?????? How intensely frustrating those first few days and weeks would be! Imagine the disruption of family life if children spoke a new language and their parents another!

Many years or decades later when meeting other families and trying to explain their origins their sign language and pictograms confused the whole issue of Noah, the Ark and animals and the tower but retained the central them of a flood and eight people and animals surviving in a boat.

As they travelled and settled they would invent new gods and religions for themselves or possibly demons would appear and convince the people they were aliens or gods with super powers to be worshipped. This still happens with cults that hit the news occasionally. Then as the stories were retold and embellished the original truth was overlaid with fiction that constant woozle made seem real – just as happened with EVONONSENSE these last two centuries. On meeting other tribes and trying to explain their own histories and new gods a monumental woozle would be built up. When writing was invented these jumbled histories and myths were edited to give glory to the kings and the gods until today practically everyone believes the Flood was just an overflowing river around Sumeria.

True Christians who insist the Flood was a divine judgment by GOD are jeered at just as Jesus was jeered at as he hung dying on his cross.

Worms and the similar layers of sediments around the world prove The Flood really was a worldwide catastrophe quite recently!

Chapter Fifty Six
Old Testament GOD versus the Israelites. Mass slaughter.

Richard Dawkins and many others have sneered at the reality of GOD by protesting that the Books of the Bible have numerous instances of GOD slaughtering masses of pagans, Israelites and others often including wives, children and animals, and many accounts of the miracles performed by GOD, Jesus, prophets and the disciples – as well as Pharaoh's magic practising priests. And GOD caused The Flood that killed all but eight people and some creatures on a boat. 'Who wants to believe such nonsense and worship such an evil god!' claim Dawkins and his ilk in their ignorance about the truth and circumstances of the deadly incidents in the books.

There is no denying there are many mass killings of Israelites and other nations in the Bible by GOD despite Him saying they are His chosen people!

Antiochus the Syrian invaded Jerusalem in response to reports the Jews were celebrating a false report of his death. He ordered the slaughter of many and desecrated the temple by offering a pig to his own god on the altar dedicated to GOD then sprinkling pig broth all over the temple and burning all the scrolls – maybe that is how we lost The Book Of Enoch. Old films show Nazis rounding up Jews from synagogues during World War 2 and casually throwing the synagogue's scrolls on fires totally unaware of the intrinsic value or that the scrolls prophesied what happens to people who disobey GOD to kill, pillage and story.

Genesis tells of GOD killing all the corrupt humans, Nephilim, hybrids and vicious creatures in The Flood. It details the destruction of Sodom and Gomorrah.

Exodus records that GOD killed perhaps two million Israelites and Egyptians in the Exodus years. It recounts how GOD slaughtered all the firstborn children and animals of the Egyptians to make Pharaoh release the Israelites then killing his entire army in the Red Sea.

Exodus's main story is the Israelite's wanderings in the wilderness for forty years until all the older generation had died off because they had been frightened of reports of the Nephilim giants in Canaan and had wanted to rebel against Moses's leadership and elect a new leader who would take them back into Egypt and slavery! Exodus 12:37 lists six hundred thousand men followed Moses and all died in the wilderness!

Exodus has many accounts of the Israelites' enemies being destroyed for worshipping idols and just as many accounts of the Israelites being destroyed for idolatry as well as doubting or questioning GOD such as the slaughter of all who joined in demanding and then worshipping the Golden Calf just days after they had sworn a blood covenant against making idols with GOD. Later GOD caused many to die because they complained about the tedium of eating nothing but the honey flavoured manna and wished they had the meat and vegetables they had while in captivity in Egypt. GOD sent huge flocks of quails to them but it seems the birds had eaten poisonous seeds and when the Israelites gorged on the birds until the meat ran out of their nostrils the poison was concentrated in them and they died in huge numbers.

Some of the other Old Testament books mention how often the Israelites would forget to keep to the worship of GOD and be seduced to the worship of various false gods – usually Baal or Moloch - and Ashorah stakes and the inevitability of GOD letting many be destroyed until they remainder stopped their idolatries for a while.

Ezekiel details how the Israelites mad a windowless room in which each man had his own idol of a animal, serpent or other creature while other Israelites turned their backs on the altar devoted to GOD and worshipped the sun while getting high on the smoke of hallucinogenic twigs. This worship of the sun is the true abomination of desolation of The Last Day.

But each incident clearly boils down the Israelites being slaughtered for turning away from GOD or from obeying Him – and just as many times GOD either directly slaughtered nations opposing the Israelites or encouraging them to idolatry.

The one billion or more killed in Genesis must be added to the masses killed in the other books to illustrate how awful life has been since Eve listened to Satan.

The last book Revelation foretells of the mass slaughtering to be carried out by Satan's slaves during the Last Days before Jesus returns with his angels to destroy all Satan's murdering slaves. The death count in the last days will surpass all of history back to Creation with only Christians and few Jews being lifted off Earth and saved during The Rapture while Earth and all its satanic people burn.

So, yes, the Bible is full of the mass deaths or armies, tribes and cities and the deaths of individuals including the innocent daughter of Jephthah as recorded at Judges 11:30-40!

It is most important to grasp the fact that most of the deaths in the Bible are due to people worshipping Satan's false and imaginary gods and idols or for directly fighting against GOD and His chosen people.

The Israelites who demanded Aaron make them a god-idol to worship while Moses was up Mt Sinai speaking with the True GOD will not be resurrected to life as just a few days previously they had sworn a covenant with Moses and he had sealed them with the blood of sacrificial oxen. When Moses stayed up the mountain some of them demanded Aaron make them an idol to be their god and taking their gold jewellery he did so. GOD saw them dancing round the Golden Calf idol and sent Moses hurrying down the mountain to slaughter the ringleaders while GOD sent a plague to kill many more. How quickly they had turned to worship false gods! Over then next centuries and until today many Israelites or Jews are quick to turn to worship idols knowing perfectly well they are signing their own eternal death warrants!

+Today and ever since before the Tower of Babel billions of people have worshipped false gods and idols invented by themselves or by Satan and his demons.

We have no tally of how many others will die during the tribulation or be resurrected to death on Judgment Day because they turned away from GOD or never sought after Him as described at Zephaniah 1:6 'And them that are turned back from the Lord; and those that have not sought the Lord, nor enquired for him'.

All these people will be resurrected on Judgement Day and many will be delighted to find they are granted a new life on the renewed Earth because they acted in ignorance but vast numbers will be resurrected to be told that they were wilfully evil and have to be destroyed instantly and permanently.

Happily Jephthah's daughter will be resurrected to life as she was blameless and did not resist or make her father sin by making him break his vow to GOD. She was a shadow ot the sacrifice of Jesus some 1,000 years later.

When we are told to fear GOD it is not an empty warning!

The fear of GOD is the beginning of knowledge.

Proverbs 2:10 'For wisdom will come into your heart, and knowledge will be pleasant to your soul; discretion will watch over you, understanding will guard you,

Proverbs 1:7 The fear of the LORD is the beginning of knowledge; fools despise wisdom and instruction.

Job 28:28 And he said to man,'Behold, the fear of the Lord, that is wisdom, and to turn away from evil is understanding."'

James 1:5 has a Prayer For Wisdom and Knowledge James 1:5 If any of you lacks wisdom, let him ask GOD,who gives generously to all without reproach, and it will be given him.

Solomon 1 Kings 3:5-12 At Gibeon the LORD appeared to Solomon in a dream by night, and GOD said, "Behold, I give you a wise and discerning mind, so that none like you has been before you and none like you shall arise after you".

Daniel 2:21 He changes times and seasons; he removes kings and sets up kings;he gives wisdom to the wise and knowledge to those who have understanding;

Psalm 119:66 Teach me good judgment and knowledge, for I believe in your commandments.

Proverbs 16:22 Good sense is a fountain of life to him who has it, but the instruction of fools is folly.

Proverbs 1:5 Let the wise hear and increase in learning, and the one who understands obtain guidance,

Ecclesiastes 7:12 For the protection of wisdom is like the protection of money, and the advantage of knowledge is that wisdom preserves the life of him who has it.

Christian Quotes About Knowledge

The ultimate ground of faith and knowledge is confidence in GOD. - Charles Hodge

Wisdom is the right use of knowledge. To know is not to be wise. Many men know a great deal, and are all the greater fools for it. There is no fool so great a fool as a knowing fool. But to know how to use knowledge is to have wisdom. - Charles Spurgeon.

Chapter Fifty Seven

Baal and 'Allah'.

The Bible is silent on the events from the scattering of the people at Babel until many years later when Abram at age 75 was told by GOD he was to become a great nation through which all the families of the Earth will be blessed. Abram lived through the last 40 years of Peleg's life and must have had first-hand accounts of how the land became split up. Secular history has many legends and ruins that all purport to be older than the Flood but obviously cannot be and are obviously all referring to the same event but thoroughly mixed up by the many new languages GOD made the Babylites speak.

Difficulties in correctly dating the reigns and relationships of kings and queens named on ruins and statues and the eagerness to deny Bible truths means an awful lot of woozle has accumulated over the years after Babel and clouded the issue of the origin of the various deities that came to be worshipped after the Flood.

Old Babylon was the centre of idol worship just as today's Babylon is centred in Rome around the pope. Many new gods had been invented in the Babylon years, perhaps demons had been materialising bodies and pretending to be deities who had the power to influence weather, crops or battles. Most of these gods had to be worshipped with sex, drugs and rock'n'roll rituals. Attempts at correctly dating and location each of these new gods is a fruitless task as there were so many and due to the difficulties of the confused language it is likely that some gods had different names in different cities in different countries.

The Bible is only concerned with the god that became developed and known as Baal and is usually depicted as a man in a short tunic with a pointed helmet with bullhorns and right arm upraised with a club. This is the one Disney gave a prominent place to in the library scene in his film Bedknobs and Broomsticks and which Jehovah's Witnesses worship and include as subliminal images in their Watchtower, Awake and books. It is sad that Baal has more followers today than at any time since he was invented by Satan or one of his demons in Babylon about 4,000 years ago!

After YAHWEH GOD had confused the language at Babel some groups of Ham's descendants had gone far south into Arabia to become the Arabs, prospered and multiplied, and had been influenced by Satan to devise an entire family of gods based on meteorites and to be worshipped as gods able to influence the weather, crops, etc. We do not know who first invented the word 'Allah' to mean god in the years after Babel but just as the Israelites could easily fool themselves into thinking a calf made of melted down golden earrings could be a god then at some stage in the years after Babel some tribe of Arabs chiselled themselves idols of stone and imagined them real gods. The Arabs who ruled the Mecca area had built a temple called the Kaaba as a windowless temple devoted to the chief idol named Baal which was a rough statue with a crescent on its chest and its hands raised palms up just as Muslims hold their hands today! This stone was to be worshipped and promoted as creator of heaven and earth. This is why Muslim's heads are so full of mixed up lies they cannot distinguish between YAHWEH GOD and the idol 'Baal-allah'. The Baal stone had been witnessed flashing across the skies and in those days the night skies would be clearer and meteorites more easily and more regularly seen that today when everyone is indoors watching television in the dark hours. The Baal idol had a wife idol called Allat and three children named Al-Lat, al-Uzza and Manat which with the 36o other idols placed round the Kaaba gave one idol to be worshipped each day of the year. Maybe it was around this time that GOD gave Ezekiel the vision of the windowless room with many idols?

By Abram's time the whole region was worshipping meteorite stones and idols of Baal-Allah with sacred trees or groves planted close. Quite often the idols were placed on hilltops. Flights of fancy expanded this whole stone worship to a national religion with temples, duties and rituals. The worship of Baal was encouraged with sexual rituals and sacrifices of live children. The temples and idols had crescent moons and stars carved on them as can be seen on Muslim temples today.

Abram arrived in Canaan, built altars and 'called on the name of GOD'. Or prayed to Him. A drought throughout Canaan forced Abram to continue on down to Egypt.

Now this drought is very strange and no-one seems to have picked up it being unusual as the

Flood had only occurred about 300 years previously. Writers of religious and secular websites all claim only the Jordan waters the Bible lands yet surely the Earth would still have many springs trickling from all the uplifted and disturbed lands? That there was a drought seems proof that the Ice Age was still locking up more and more water on the new poles and mountain ranges. GOD had promised the rainbow after the Flood and rainbows need water droplets and clouds. If both were lacking allowing the sun to bake the land dry and parched than there really must have been fantastic amounts of ice and snow on Earth during the exile in Egypt.

Abram and Lot left Egypt to the south of Jordan where they decided to separate to ensure sufficient grazing. Abraham moved into the lands of the Canaanites but Lot went to the lands extending down to Sodom and Gomorrah.

The Old Testament tells of many occasions when GOD allowed the attack, slaughter and enslavement of the Israelites because as they travelled across Canaan they had turned to the worship of various local gods and idols, mainly Baal and the accompanying Ashorath stakes, and even burned their children in fires to Moloch – that really astounded GOD who had never even thought of such a thing! The Israelites just refused to learn that GOD would punish them for this evil worship.

2 Chronicles is full of accounts of the Jews making and worshipping idols of Baal/Molech with Ashtoreths and of taking them into the temple to worship in place of the worship of GOD as laid down in the Law. Each time they did so GOD let them be attacked and slaughtered until they destroyed all the idols but inevitably they would soon begin to lust after false gods again.

Ezekiel 8:10 is a vision given to Ezekiel in which he sees Israelites in a windowless room worshipping all sorts of idols of animals and other creatures and waving incense over them. Then he is shown women 'weeping – the false keening pagan women do – over the supposed death of Tammuz aka the son of Ishtar/Ashtoreth. Next he is shown about 25 elders with their backs to the alter and their faces to the sun – the East just as Muslims do- and the were intoxicating themselves with the smoke of aromatic twigs. This worship of the sun is the real abomination of desolation!

Despite all the lies and claims made by the devilworshipping followers of Baal the whole religion is based on nothing but silly myths and fairy tales easily shattered by any intelligent person.

Many people have concocted elaborate theories about the origins and rise of Islam introducing Jesuits, Catholics, Illuminati as the agents responsible but this is all due to the fact they do not accept Satan is a real being with immense powers and has an army of demons at his beck and call. If that can be realised then all the silly conspiracy theories fade away and Satan steps in to possess a human just as he and his demons had been doing since The Flood – which of course none of the conspiracists believe in anyway!

Mud demonised while meditating. Takes Mecca, Starts Islam. Koran mishmash.

In about 610 AD an Arab descendant of Ham's named Muhammed used to go to a quiet cave to pray to imaginary gods. The cave is on hills overlooking Mecca and from it he would be able to see the kabaa with its 365 idols and the chief one called for centuries 'Baal-Allah'.

One day Satan was wandering Earth looking for opportunities to fool and enslave the humans and came across Muhd and decided to have some fun with him by materialising a wondrous body and claiming to the the angel Gabriel and sent as a messenger from GOD. He told Muhd that he had been given the special honour of giving YAHWEH GOD's latest revelation to the world. Obviously this was Satan's lies as Jesus had prophesied the future in his Sermon on the Mount and later, in Revelation 18 he told John: "For I testify unto every man that heareth the words of the prophecy of this book, If any man shall add unto these things, God shall add unto him the plagues that are written in this book". Moses had spoken with GOD for just forty days to receive the Ten Commandments of Christianity but it took

Satan 23 years to give Muhd all the nonsense that is the koran. Most of this interchange was given while macho Muhd was crossdressed in his wife's dresses. Is there a weird deep link between crossdresser Muhd and the Muslim insistence on their women being invisible in burkas?

Muhd then convinced his family and many others that he had been appointed to lead the world to worship the stone called 'Allah' by means of preaching the lies of the Koran or by war and the sword. He became ruler of the area round Mecca, threw all the other stones gods out of the kabaa and retained just the meteorite worshipped as 'Allah' for hundreds of years by earlier Arabs. That stone is now set in the kabaa and all Muslims are obliged to travel there to try kiss it once in their lifetime. All Muslims strenuously deny they are worshipping the stone or Muhd but translation of the prayers shows they are praying to the dead man Muhd – yet another evidence of Satanic contradiction of GOD's prohibition on necromancy as written in Deuteronomy 18:11 tells us that anyone who "consults with the dead" is "detestable to the Lord." Muslims and Catholics all pray to the dead and all are doomed.

(See Lourdes, Mormons etc for other instances of Satan's impersonating angels or Mary.)

Poor stupid Muslim women have to be completely covered and have their genitals mutilated so they cannot enjoy sex because it was Eve who Satan used and it was lusting after sex with beautiful human women that caused the fallen angels to come to Earth and sin. Satan knows his children were drowned during The Flood and knows he is going to be destroyed by Eve's descendant. It was a masterstroke in making Muslim men believe their women should be covered and not enjoy sex and of course the men fell for it. Compare the shapely well dressed figures and happy faces of non-Muslim women with the sullen subservience of obese Islamic women to see Satanism in effect.

Unfortunately today over one and half billion Muslims worship or revere the meteorite 'Allah' as their god and carry out jihad and atrocities towards anyone who refuses to kneel and worship the stone. They are all walking dead.

Until the start of the 21st century Muslims and Islam were seen as just another religion but the hijacking of four aircraft on 9/11 and the subsequent proof of the hijackers being Muslims intent of forcing their devilworship onto the entire world or be beheaded made many Christians realise that Muslims worship Satan and are a manifestation of his plan to destroy the woman's seed.

In the years since 9/11 we have seen more and more Muslim atrocities which all prove that it is a Satanic religion and not the peaceful one it's adherents and apologists claim. Islam's Satanic origins can be seen in the way Muslims are told to lie, cheat, steal and murder in direct contradiction of GOD's Ten Commandments. YAHWEH GOD who raises up and puts down kings and rulers allowed Barack Obama to become President of the United States for eight years during which he claimed to be Christian and legally entitled to be president but now in 2017 a mass of evidence has arisen showing he is a fullblown Muslim with a mission to turn the US to worship the imaginary god Baal-Allah. GOD raised up Donald Trump to reveal the lies of Obama just as He arranged the veil of the holy of holies to be ripped open the minute Jesus died to reveal that Caiaphas and previous high priests were liars, hypocrites and therefore the offspring and followers of Satan.

GOD abandoned Great Britain due to the apostasy and ecumenism of the supposedly Christian leaders. They had forgotten GOD said He is a jealous god and hates all idolatry and idolaters. The leaders of the Christian church are photographed kneeling to idols of Mary with the pope in Rome instead of warning him that his idolatries will ensure Jesus slaughters him when he returns.

GOD raised up a Muslim mayor for London to show what would happen to all who allow idolatry in their countries. Shueb Salar, the advisor to Muslim mayor Saddiq Kahn, was suspended after he had posted evil comments on his Twitter account about homosexuals and women including a tweet saying "currently hating on all you faggots who have finished uni", and the next month he tweeted "**** all you hoes". 'Hoe', meaning 'whore' is a derogatory term for women popular in some subcultures. Salar also called women 'bitches'. Only when footage surfaced of Salar on a shooting weekend was he actually sacked, ostensibly for tweeting he could be a hitman. He also posted a video online showing him firing a rifle and bragged about spending a weekend "shooting stuff with real guns, knives, crossbows and bow and arrows". He thought it was funny to add the comments #I'mASecretHitman, #ShoutMeIfYouWantMeToTakeCareOfSomeone and #I'llMakeItLookLikeAnAccident. Like most Muslims extremists he feels it necessary to slaughter non-Muslims or 'kuffars' as we are seen as cattle. This is the reason there are human slaughterhouses in Muslim Syria and Turkey in which humans are butchered for the meat trade as the Muslims do not see any harm in eating what they see as cattle. His homophobia is no doubt fuelled by the pedophilia and homosexuality he indulged in as a boy and teenager.

Salar's views about women, let alone homosexuals, are routine among young Muslim men and is due to far too many of them growing through their adolescence in a hotbed of consensual promiscuous homosexuality and bestiality. The koran says a Muslim male must not have sex with a Muslim girl before marriage but if he feels the need for sex it is acceptable to have sex with boys, men, donkeys or goats. Naturally early sexual activities imprint on the mind and make the person addicted to that type of sex which is why so many Muslim men are closet homosexuals. This closet homosexuality breeds self-loathing which outlets as contempt for homosexuals who are free to pursue their perversion. Many Muslim male's heads are absolutely messed up over repressed homosexuality, pedophilia, bestiality and mysogyny.

One case in point is the murder on Palm Sunday 16 April 1995, of Iqbal Masih who was returning from mass with his two cousins when he was shot in the back with a shotgun. He was killed instantly. The official police report states that a local man, Ashraf Hero, admits killing Iqbal accidentally by firing of the gun in panic after the boys witnessed him engaging in bestiality with a donkey. The police extracted this bizarre confession from Hero after hours of torture. The Pakistani Human Rights Commission investigated because of Iqbal's many enemies in the carpet business, but eventually agreed with the police report. Controversy still surrounds his death and many believe that Iqbal was assassinated on the orders of the carpet manufacturers and that Ashraf Hero was a convenient scapegoat or contracted to kill.

On 15 August 2017 in Morocco fifteen Muslim boys took turns having sex with a donkey only to discover themselves dangerously ill with rabies! Satan's koran says sex with animals is acceptable but GOD's Bible says it is evil and punishable by death.

The koran and hadith include these quotations:

Muhammad let young boys watch him clean his private parts (Bukhari1:4:152)?

He sucked the tongue of a boy (Musnad Ahmad 16245)?

He dressed up in little girl's clothes and women's clothes (Bukhari 2442)?

He slept with or had sex with a dead female body in a coffin or grave (Kanz al Ummal 34424)?

He was a pedophile (Bukhari: 5:58:234, 236; 7:62:64, 65, 88)?

He was said to consume house flies (Bukhari: Vol. 4, Bk. 54, No. 537)?

He commanded to drink camel urine (Bukhari: Vol. 8, Bk. 82, No. 79)?

He admitted to having a devil (Muslim: Ch.14, Bk. 39, No. 6759)?

He confessed to being demon possessed (Ishaq: 106)?

He told false revelations (Tabari VI: 107, 110; Ishaq: 166)?

He encouraged adherents to lie and break oaths (Ishaq: 365; Tabari VII: 94; cf. Ishaq: 519; Bukhari: Vol. 7, Bk. 67, No. 427)?

He had sex with animals (Abu Dawud38:4450)?

When the Muslim holy book includes such perversions is it any wonder so many trials of Muslims involved in pedophilia and their sexual crimes in general have revealed a deep contempt for women, particularly white women, who are seen as fair game, or as 'hoes'. The mass sexual assaults in Cologne on New Years Eve illustrate the point exactly.

Just this week Erdogan of Turkey has passed a law demanding worship of Baal-Allah and demanding Muslims everywhere try to go to worship in Jerusalem's Muslim mosque as a way of stirring up racial tension. The spirit of the serpent still permeates all Muslims.

This wild behaviour by Muslims is directly traceable to Satan twisting practically everything GOD had said in the Ten Commandments. They insist on ignoring planning laws and deliberately build houses and mosques bigger than allowed and protest like crazy wild animals when their illegal acts are pointed out.

The koran and Islam's leaders tell Muslims to emigrate around the world and overwhelm the local population and the school, health and housing by breeding many children and demanding Islamic laws and policies must be applied. They ignore planning laws and build giant mosques to make Allah seem bigger than GOD. They destroy Christian churches and kill Christians for fun. This pleases them as the evil smirks on their faces show. But they have an awful fate awaiting them as just like King Zedekiah they will all have to watch their children being slaughtered before their eyes – as written at Jeremiah 52:10 'The king of Babylon slaughtered the sons of Zedekiah before his eyes, and he also slaughtered all the princes of Judah in Riblah'.

On his return Jesus and his angels will slaughter all Muslims who are marked with the moon and stars from worshipping Satan's false god Baal/Allah.

Chapter Fifty Eight
Jesus's Ministry, Crucifixion, Resurrection, Witnesses.

It seems probable that Jesus had to delay his ministry until aged about 30 due to the need to look after his mother brothers and sisters after the death of his stepfather Joseph who was considerably older than Mary. It was a loving gesture of Mary to refer to Joseph as Jesus's father when she and he found the boy sat in the temple questioning the teachers of the law.

As we are told nothing about him from being 12 until he got the call to begin his ministry at age 30 when he travelled to seek John baptising in the Jordan. Then he spent 40 days in the wilderness having Satan tempt him. As Satan stormed off in defeat angels came and ministered unto him. How amazing that must have been as they worshipped him and he greeted them as old friends! HALLELUJAH!

Jesus travelled around the towns and cities of Galilee to teach in the temples on the Sabbath and was well received and glorified. He gathered the twelve disciples and together they baptised thousands of people who believed him. On returning to his home town Nazareth he attended the synagogue on the Sabbath as was his custom. But this time as a teacher! Requesting the scroll of Isaiah he read what we call chapter 61. The listeners were initially delighted with him but when he spoke of Syrians being cleansed by the prophet Elias they perceived he was linking themselves to the prophecies and also crediting himself with being the expected Messiah and anointed with the Spirit of GOD. Despite the earlier praise they rushed him out of the temple to a high cliff and wanted to throw him to his death – another instance of breaking Commandment Six! - but he 'walked through them'. This must be an instance of his having angelic help as how could he slip through a group of men with murder in their minds!

For the next three years he travelled about preaching, performing the miracles written in the Gospels and driving out devils. No-one in the entire area could fail to hear about him, and his works in casting out demons, making the blind receive their sight, the lame walk, the deaf hear, cleansed the lepers, raised up the dead, and constantly preached the gospel to the poor. Caiaphas would kept informed of Jesus and how he was fulfilling the many prophecies in the books that all synagogues possessed and all priests were familiar with.

Jesus went to the temple on the day when cattle, lambs and doves had to be bought and slaughtered for the burnt offering sacrifice commanded by GOD many centuries earlier. There he found the sellers and their birds and lambs were making the outer Court of the Gentiles into a noisy dirty market with other people using it as a shortcut across the town. This court had been made in the days of Herod the Great and it is thought Caiaphas took a commission on sales and rent from the traders. Worse were the greedy moneychangers who sat at their tables to exchange the Roman, Greek and other daily coinage for shekels from Tyre as the chief priests had deemed the daily coinage with Ceasar's or another human's head to be idolatrous. Making a whip of cords Jesus ran about driving out all these traders and the ones with no real reason to be in the courtyard. Caiaphas would be furious that Jesus was contradicting him as well as making him lose money. This loss of face, authority and money on top of the fear of the Lost Ark Scandal being exposed made Caiaphas and Annas see Jesus as a threat that must be eradicated. They schemed and plotted about how to entrap him and have him executed.

The next morning Jesus and the disciples walked back to Jerusalem. Seeing a fig tree by the roadside Jesus went to see if it had any fruit but found none. Unexpectedly he cursed the tree by saying "Let no fruit grow on you ever again!" Immediately the tree withered away. The disciples marvelled at the sudden withering and asked Jesus about it. He told them that if they had faith they could do the same, and more, they might tell a mountain to move!

Actually of course there is much more to this speedy withering – it was a triple lesson from Genesis and Numbers and the coming judgment if the disciples had known the scriptures! Lesson one was Jesus himself who caused the Earth to send grass and fruit trees bursting out of the ground bearing flowers and fruit on Day 3 of Creation so for him to send a tree withering back to the soil in minutes would be quite easy! Lesson two was the fact that the Jews should have known of Aaron's rod that overnight went from a dead dry old walking staff to being brought back to life complete with leaves and ripe almonds! Lesson three was that the unproductive ones will be cut down and destroyed. On one level both lessons seem to show GOD's time and Jesus's powers are unfathomable to we humans – but surely, both lessons are parables about what can happen to people if they do not produce fruit as the fig tree did or if they are spurred into producing fruit by GOD like Aaron's dry staff? Caiaphas and all unbelievers are the fig tree - Jesus was Aaron's staff? EUREKA!

Two days before his arrest Jesus was questioned by the Pharisees who tried to incriminate him over the question of paying the Roman tax. He called them hypocrites and replied that as

Ceasar's head was on the coinage the tax had to be paid. Next the Sadducees asked him about which of seven brothers would get the wife they all married when they were raised to life on Judgment Day. He replied that the resurrected will not marry but will live chastely like angels. He should have called those hypocrites too as they believe there is no life after death!

Every time Jesus spoke a parable that stung the consciences of the Pharisees and Sadducees they were more determined to arrest him but feared causing a riot during the coming Feast of Passover. Their chance came when Judas turned traitor for money. He was suspected of stealing the disciples' money while carrying the bag and now Satan lured him with the promise of wealth. His hopes for real riches were dashed when all chief priests only offered him thirty pieces of silver. Filled with remorse and guilt Judas threw the money at the priests feet, went away and hung himself. The actual coins are unknown so the total value cannot be decided other than knowing the money was sufficient to buy the potter's field.

The Gospels of Matthew, Mark, Luke and John all tell of the arrest, trials, vilification and abuse of Jesus and his being nailed to the cross on the fifth day or 'Thursday' at mid morning.

He died at three in the afternoon and was put in the tomb before six o'clock.

Next day Caiaphas and the Pharisees went to Pilate and asked him to make the tomb secure and guarded in case the disciples came in the night to take the body to make it appear he really had risen. Pilate agreed. The tomb was sealed with a great stone spiked to the rock and a guard was posted.

Jesus's body was in the tomb about 60 hours until the 'Sunday' or first day of the week in order to fulfil the prophecy and his own words.

Before dawn broke Mary Magadalene and the other Mary went to see the tomb. A great earthquake rumbled as an angel came and rolled away the great stone. The guards were petrified with fear. The angel told the women Jesus had risen but to look to see where he had lain and then go tell the disciples. They saw the body wrappings and ran to the disciples only to have Jesus appear and stop them and tell them to rejoice that he had risen exactly as he said he would.

The guards recovered and ran to tell the chief priests what had happened. A meeting was rapidly called with temple elders and it was decided to give the guards lots of money to bribe them to say the disciples had come by night and stolen the body.

Though at the time and still today black magic practitioners are known to use bits and pieces of dead bodies in their devilworship there isn't much evidence it as a common occurrence in the Jewish world.

Muslims are a different matter and they have been so brainwashed by silly stories that they are all convinced that not only was Jesus just an ordinary man who preached but that he had been taken down off the cross before death and a dead man hung on instead! All Muslims are walking dead for that blasphemy.

Any person claiming to be Christian but doubting Jesus did returns to life and go to heaven is no Christian!

After his resurrection Jesus was seen by many hundreds of people around Galilee and on the road to Damascus. The Bible says that Jesus made a number of appearances after His death. They were to a number of different people over a forty-day period. The Bible specifically says that Jesus appeared to Mary Magdalene, the women that came to Jesus tomb (Mary the Mother of James, Salome, and Joanna), Peter, and two disciples on the Emmaus road. He also appeared to the remainder of the Twelve Disciples with Thomas absent. Later he appeared to them with Thomas present. There was also an appearance to seven disciples on the Sea of Galilee. On another occasion he appeared to over five hundred people at the same time. There is also an appearance to James. Finally Jesus appeared to Saul of Tarsus - the man who became the Apostle Paul. These appearances convinced His disciples, beyond any doubt, that He had risen from the dead.

We are not told of what else he was doing or where, during the 40 days between his resurrection and ascension to heaven. Could he have been appearing to believers in distant cities? Or did he eavesdrop on how Caiaphas and the others were dealing with the problems of the Lost Ark and the rise in Christianity?

Though Jesus was crucified and after his last appearance to Saul was not seen again he didn't disappear without trace.

His disciples kept of preaching and living as he had taught them. The manner of his death and growth of his disciples were mentioned in various letters such as Lucian of Samosata who wrote of the early Christians as follows: "The Christians ... worship a man to this day – the distinguished personage who introduced their novel rites, and was crucified on that account."

The Talmud writings completed in Babylon might seem an unlikely source of truth of Jesus but he is mentioned as Yeshua and was being watched with the aim of stoning him to death. 'On the eve of the Passover Yeshua was hanged. For forty days before the execution took place, a herald ... cried, "He is going forth to be stoned because he has practiced sorcery and enticed Israel to apostasy."

Jewish historian Josephus wrote: 'About this time there lived Jesus, a wise man, if indeed one ought to call him a man. For he ... wrought surprising feats.... He was the Christ. When Pilate ...condemned him to be crucified, those who had . . . come to love him did not give up their affection for him. On the third day he appeared ... restored to life.... And the tribe of Christians ... has ... not disappeared.'

On the other hand there are many who deny Jesus existed such as Robert Price who wrote a pretentious article titled 'Deconstructing Jesus' in a vain attempt to deny Jesus. When we read Price is a 'former Baptist pastor' we can know that he let Satan whisper that 'surely he didn't think Jesus was real?' Once the seed had been planted Price was 'drawn away by his own desires and enticed' exactly as Eve had been. Price's desire was the infamy of being lauded as a Jesus denying author.

Bart D Ehrman is the present poster boy of the Jesus deniers for his many books, articles, seminars and television appearances in which he takes pride in saying he was raised a liberal Christian but became an agnostic atheist as he addled his brain with philosophical claptrap about the problems of good and evil – another case of Satan whispering plausible lies. In one book he repeats the ridiculous fairy tale that Jesus was just an itinerant preacher who died and whose followers all hallucinated they had seen him after he had died! He pretentiously claims to be both humanist and agnostic by thinking himself evolved from monkeys. I always think that anyone who has an initialled name like Bart D is a bit suspect – a bit of a poser as we say. Does he have anything relevant to say about the problems a Christians faces today as we run up to the End Times? No! But he's sold his soul to Satan and no doubt is living high on the hog until Jesus returns.

One aspect of the four Gospels and indeed the whole Bible is the constant mention of angels – does Ehrman, Price and their ilk think the actions of angels are just fairy tales like EVONONSENSE?

Every denier of any part of the Bible and especially mention of GOD, Jesus and Holy spirit ought to know that calling GOD and Jesus liars is a guaranteed way to be destroyed on Judgment Day!

Chapter Fifty Nine

Why Jesus's Blood Had To Be Spilled On The Mercy Seat!

Before I became a Christian the American evangelists and preachers used to baffle me with their demands that Christians be bathed in the blood of Jesus! Their sermons include an awful lot of mentions of Jesus' blood. I just didn't get it because UK ministers or pastors I saw on television seldom spoke much about Jesus's blood.

The UK pastors concentration on Jesus's death on the cross gives the vast majority of believers the impression that just being crucified and dying from loss of blood was sufficient atonement for humanity's sins. Obviously having huge rough spikes hammered through hands and ankles would start a lot of bleeding in addition to any bleeding cuts and bruises from being beaten, flogged and being crowned with thorns. Surgeons have determined that being crucified causes death from lack of oxygen as the lungs and chest cannot expand to get air in and eventually the victim dies from a combination of lack of oxygen, fatigue, loss of blood and pain. Death can be speeded by breaking the legs so causing more constriction of the chest and lungs.

This spilling of his blood by blows, lashes, thorns and nails and its dripping on the ground is what the world thinks is important about Jesus's crucifixion.

They are wrong out of ignorance thanks to Satan blurring the Biblical accounts.

Matthew 1:21 records Joseph the carpenter being told by an angel in a dream to not break off his betrothal to Mary because he had realised she was pregnant but to marry her 'And she shall bring forth a son, and thou shalt call his name Jesus: for he shall save his people from their sins'. Which were his people – the Jews?

The Jews considered themselves sinless as they meticulously followed their interpretations of the Ten Commandments in the tiniest detail and the high priests and elders made sure everyone saw them doing so by choosing the most prominent seats in the temple and wearing the longest fringes to their garments. What sins could they be guilty of? Caiaphas the high priest and his father-in-law Annas kept a great terrible Jewish secret: a huge mountain of unforgiven sin! It was this mountain that Jesus was sent to wash away.

Jesus had to die not only because he claimed to be the Son of GOD but because he knew the Jews were being systematically fooled, cheated and betrayed by their lying, hypocritical chief priest Caiaphas and the elders led by Annas who governed the temple in Jerusalem! According to the accounts of Matthew and John Annas still wielded considerable authority and took a leading part in plotting to have Jesus killed as recorded at Matthew 26.

Caiaphas had been ordained high priest by his father Annas and like him had appeared to follow the Passover ritual of sprinkling the blood of an innocent lamb on the Mercy Seat of the Ark to obtain forgiveness of the previous year's sins. Each Passover when the high priest solemnly entered and exited the Holy of Holies all the watching Jews would think their sins of the previous year had been forgiven.

However each high priest and the elders knew the Ark of the Covenant was not in the Holy of Holies and no atoning blood could therefore be spilled on the Mercy Seat! Nor could GOD come and display his presence by making a cloud over the Mercy Seat. The Ark contained the two stone tablets on which GOD had personally engraved the Ten Commandments and it should have been in the Holy of Holies but it was lost, presumed stolen by the Babylonians in about 587 BC! After returning from exile and building the new temple the high priests had decided to carry out a pantomime each Passover and pretend to take the lamb's blood and sprinkle it on the Ark! They were lying and cheating the people! Annas has revealed the secret of the lost Ark to his son Caiaphas when making him high priest just as a previous high priest had initiated Annas into the secret. What smug satisfaction did the high priests derive from this empty ceremony they had been performing for over 600 years since the time the

First Temple had been ransacked and destroyed by Nebuchadnezzar's army? The invading army had stripped the temple of all the gold decorations, motifs and fittings installed by Solomon and carried them all off to Babylon.

Fearing loss of the Ark to Nebuchadnezzar's army King Zedekiah had hidden it and other sacred items in tunnels beneath Jerusalem. His death in Babylon and the deaths of others who knew the secret ensured the location of the Ark and its tablets of stone on which GOD had personally written the Ten Commandments was unknown. The new Second Temple was a shadow of the former one and what no-one but the priests knew was that the Holy of Holies was empty and GOD did not come fill it with his glory.

In the intervening centuries the temple remained the centre of religious ritual teaching and doctrine. The high priests and elders became ever more greedy and hypocritical as they devised new laws and rules covering every aspect of life. The Jews had once again become sinful and forgetful of keeping faith with GOD as the prophet Malachi details in his book written about 460 BC. Jesus illustrated the hypocrisy of the Jews when they claimed his disciples had sinned by plucking grain but he silenced them by reminding them they would not hesitate to work on the Sabbath if their sheep fell in a pit.

By the time Jesus began to preach in 30 AD the Ark had been missing for over 600 years and the Jews had accumulated a vast burden of unforgiven sins. Caiaphas knew that Jesus would know this and fearing exposure he conspired with his father Annas and the elders to have Jesus falsely accused and killed.

Matthew 23:37 seems to tells us that Jesus had been regularly directing affairs on Earth and directing men to become prophets to try inform and advise the humans of the correct way to worship GOD and turn away from Satan only to have Satan foment hatred of the prophets to have them killed just as the Old Testament records so often. Jesus joined the three in the furnace. He argued with Abraham about destroying Sodom if just a few good people could be found there.

EUREKA!

Reports of the preachings, baptisings and prophecies of John the Baptist reached Jerusalem during the period while Jesus was still working as a carpenter. John had been born by a miracle after the angel Gabriel had visited Zacharias while he was offering incense in the temple in Jerusalem. Gabriel struck Zacharias dumb for questioning GOD's promise. He remained speechless until his son was born when the moment he wrote the boy's name on a tablet his speech returned and all who witnessed it spoke about it all over the hill country of Judea.

Caiaphas, a Sadducee, was high priest at that time and must have known of Gabriel's visit, Zecharias losing and gaining his speech and the promise of John being filled with Holy Spirit to preach of the coming of the Messiah. John went to live in the wilderness for some reason until the word of GOD came to him and he began preaching the need for repentance of sins with baptism being the sign of the inner cleansing. One day as John was baptising he saw Jesus approaching and instantly knew who he was. Jesus asked John to baptise him. John demurred initially but Jesus said they needed to follow the correct procedure whereupon John baptised him.

As he came out of the water a dove came on him and the voice of GOD thundered down from heaven: "This is my beloved Son, in whom I am well pleased".

This could hardly fail to impress all witnesses and the incident would quickly spread across the land. Caiaphas would hear the report and realise Jesus fulfilled the prophecies about the coming of the Messiah who would root out the idolaters and lazy priests who have fallen into a rut of complacency more concerned with gathering money and quibbling over trivialities than keeping the Jews holy. Caiaphas was well aware that Jesus would find the prophecy of Jeremiah 7:11 fulfilled: 'You have made my house into a den of robbers' by allowing the selling of animals and birds for sacrifice in the courtyard and the tables of the usurious money changers who gave Jewish coins for the Roman or Greek coins Caiaphas prohibited as tithes.

In addition, being a Sadducee who denied resurrection of the dead Caiaphas would be angered that the Messiah was clearly going to raise the dead as written at Daniel 12:2 which reads: 'Multitudes who sleep in the dust of the earth will awake: some to everlasting life, others to shame and everlasting contempt'. Caiaphas had many reasons to want Jesus killed before he revealed the Ark was lost and the Jews had a great accumulation of guilt.

Caiaphas could not fail to be kept aware of all the miraculous cures and resurrections Jesus and his disciples were performing around Jerusalem and the surrounding cities. No Jew in a position like Caiaphas's could be ignorant of the scripture's multiple prophesies about a Messiah and what that Messiah would do to the religious leaders. Caiaphas had failed as high priest in guiding the Jews in their worship and beliefs and also had the secret of the Lost Ark hanging over him. There was only one course of action open to him: Jesus must be stopped from exposing Caiaphas's sins!

Eventually as written in the Gospels of Matthew, Mark, Luke and John the high priests succeeded in having Jesus betrayed by Judas, arrested and taken before Pilate, Herod and then Pilate again.

Caiaphas pressured the Roman governor Pilate to pronounce Jesus be crucified rather than the criminal Barabbas and Jesus was nailed to the cross at 9am.

There is a silly belief among some cults that because Jesus was silent on the cross he wasn't really human and his body was a sort of shell! This is the typical sheer stupidity that cultists like to indulge in. They forget that by the time Jesus was nailed to the cross he had missed a night's sleep, probably not eaten, had been viciously whipped – extra viciously? - with whips intended to rip flesh off, had lost lots of blood, had a crown of thorns pressed into his scalp, had had to carry his cross for some distance and then had the agony of the nails and the cross being raised so all the weight fell on the nails! No wonder he had no breath to cry out until the end! But then as a sacrificial lamb is silent as it faces the knife so Jesus was silent.

Modern atheists and others put the blame for Jesus's death on the Romans such as in this BBC article: 'In the end the Roman authorities and the Jewish council wanted Jesus dead'. This is clearly wrong as Pilate had not found Jesus guilty of anything and wanted to dismiss all charges as a mere religious squabble among Jews. It was the chief priests who demanded Jesus be killed. To appease them Pilate offered them the choice of Barabbas, jailed for murder during a recent rebellion, or Jesus. The chief priests urged the people to yell Barabbas! Pilate knew that the mob may have rioted again if he released Jesus and though he washed his hands of the matter he did order Jesus to be flogged and then crucified.

Matthew 27:26 records the Jews boasting that Jesus's blood would be on themselves and their children! This is a perversion of the sacrifice of the innocent lamb to wipe away the sins – Jesus's blood added more sin onto the Jews who denied and still deny he is the son of GOD but it washed away the sins of all people who accept he is the son of GOD!

During the first few hours as he hung on the cross the chief priests and temple elders gathered around and joined others in jeering and taunting Jesus about why he could not save himself if he was the son of GOD. Caiaphas and Annas and others who knew the secret of the missing Ark would be gloating inwardly that their secret would be safe now that Jesus was dying. They may have felt uneasy as from noon the onlookers had to light torches as an unnatural darkness came over the Earth until Jesus died at 3pm.

Jesus's head fell forward in death and Jerusalem was shaken by an earthquake. The 3inch thick veil hiding the Ark was ripped open by GOD's angel to expose the empty sanctuary – there was no Ark. Caiaphas was caught in full embarrassment as his guilty secret was evident to everyone! How did he wriggle his way out of that predicament?

The torn curtain showed that Jesus's death had provided everyone with direct access to GOD but it is likely few realised that as their immediate concern was directed to how could they have been forgiven if blood could not be sprinkled on the mercy seat. However as the Holy of

Holies was in the innermost part of the temple it is likely only the high priest and elders witnessed the event and realised their lies had been revealed but as they were all bound together to preserve the scandal then the lack of an Ark would go unnoticed by most of the population of Jerusalem.

The onlooking Roman centurion and his soldiers saw the earthquake crack open the rock under the cross and they realised Jesus was truly GOD's Son. This same earthquake cracked open the graves of many of those saintly faithful prophets who had been put to death in earlier actions and on the day Jesus was resurrected these saints climbed out of their cracked graves and went walking round the holy city and appeared to many as further proof of GOD's ability to raise the dead. Jesus referred to those dead saints at Matthew 23:37 "O Jerusalem, Jerusalem, thou that killest the prophets, and stonest them which are sent unto thee, how often would I have gathered thy children together, even as a hen gathereth her chickens under her wings, and ye would not!" This indicates he had also been coming to Earth before being born to Mary.

According to the Jewish tradition a dead body must not remain on the cross overnight into the Preparation Day of the Sabbath causing the soldiers to break the legs of the two criminals either side of Jesus to make them die faster. On approaching Jesus and seeing he had died a Roman jabbed his spear into Jesus's side and out gushed blood and water.

The reason GOD caused an earthquake was to allow Jesus's blood to fall down onto the Mercy Seat of the Ark of the Covenant as the greatest and final blood offering for sin for all the human race! It had been hidden there about 620 years previously when Nebuchadnezzar's Babylonian army besieged then destroyed Jerusalem and carried off King Zedekiah and most articles of the temple. They did not take the gold covered Ark of the other items from the Most Holy as they had been hidden in tunnels under the hill Jesus would be crucified on during the 30 months siege.

Jeremiah 52 records this ransacking of the temple but no mention of the Ark shows it was not there to be stolen!

The Ark contains the replacement pair of stone tablets etched with the Ten Commandments that Moses brought down from Mount Sinai at the end of the second 40 days he spent up there talking with GOD after he had smashed the first pair on seeing how quickly the Israelites had turned away from GOD and made themselves a golden calf to worship.

During the long silence of the years of captivity in Babylon, and the return and rebuilding of the temple and its subsequent destruction the Ark has stayed safely hidden underground guarded by four angels. Whether these angels stay on duty constantly or maybe were allowed 'periodic leave' to return to the joy of God's presence is not known.

In 70 AD Jerusalem was again attacked and the temple destroyed and then in 135 AD the city was practically razed and lost most of its Jewish structures.

The Jews were dispersed and stayed dispersed and mostly persecuted around the world until eventually Israel was declared a state in May 1948 since when millions of Jews have returned there in the mistaken belief that they are the Chosen Race and will rule the world when Jesus comes again.

Unfortunately they lack the understanding and wisdom to know that as they rejected Jesus and had him killed they lost the right to rule alongside him and most will be killed during WW3/Armageddon and 'only a remnant will be saved.' Romans 9:27, Isaiah 10:22.

Jewish Blood Guilt – Blood Libel Belief.

Matthew 27:25 tells of the Jews accepting responsibility for the death or blood of Jesus. 27:**25** - Then answered all the people, and said, 'His blood be on us, and on our children!'

This is the worst root of the 'Blood Guilt or Blood Libel Belief' that is blamed every time anti-semitism is seen and which led to many pogroms against Jews in medieval times as well as the WW2 Holocaust but it is not the only Blood Guilt of the Jews. The 'Holocaust' is popularly thought to mean Hitler's destruction of the Jews during the years 1933 to 1945 but actually the word means the Jewish rites of burning babies in idols of Baal, Moloch, Molech or Malcham. The idol had a bulls's head and horns, two arms and fire in its great belly. A baby was put on the hot arms to endure awful pains before the arms were raised by pulling chains to make the infant fall inside the belly and be burned up. Many times the Jews resorted to this sacrificing to Moloch. Usually the idol was set up on a hill with a pole or tree –an Ashorah or Ashtoreth beside it. Hitler's burning of millions of Jews seems to be divine retribution?

Ben Shapiro is a Senior Shillman Journalism Fellow at the David Horowitz Freedom Center who condemns attacks on Jews with false statements such as 'the labeling of Jews as bloodthirsty villains led to the Holocaust in the first place' but accepts there is a blood libel based on popular belief that Jews abduct children for ritual slaughter and to provide blood for drinking. This might seem the idiocies of primitive people who dislike Jewish separateness, their curious clothing and rituals as well as the belief that Jews in business will cheat their customers. (Curiously I found the very opposite when trying to sell two lovely gold rings which had proved so fragile they had both broken and so needed selling for scrap – a Jewish dealer offered me about 40% more than British dealers!)

The blood guilt/libel is obviously the Jews accepting they killed Jesus's and his blood would be symbolically splashed upon them. Then there is another little discussed matter of Jews and children's life - or blood as in the context of Jesus's crucifixion - to be considered. Their Talmud, written during the years of exile in Babylon, does quite matter of factly discuss whether sleeping or lame children could be sacrificed by burning on altars dedicated to Baal as so shocked GOD as written in Jeremiah 19:4 "because they have forsaken Me and have made this an alien place and have burned sacrifices in it to other gods, that neither they nor their forefathers nor the kings of Judah had ever known, and because they have filled this place with the blood of the innocent, and have built the high places of Baal to burn their sons in the fire as burnt offerings to Baal, a thing which I never commanded or spoke of nor did it ever enter my mind'. Jews had been burning their children to Molech and Baal for many hundreds of years! While burning children to Moloch is an awful crime and surely sufficient to condemn the perpetrators to eternal damnation there is another little known 'blood crime' Jews are guilty of – the draining of the blood of kidnapped children sacrificed to their idol Baal? Over the centuries many children's corpses have been found empty of blood and covered with puncture wounds. The blood is used in secret worship of Baal and also used to bake the cakes offered to the idols. At 1 Kings 18:28 we read of Elijah challenging the Baal priests of King Ahab to have Baal send fire to burn a sacrificial bull on Mount Carmel. The priests called upon Baal to send fire but none came no matter how loud and long they called and even when they resorted to dancing wildly and stabbing themselves with 'lancets as was their custom!' Jews link the blood of Gentiles with their worship of Baal. GOD has long ordered the Jews to not cut or pierce their skin: Leviticus 19:28, Deuteronomy 14:1. Curiously while the original Kong James Bible translated 'lancets' as lancets – meaning the sharp pointed needle surgeons use to open veins or drain various swellings - the New and most other modern bibles write 'lances' as in spears – but would priests carry spears about with them?

Travellers in Syria and India have reported and filmed similar wild dancing and self-cutting by various pagans in their demonic worship trances.

Another confirmation of the Jews having Jesus's blood on their hands is Peter's speech in Acts 4:12. After being filled with the Holy Spirit at Pentecost Peter and John found a lame beggar in the Beautiful Gate of the temple. They healed him and he was so delighted he ran into the temple, leaping and praising GOD so much the worshippers gathered to see who had performed the miracle. Peter addressed them: Acts 4:12 "You Israelites, why do you wonder at this, or why do you stare at us, as though by our own power or piety we had made him walk? 13 The God of Abraham, the God of Isaac, and the God of Jacob, the God of our ancestors has glorified his servant Jesus, whom you handed over and rejected in the presence of Pilate, though he had decided to release him. 14 But you rejected the Holy and Righteous One and asked to have a murderer given to you, 15 and you killed the Author of life, whom God raised from the dead. To this we are witnesses".

As usual the Jews ignored ignored the First, Second and Sixth Commandments prohibiting false gods, idols and murder as they took the incineratings in the Molech to the abstract with arguments about could a sleeping child be burned! Ezekiel 23:37 also lists the sins of the young king Manasseh as sacrificing one of his sons to the idol and also setting up an Asherah stake in the temple. (Jehovah's Witnesses worship Asherah as they celebrate Jesus being sacrificed to to Baal on an Asherah stake.)

Add to this the fact that despite GOD's prohibition on blood a rabbi who circumcises a baby does suck the child's penis and blood after making the cut!

Jews never mention Jesus except to ridicule or deny he was the Messiah. One extreme belief many Jews have is that Jesus is boiling in excrement for daring to claim to be the Son of God! How curious are the ways in which Satan's followers twist the Commandments!

The Ark of the Covenant is found safe and well guarded.

In 33 AD Jesus's blood dripped down a crack made by a medium earthquake that had rocked Jerusalem at the instant Jesus died. For 1,949 years the blood of Jesus stayed safe and dry on the Mercy Seat of The Ark and on the cracked wall above. The blood was a covenant between GOD and humans that anyone who believed in Jesus could take advantage of to have their sins washed away and gain eternal life.

During those 1,949 years and onto today, and until the moment Jesus returns to destroy all who have refused the offer of the blood, a great number of humans have taken up the gift and gained eternal life – their sins washed away out of GOD's memory. Those people are all ages, races and colours in all the world and only GOD knows every detail about them. In these last days until it is too late many millions more will accept the message and the gift of the blood and be saved from the grasp of Satan and the coming wrath of GOD as His angels pour out their seven bowls.

After 1,949 years undisturbed the Ark was lit in the beam of Ron Wyatt's flashlight. What a EUREKA moment that was!

Ron had prayed long and hard and had contact with angels and perhaps Jesus himself as he studied and then traced the route of the Exodus to the Red Sea and found commemorative pillars on each side with Egyptian chariot remains including gold wheel covers. He found the two rocks that Moses split to get water for the complaining Israelites.

While walking round Jerusalem trying to pinpoint the actual site of the crucifixion and the hiding place of the Ark he had his arm suddenly point to a pile of rubble and his mouth said: 'That is Jeremiah's grotto and the ark is down there.'

Working spasmodically during the short holidays he managed to afford he dug down and followed the tunnels and eventually found a cave entrance blocked with cemented rocks and knocking through he found the Ark and directly above its east side was a crack in the rock and dark dried blood stains and some had dripped on the Ark – Jesus was the perfect unblemished sacrifice!

When Ron told the Jewish Archaelogy Department about it they quickly banned him from the site. He started packing to return home only to next day get a call to come help the Jews as four of their men had gone into the cave and not returned and their walkie-talkies had given the sounds of panic and terror.

Ron went down and found the four dead with both eyes crossed – they had been killed by massive bilateral strokes.

When Ron enquired of the angels why, they said it was because the Ark has to stay hidden until Jesus comes again to save remnant of faithful during the worst days of World War Three, or as it is popularly known, 'Armageddon'.

THIS IS THE REAL MESSAGE OF JESUS'S BLOOD BEING SPILLED FOR US – HIS BLOOD HAD TO BE SPLASHED ON THE MERCY SEAT TO TAKE AWAY OUR SINS – NOT JUST TRICKLE FROM HIS WOUNDS TO DRIP ON THE GROUND AS PROOF OF HIM BEING DEAD.

Chapter Sixty

Jesus went to preach to the fallen angels?

1 Peter 3:18 tells us: 'Because Christ also suffered for sins once, the righteous for the unrighteous, that he might bring us to God; being put to death in the flesh, but made alive in the spirit; in which also he went and preached unto the spirits in prison, that aforetime were disobedient, when the longsuffering of God waited in the days of Noah, while the ark was a preparing, wherein few, that is, eight souls, were saved through water'.

Multitudes of pastors, ministers, preachers and scholars have tried to explain this verse and failed! They tie themselves up in metaphysical claptrap unable to understand that while Jesus definitely did have a human body he was also perfectly aware that he was a spirit too! At twelve he knew he was the son of GOD! He demonstrated it for the first time by living without food and water in the desert being tempted by Satan. He demonstrated it again with his first miracle of changing the water into wine! How that must have astounded all the wedding guests! What questions would they all put to his mother about this amazing son of hers! Did anyone know about Jesus's conception except her and Joseph – was Joseph still alive? Did Jesus delay his ministry because of need to work in the family carpentry business until his brothers were old enough to support the widowed Mary? Or did he need to study the scriptures until age 30 to be able to counter all false doctrine and charges about GOD, himself and his ministry?

During his ministry he had to show he had the most perfect grasp of the scripture and their real meanings – but did he have to learn them by intensive study or did he have first hand knowledge of all the Old Testaments events from being both a participant and an observer? If it was he that joined Shadrach, Meshach and Abedego in the furnace or it was he who bargained with Abram before the two angels went on to destroy Sodom and Gomorrah he would surely know all the details and what was spoken by whom?

If the early Bible translators had not used LORD for GOD, Jesus and angels on missions from GOD the confusion over trinity would not have arisen and many Bible verses would be less confusing. We read that no-one can see GOD's face and live yet we read of 'GOD' speaking to humans when clearly the context of the passage shows that it is an angel that is conveying a message or acting as intermediary between GOD and the human.

Jesus died on the cross and being taken down was wrapped in linen and laid in a new unused tomb leaving family, friends and disciples to mourn. No doubt Caiphas and Annas gloated over their victory unaware that it had been a pyrrhic victory they will bitterly regret on Judgment Day. Jews still gloat over killing Jesus!

The verse in 1 Peter reads: 'being put to death in the flesh, but made alive in the spirit; in which also he went and preached unto the spirits in prison, that aforetime were disobedient, when the longsuffering of God waited in the days of Noah,'

Jesus's human body had suffered and died and in the quiet darkness of the tomb it would possibly but not necessarily pass through the normal stages all dead bodies pass through: cooling, rigor mortis, the downward draining of fluids, the first stages of internal decomposition. This is the usual progress but did this happen to Jesus's body? GOD and he knew that in just about 60 hours time the body would need to be resurrected as prophesied. It is possible that GOD had angels somehow keep the body in perfect condition though Jesus had proven with his resurrection of Lazarus that the effects of death can be reversed instantly – just as the Bible promises at Revelation.

Regardless of the physical body the spirit of Jesus could not die as it had had a pre-mortal life with GOD in heaven as confirmed by John 17:5 in which Jesus appeals to his father: **5** 'And now, O Father, glorify thou me with thine own self with the glory which I had with thee before the world was' – and in Hebrews 1:5 **5** 'For unto which of the angels said he at any time, Thou art my Son, this day have I begotten thee? And again, I will be to him a Father, and he shall be to me a Son?'

Jesus knew perfectly well he had lived in heaven with GOD!

His spirit was released at death but it/he also had the power to rematerialise the human body again as many verses illustrate. As a spirit he could enter the locked upper room on two occasions; he even ate fish to convince the disciples his body was real. Then moments later he disappeared again! He could also vary his appearance – or dim the eyes of others – as on the road to Emmaus when two disciples failed to recognise him. Luke 24:13.

In this spirit mode he was able to pass through solids as easily as today we know xrays and gamma rays can pass through metal or our luggage at the airport check-in.

In spirit mode he was able to fulfil the 1 Peter verse by travelling to the place in which the fallen angels have been bound from The Flood until Judgment Day. Just where that place is we cannot know apart from believing it to be a literal hole inside Earth as some claim are the interpretations of Jude 6: 'And the angels who did not stay within their own position of authority, but left their proper dwelling, he has kept in eternal chains under gloomy darkness until the judgment of the great day', and 2 Peter 2:4 'For if God did not spare angels when they sinned, but cast them into hell and committed them to chains of gloomy darkness to be kept until the judgment'.

The pope's new throne seems to show Satan and the fallen angels bursting out of a pit in the short time of Satan's 'Little season' as they are freed after The Millennium as prophesied at Revelation 20:7.

The problem with 1 Peter 3:18 is the original Greek has been translated as 'preached' – as in preached a sermon intended to stir the mind. If Jesus had gone to them to preach their error and how his death had fulfilled the gospel it could be argued that the preaching might bring on genuine repentance in some of those angels less evil than Satan. Could this be what Paul alluded to in 1 Corinthians 6:3 'Do you not know that we will judge angels?' What other angels have sinned or neglected their duties so as to need judging?

Other translators say the verse should be 'proclaimed' – as in making an announcement. This seems to give a more logical meaning to the verse as the bound angels had known Jesus in heaven and would have been aware of the trouble that following Satan would bring on them. A proclamation from Jesus about how the divine plan was proceeding would make the fallen angels regret their actions and cause them agony such as Satan will receive when he is cast into the lake of eternal fire.

Also of course it is possible Satan is still able to go visit these bound angels and boast about how he is continuing the awful crimes he led them all into - though obviously he is unable to free them to cause more trouble. What a shock they would have had if Satan had visited them to crow about how he had had Jesus reviled and killed only for Jesus himself to go visit them minutes later and prove that he is very much alive and they and Satan will totally perish when GOD decides the time is right?

Many people have tried to explain the verse as meaning Jesus went to preach to the spirits of the humans who lived, sinned and drowned during The Flood, or to all the spirits of all the dead who have died since Creation, or more ludicrously: that it was Jesus who was actually Noah who preached the coming Flood! We can be assured Jesus was never Noah – and he did not preach to the souls of any drowned humans as the Bible tells us that 'at death their thoughts perish.' Thus what would be the point of preaching to a disembodied soul which would know nothing of its former life?

Also again: Scripture tells us the dead have all perished with a small number recorded in GOD's Book of Life but most known only from an entry in the Books of Death.

Can a name in a book be preached to? Obviously not!

Much information about the fallen angels Jesus and his dealings with their demon offspring is found in the Luke 8.30 story of Jesus meeting a demon possessed man in the tombs and casting out a legion of demons that then entered a herd of pigs, while the Book of Enoch is mentioned in the Bible but was kept out of the collection that eventually became the Bible in Nicea in 325 because it tells of the origins of the fallen angels and their fate.

We need to see through the smokescreen the Scarlet Woman of Rome put up about just why Enoch and Jasher were kept out of the Bible - it was not because they were non-scriptural but because they tell the truth of the Scarlet Woman and Satan. What better way to have them dismissed than by claiming them to be fiction? Yet the truth in them and their quoting proves they were seen as legitimate books at the time of Jesus.

Enoch tells of Satan inducing a group of 200 angels to rebel against GOD and come to Earth to enjoy sex with the lovely human women with tales of the beauty of Earth women. The angel Semjaza made himself leader of the group after making them all swear an oath to be bound together in case GOD punished them. He and they descended from heaven onto the top of Mount Hermon, created handsome bodies and went wandering about seducing the women and impregnating them. A link with Nephilim and Mount Hermon is at Joshua 12:4: 'And the coast of Og king of Bashan, which was of the remnant of the giants, that dwelt at Ashtaroth and at Edrei, 5 And reigned in mount Hermon, and in Salcah, and in all Bashan, unto the border of the Geshurites and the Maachathites, and half Gilead'. There were temples to Baal and Ashtorath on Mount Hermon.

It is strange that Enoch wrote of the fallen angels coming to Mount Hermon as now in these last days we can read of a very strange occult connection with Mount Hermon! The Lucis Trust – a nice way of avoiding using the full name of Lucifer – was founded by a woman named Alice Bailey who had become demon possessed after listening to HP Blavatsky's demonic utterings. Bailey then claimed to have met and been inspired by a 'Tibetan master' and having received enlightenment and since then spoke a lot of One World ecumenism. Why do fools think men living in poverty in the high barren icy mountains of Tibet have a direct line to GOD? Those men are like puppets on the end of strings jerked by Satan - if they are real humans and not just demons materialising bodies for fun as was the 'Prophet Samuel' conjured up by the witch of Endor – to the witch's shock! Bailey is just the same as AR Wallace who also became demon possessed after attending a spiritualist's meeting. She founded a publishing company named it Lucifer Publishing Company to publish the satanic lies spouted by the 'Tibetan master'. Bailey later managed to have the Lucis Trust accepted by the United Nations and has a chapel in the UN Building and is consulted over various UN programs. And most strangely the Lucis Trust has a chapel in the UN complex on top of Mt Hermon! The Lucis Trust's website has this piece of New Age nonsense: **SERVICE AND THE DIVINE PLAN.** 'The time is long past when a line of demarcation can be drawn between the religious world and the political, economic, social or scientific. Spirituality as it is practised today in its many forms bridges between higher dimensions of soul and all that is human: it is essentially concerned with the establishing of right human relations. In seeking to live a spiritual life and express higher values through all relationships every sincere person of goodwill serves this wider evolutionary process. And so the server is organically connected with those great Serving Lives, the Rishis, Saints and Bodhisattvas of all faiths, the Spiritual Hierarchy. In Alice Bailey's writings the Hierarchy, is presented as one great community of consciousness'. Further Satanic claptrap is: 'Twelve Spiritual Festivals: The time of the full moon is a period when spiritual energies are uniquely available, facilitating a closer rapport between humanity and the spiritual Hierarchy'. This is as clear a clue to the occult origins of Bailey's group as can be!

To have at least 7 links between Mt Hermon and Jesus, Lucifer, Fallen Angels, Nephilim, Baal and mentions in Joshua and Enoch is very strange!

As the UN has a base on Mt Hermon and as Satan took Jesus to a high mountain from which he could see all the kingdoms of the world - although due to curvature the view could only be about 200 miles maximum so only the Biblical kingdoms could be seen - I rather think that

Jesus' transfiguration and Satan's and UN using Hermon might be the link to the NWO and mark of the beast of these End Times.

EUREKA!

Chapter Sixty One

Christians persecuted over The Truth. Arius v Athanasius.

Following Jesus's return to heaven the new Christian faith gathered followers until the apostle Stephen angered the Synagogue of Freedmen in Jerusalem so much they conspired to have him charged with blasphemy against Moses and GOD. He was hauled before Caiaphas, sentenced to death and stoned while a zealous young Jew named Saul looked on approvingly.

Saul immediately began searching the city for Christians and dragging any discovered off to prison and probable execution. This caused the Christians to flee to other cities in Judea and Samaria. Saul then requested letter from Caiaphas to the temples in the Damascus area to capture Christians for trial and punishment in Jerusalem. On the way there Jesus came to him and converted him to become one of the most ardent apostles until eventually in 67 AD Nero had him tortured and beheaded.

At the same time Peter was crucified, upside down at his request, since he did not feel he was worthy to die in the same manner as his Lord.

Andrew preached in Greece, where he is said to have been crucified.

Thomas preached as far east as India, where the ancient Marthoma Christians revere him as their founder. They claim that he died there when pierced through with the spears of four soldiers.

Philip was arrested and cruelly put to death by a Roman proconsul.

Matthew is reported to have been stabbed to death in Ethiopia.

Bartholomew has several accounts of how he met his death as a martyr for the gospel.

James was stoned and then clubbed to death.

Simon was killed after refusing to sacrifice to the Persian sun god.

Matthais is thought to have died with Andrew by burning.

John is the only apostle to die of old age but was saved escaping unhurt after being cast into boiling oil at Rome.

This set the pattern for what has been 2,000 years of persecution of Christians by Satan with the aid of his Catholic, Muslim, Jewish and pagan slaves in Judea, France, Britain, Italy, Russia, Spain, Turkey, Syria, Africa and elsewhere – wherever Christians have travelled to preach the Gospel.

While all persecutions are sad it is well to remember that people who continue to worship pagan deities or idols after hearing the Gospel are doomed to destruction either when Jesus returns or on Judgment Day. He will judge the quick and the dead.

History books and the internet are full of massacres of Christians although quite often the reports confuse Christians with Catholics or others who seem to worship GOD and Jesus but maintain the worship of pagan gods in the guise of GOD, Mary and the infant Jesus. This confusing make it impossible to compile an accurate body count of True Christians – though GOD is keeping a perfect tally.

As ever, Satan continues striving to obscure and lie about the Truth of GOD and Jesus and is aided in this by his faithful slaves such as the one who compiled the entry for 'Christianity' in the online Britannica.com which asserts: 'Christianity's largest groups are the Roman Catholic Church, the Eastern Orthodox churches, and the Protestant churches'. Satan's deception in that sentence is that the Catholic and Orthodox churches both fail the litmus test for a truly Christian church as the leaders of both cults make and pray to their idols and the dead – their gods!

The Eastern Orthodox church fails the Christianity test by its idolatry of icons of dead, by its lies in claiming Mary was perpetually virgin, by claiming a person can obtain salvation after death. Orthodoxy is practically Catholicism.

Catholicism is actually based on the Babylonian pagan worship of the sun which was claimed to be the daily reincarnation of the dead Nimrod, and Astarte Queen of Heaven, wife of Baal or Molech, and their son Tammuz. Astarte and Tammuz are the real 'Madonna and Child' worshipped by all Catholics. Following the downfall of Babylon by King Cyrus and his Persians some surviving priests escaped to Pergamos in Turkey and then onto the growing city of Rome. The Romans had a full panoply of gods and the new Babylonian ones eventually were accepted a century before Jesus was born. Sixty years after Jesus was crucified Pergamos was still referred to as Satan's Seat for its continued sun worship idolatry.

Astarte is also known as Ashtoreth as used with the idol of Baal in the occult Disney film 'Bedknobs and Broomsticks'. Jeremiah 7:18 tells of how the Israelites baked cakes for the idols of Astarte and ignored Jeremiah's warnings that they were idolatrous and doomed to extinction. Pope Callistus is said to be the first pope to claim the title Pontifex Maximus and today's Catholics dedicate cakes to Astarte and Tammuz, pray to the dead Mary and other dead people, and ignore the warnings they are doomed to death when Jesus returns.

Catholics align their churches to the east to pray to the sun as it shines through the huge sunburst windows they favour, and raise their cakes to the sun before breaking and eating them as they drink the wine they are told and insist is blood. Today various people rise early on the day they dedicate to Easter -Astarte - and go worship the rising reborn sun. The pope also preaches pondslime-fish-monkeys-men EVONONSENSE and proclaims Lucifer is actually Jesus. He is in favour of Satan's One World Government and its Mark of the Beast stamp. His new throne shows Satan and his fallen angels being released from the pit during the 'short season' of release at the end of the Millennium when they will opportunity to try find some humans willing to follow them in rebelling against GOD for one last time only to have them all totally destroyed and end all sin on Earth!

After the last apostle died the various Christian groups or churches began to drift away from the purity of the Gospel message and add their own interpretations to the words of Jesus and the apostles. Often they were infiltrated by Satan's slaves who were determined to incorporate pagan beliefs and practice – just as the Israelites refused to cease worship of Baal, Astarte, Tammuz, Molech and the sun so many times in the Old Testament. Human nature caused various dominant people to lead the churches in direction that seemed most accommodating or most pleasing. It seems many leaders found and continue to find it easier to cater to human desires and fancies than hold strictly to the Gospel or the whole Bible. This is seen today in Archbishop Sentamu of York who claims himself a Christian and should be upholding the truth of Creation but instead panders to the Evoquacks and says he evolved from pondslime over hundreds of millions of years.

As Christianity spread across the Roman Empire attempts were made to force Christians to bow down and sacrifice to idols and a mythical trinity but of course they refused and were martyred by the sword, fire or even worse. The truth is that nowhere in the Bible is there mention of a Holy Spirit being seen in heaven or accompanying GOD or Jesus in their frequent appearances before humans! The desire to have Holy Spirit made into a real being is to openly worship the Babylonian trinity.

By 325 AD the different styles of worshipping GOD was causing so many claims and counterclaims with strife between the various groups that Constantine, Emperor of the Roman Empire, convened a conference of about three hundred Christian bishops and deacons from the eastern area of the Mediterranean in Nicea, a little town south-east of Istanbul where the Bosphorus Straits flow between the Black Sea and the Mediterranean. Constantine told the assembly they must agree on some common doctrine for their churches to put an end to the constant disputes between all considered themselves Christian. What Constantine wanted was the three hundred bishops and deacons to arrive at a compromise of their faiths for the greater good of peace and stability of the Roman empire.

There could be little chance of the group abandoning some of their central beliefs. Pastor Arius from Alexandria in Egypt had many years earlier put forward and preached the only logical doctrine given by the Bible – that Jesus was not GOD but GOD's son - and Holy Spirit was GOD's gift of understanding and truth and not an actual being or spook. While memory of Jesus and the apostle's preaching was fresh in the minds of the living Arianism was accepted as the True Christianity – which it is! But as the older generation died off the younger ones began to fall away seduced by the pleasures offered by the other churches that had perverted the reality of GOD, Jesus and Holy Spirit into the persons and imaginary gods of the Babylonian trinity: Nimrod/Baal, Ashtoreth/Istarte, Tammuz/Jesus. Part of the Nicea group wanted to remain true to the Gospel, others wanted to cling to their pagan beliefs and rites but with a gloss of Christian Truth, another segment wanted Athanasism trinity made the official doctrine. For the sake of expediency all but two attendees agreed to sign what became named the Creed of Nicea and its central belief of GOD and His son Jesus.

The Creed of Nicea as with all other compromises did not satisfy those cults that cling to pagan beliefs and practices. Arians were quite happy with the truthful creed but the Trinitarians were not content but had to adopt Araianism while openly adding their own interpretations and paganisms.

In the years after Nicea the various factions became ever more hostile to the simple truth of Arianism and the churches that based their services and beliefs upon it. Reports of murders, church burnings, excommunications by Satan's slaves were widespread so much that another meeting was convened at Seleucia in 359 AD. This time after long delay Constantine prevailed upon the assembly to once more accept Arainism and it was made the official doctrine – as it should be!

Arius lived just eleven years after his success and died of some awful stomach disease or poison in Constantinople. He is reviled as an heretic by those who worship Satan's trinity. But unlike all those he is guaranteed a new life with Jesus in the Resurrection.

One perfect example of Satan's slaves preaching his lies is found on the website GotQuestions.org whose mission statement is: 'GotQuestions.org is a ministry of dedicated and trained servants who are Christian, Protestant, conservative, evangelical, fundamental, and non-denominational'. Sadly in their article on Arius and the battle between Arianism v Trinitarianism they support Satan with their slanders: 'As Arius began promulgating his heresy,' and 'Upon hearing Arius's false teaching,' and 'Athanasius was able to identify the wolves in sheep's clothing that were infiltrating the church'. Satan loves GotQuestions.org!

In 381 AD another conference was called to try arrive at a better creed and this time as the Arians had been suppressed and dismissed as heretics whose opinions counted for nothing Satan succeeded in having his slave Athanasius put forward his rambling statement of Trinitarianism now known as The Athanasian Creed. It includes this blasphemy: 'Thus the

Father is God; the Son is God; the Holy Spirit is God: And yet there are not three gods, but one God. Thus the Father is Lord; the Son is Lord; the Holy Spirit is Lord: And yet there are not three lords, but one Lord.' It finishes with another false doctrine: ' This is the catholic faith. One cannot be saved without believing this firmly and faithfully."

The Bible states that GOD's first commandment is: You shall have no other god'. Jesus said the only way to be saved his through him alone!   Satan loves those who follow the Athanasian Creed all the way to Sheol's permanent oblivion!

Athanasius's creed is still the official creed of most of the world's Christians but all ministers and pastors who preach it are sentencing their congregations to death for lying!

It was also at Nicea in 325 AD that the present Bible was agreed upon with 39 books dealing with the time from Genesis to Malachi written about 450 BC and a further 27 books starting at Matthew and covering the period of about 4 BC to about 100AD when John died after being given the vision of Revelation.

The Nicea council kept the books of Jasher and Enoch out of the collection to be called the Bible because Jasher and Enoch give details of how Satan became evil and revealed secrets of the Earth's humans had no need to know.

The Book of Enoch has: 6:1-6.And it came to pass when the children of men had multiplied that in those days were born to them beautiful and fair daughters. And the angels, the sons of heaven, saw and lusted after them, and said to one another: 'Come, let us choose us wives from among the children of men And have children with them.' And Semjaza, who was their leader, said to them: 'I fear you will not agree to do this deed, And I alone shall have to pay the penalty of this great sin.' And they all answered him and said: 'Let us all swear an oath, and all bind ourselves by mutual curses so we will not abandon this plan but to do this thing.' Then they all swore together and bound themselves by mutual curses. And they were in all two hundred who descended in the days of Jared in the summit of Mount Hermon'.

Naturally many religions sneer at Enoch and claim it is just a fairy tale but in fact as Jude quoted Enoch and as it makes perfect sense as Jesus dealt with multitudes of demons at Luke 8:30 Jesus then asked him, "What is your name?" And he said, "Legion," for many demons had entered him'. As we rush up to the End Times that legion of demons is still alive and constantly entering into willing or unwilling humans to make them sin or behave like the man in the graveyard.

Maybe the reason Enoch's description of Satan and his demon army was kept out of the Bible is because the Catholics know that is really who they worship? How many Catholic shrines have been erected where apparitions claiming to be Mary appeared? Each of those was a demon or perhaps Satan himself.

Creation is covered in Genesis though all Evoquacks sneer, and Malachi covers the Jew's neglect of the duty to follow the orders and laws of GOD- they had been offering stolen, blind, lame creatures for the sacrificial offerings and withholding tithes as they had little regard for GOD. This may have been due to the High Priests being Sadducees who denying resurrection have no fear of dying as to them that is the end of everything – exactly as so many pagans and fallen away Christians do! The New Testament continues revealing the Sadducees were sinning and ignored the warnings Jesus gave them and eventually killed him to hide their sins. While alive a Sadducee could live well and ignore GOD and laugh at being told there is hope a new wonderful life on Earth ruled by Jesus and die in full knowledge of their many unforgiven sins – but what a shock Judgment Day will be for so many billions who deny GOD and Jesus!

All Christian churches that join ecumenical movements and take part in Catholic or other idolatrous church's services are sentencing their congregations to death.

Photographs show various Christian leaders bowing down to Mary. I just watched the 'Patton' movie and in one scene his army has to invade Sicily in World War Two and he has to go meet the local cardinal who sits in majesty at the top of a long flight of stairs with a red carpet for Patton to walk up. Clearly Catholic leaders think themselves above mere mortals. Patton should have taken a chair at the bottom of the steps and beckoned the cardinal to come down and see him! Politicians always fawn over the pope although President Putin to his credit refused to and stated that he thought the pope was not a Christian! Donald Trump's wife fawned over the pope and clearly thought him the most wonderful person on Earth.

It is impossible to know how many True Christians have died at the hands of false Christians and all the devilworshipping pagans and Catholics.

When the Western half of the Roman Empire collapsed in 476, the popes took on the title that had previously belonged to the Roman emperors—Pontifex Maximus.

The legally recognized supremacy of the Pope began in 538 AD, when Emperor Justinian elevated the Bishop of Rome to the position of Head of all Churches. This is known as the Edict of Justinian. The Papacy would persecute the saints (God's people) for 1260 years.

The Roman Catholic desperately tries to present itself as the only true and faithful church of GOD but the god it serves is none other than the sun god Apollo-Baal of Babylon – Satan's imaginary god – along with Baal's wife Ishtar and her son Tammuz. The same Tammuz that foolish Jewish women were weeping over as GOD told Ezekiel in Ezekiel 8:14. Ishtar was also known as The Queen of Heaven and was passed off as Mary, mother of Jesus.

Thanks to Athanasius and all others who persisted in officially enforcing by threat of death the doctrine that Jesus was also GOD and vice versa it was easy for the Trinitarians of Rome to promote the idea that being the mother of Jesus Mary must also be mother of GOD. In 1950 Pope Pious made the dogma of Ishtarism official, declaring that "the Immaculate Mother of God, the ever Virgin Mary, when the course of her earthly life was run, was assumed in body and soul to heavenly glory." This dogma is so full of lies and false doctrine as to be laughable! Every Catholic who has ever uttered it or accepted it is calling GOD a liar on several points.

Catholics like to claim Apostle Peter was the first pope of Rome but this is a more Satanism as the popes are fully Catholic sun/Baal worshippers while Peter and his successor Linus were simply Christian evangelists charged with teaching The Gospel and encouraging the True Christians in Rome. Peter and Linus were followed by Clement and all took the leadership of the growing number of Christians – the 'church'.

The Romans meanwhile were worshipping their various idols of their own imaginings or imported in by travellers who had had contact with Romans around the Mediterranean. The idol Cybele is an interesting case and a good candidate for what became the cult of Mary Mother of GOD. Cybele was worshipped in Phrygia, part of Turkey and had many different names as her cult spread out across the lands. By the time Cybele had been adopted into the mass of gods that made up the Roman religious life such as Mars and Venus – Venus being another name for Ishtar. The Romans had a full system of pagan gods with their idols being worshipped at shrines all over the empire.

Catholics ignore what GOD said at Exodus 20:3-6 "You shall have no other GODs before me. "You shall not make for yourself a carved image, or any likeness of anything under the heavens'.

Every Catholic church and cathedral falls foul of the warning of Revelation 9:20 'The rest of mankind, who were not killed by these plagues, did not repent of the works of their hands nor give up worshipping demons and idols of gold and silver and bronze and stone and wood, which cannot see or hear or walk'. Right until Armageddon the pope will be leading Catholics and others in worshipping the sun and idols of Ishtar and Tammuz.

The papacy then came out of the closet of Satan and set to instituting a full system of worship of him with all the old pagan symbols and rituals they had used in Pergamos and Babylon.

One very obvious of these symbols is the hat on the pope's heads which is an obvious image of Dagon the fish god so humourously humbled by an angel of GOD at 1 Samuel 5:4 'And when they arose early on the morrow morning, behold, Dagon was fallen on his face to the ground before the ark of the LORD; and the head of Dagon and both the palms of his hands were cut off on the threshold; only the stump of Dagon was left to him'. British Christian bishops such as Sentamu and Welby also show their allegiance to Dagon when they wear their fish hats.

The pope's hat has the hexagram to show allegiance to the pagan imaginary god Remphan. The sun burst shows loyalty to Baal the sun. A Maltese cross hangs on a neck chain in homage to the Babylonian sun god. Cybele is commemorated by the popes as well as Jews and Muslims wearing the skull cap. Ishtar aka Venus is openly remembered by the numerous 8 point stars.

The popes new throne has Satan leading his fallen angels out of the pit for their little season at the end of the Millennium. The new palace has a design like a serpent head – obviously to emphasize that the ruler of the Catholics is actually Satan the serpent from the Garden of Eden.

The widespread pedophilia and homosexuality of most Catholic priests is no doubt a manifestation of Nephilim DNA and a continuation of having male temple prostitutes with the sex act between priests and boys being some sort of perverted fertility rite.

Satanic rituals include the wafer eaten in the mass – not a plain bit of bread as Jesus used and all True Christians use – but a specially baked wafer complete with Maltese Cross symbol offered up to the sun before being broken. I once went to an ecumenical service at an Anglican church and watched in amusement at the old minister lifting the wafer to the sun – naturally I didn't go forward to take a fragment.

Catholic and Muslims using rosary beads for constant repetition is contrary to Jesus's words at Matthew 6 and is like the chanting of magic spells intended to lure an evil spirit – this is why Pentecostal churches have happy clappy music with a very pervasive back beat of monotonous drumming.

Adding 1260 years to 538 AD brings us to 1798, which is the year the Pope was deposed when the French General Berthier, under Napoleon, led him into captivity. Napoleon apparently tried to crush the Papacy, and about 18 months later the Pope died in exile in Valence, France. This act ended papal power in terms of enforcing papal decrees.

During those 1260 years the Catholics perpetrated vast massacres of Christians and pagans alike. The popes declared that all Bibles and church services must be in Latin in order that the ordinary people would not know to what false god they worshipped and what lies and false doctrines they were being taught. Some estimates are 75 million died for refusing to accept Catholic idolatries. Satan helped the pope's killers devise ever more awful methods and machines of torture and death while the popes and bishops sat in bejewelled robes of purple and scarlet and supped with the kings and queens of all the world.

Since 1798 the Catholic church has freedom to once again spread like a cancer and has grown immensely wealthy and added vast numbers of crimes to the tally. Papal edicts about birth control, pedophilia, sectarian strife and psychological manipulation have made life awful for countless people around the world .

Today the pope is aided by Jesuits and all manner of secret societies and agents in all the seats of power around the world in a desperate mission of trying to have the pope elected head of a One World Government system. This is often promoted as being an essential way to ensure peace and prosperity by having one common currency and international laws. But the pope stays true to the teachings of his father Satan and insists on Sun-day being made the official day of worship of the sun.

Jesus revealed to the aged Apostle John the true nature of the Catholics and how they would once again embark on an orgy of bloody murder in an attempt to stamp out Christianity in favour of their own devilworship of Baal, Ishtar and Tammuz.

The Bible reveals the Truth: The antichrist sea beast of Revelation 13 is the Roman Catholic

Church. The position of the Antichrist is not just one man, but it is an office, like the President of the United States or the Queen of England, so it is not just a one-man Antichrist. The beast's activities reach into every corner of the Earth aiding Satan in ensuring humans die from innocently following it while thinking it is doing GOD's work.

Every Catholic needs to know that they will be slaughtered when Jesus returns no matter how sinless a life they think they have lived.

Multitudes of books have been written about the evil Catholics and popes.

As we sprint to the Great Tribulation the news constantly reports Satan's desperate persecution of True Christians and the efforts of his many slaves to promote his lies and deflect all knowledge of GOD and Creation by a torrent of EVONONSENSE, occultism, idolatry and sex,drugs and rock'n'roll.

Chapter Sixty Two

Worst mass murderers in all history. Do they carry Satan's seed? Nephilim DNA?

Any list of the worst mass murderers contains the usual ones: Hitler and his world war and genocide, Stalin and his world war, the genocides in Ukraine, Cambodia, Mongolia and Rwanda.

The total deaths of these major catastrophes seem shocking but in fact they fade into insignificance beside the deaths chargeable to Hawking, Dawkins, Attenborough, Nye, Darwin, Sagan and all similar Evoquacks! These men are all teaching there is no GOD and all life evolved from wet rocks. But even their efforts are as nothing compared to the death toll attributable to Protestant churches, Islam, Buddhism, Hinduism, Jehovah's Witnesses, Mormonism and other cults and religions. Worst of all must be the Catholic Church.

Why is this? Quite simply they have collectively led billions of people away from the worship of THE TRUE GOD YAHWEH into the deadly embrace of EVONONSENSE and the lies and idolatrous worship of Rome, Mecca and the locations, graves or birthplaces of other major cult founders.

Who or what is the absolute worst mass murderer can be a bit confusing but probably this is a fairly accurate list:

Charles Darwin! All who believe in and promote monkey-to-man EVONONSENSE and fell away from GOD centred Christianity sentenced themselves to death although vast numbers of Evolutionists are alive at this moment. He has billions of deaths on his balance sheet and many more will be added to his account during The End Times to come! For the last three years I have been studying Darwinism and its followers and do not cease to be amazed at how tenaciously the lies of EVONONSENSE are held and all mention of Creation is ridiculed despite his research proving Earth is young and The Flood really occurred!

Rome must easily rank next as its street have run with blood for almost 2,500 years. All living Catholics are doomed to destruction to add to the innumerable total in countries that have been tainted with Catholicism's idolising of Ishtar and Tammuz and Baal-Molech the sun and the fallacy of making dead people into 'saints' to be prayed to. The present pope is filling GOD's Book of Death with the acolytes who preach lies and attack the children in their churches and care homes. Popes claim to be able to forgive sins and make new laws despite GOD saying otherwise. His latest trick is declaring Lucifer is GOD! One Catholic home for unmarried mothers had 800 murdered babies buried in pits while another had 500 and how many other pits are waiting discovery we cannot know about thanks to the evil doctrine of Omerta causing the unfortunate mothers to not breathe a word of their pregnancy or else be blocked from heaven and suffer eternal damnation in the fires of a real hell – another Satanic lie! Catholic lies and brainwashing runs so deep it is impossible for the average person to comprehend how evil and pernicious its effects can be. All 1.2 billion Catholics are doomed.

Next will perhaps be Muhammed on account of the idolatry and blasphemies he forced onto all who claim to be Islamic and as the fecundity of that evil religion's followers who are presently said to number about 1.6 billion the total who have lived and died as Muslims is unknowable but all will be resurrected to be judged and destroyed for idolatry. Many Muslim kings and presidents share Muhd's guilt and have their own part in the toll Islam has taken on the human souls over the centuries.

Hinduism has billions of deaths to its accounts but no real leader can be named as it seems to have started by combining the worship of many local pagan gods and the utterings of demons operating through 'holy men'..

Stalin and other communist leaders and thinkers are mass murderers as their atheism has spread over the world for many years.

Mao is well up the list for the atheism he enforced and the mass deaths his policies caused.

David Attenborough will have hundreds of millions of soul deaths attributable to him and his GOD denying lies about humans being evolved from pondslime, fish and monkeys. How many people have watched an Attenborough nature documentary and now laugh at those like myself who preach Creation? How many teachers have learned the lies of EVONONSENSE from Attenborough and taught many students that it is the truth and thereby made it unlikely to even bother reading any truth of GOD, Jesus and their Bible? How many Education Chiefs have banned Creationism and imposed deadly EVONONSENSE on all children in school from first grade to graduation?

After those few there are many mass murderers who have put hundreds of millions of names in GOD's Book of the Dead.

Obviously Hitler, Genghis Khan, Hawk'n'Dawk and so many other deathmongers could be listed but their victim tallies and depressing lives and actions are known only to GOD. Also on the list is Buffoon Bill Nye whose GOD denying television programs targeted at children over many years has made EVONONSENSE the only accepted explanation of how humans are the highest living creatures on Earth? His lunacy or demon is now claiming that parental teaching of Creation is akin to abuse of their children!

Are these mass murderers any more demonic than the man who in November ran into a

Baptist church in Texas and slaughtered 27 innocent Christians enjoying themselves worshipping GOD? Or the Hillside Strangler, The Green River Killer, Brady and Hindley, Fred West...the list goes on and on, making the headlines then fading and leaving a nasty taste when the judge is blocked from stringing the killers up.

All these people sold their souls to Satan – just as Mary Schweitzer and Alice Roberts have done: both these women actually started as believers but let themselves be seduced by Satan's lies and promises of fame and the worship of the impressionable minds enthralled by Planet of Apes movies.

The real reason why these mass murderers have carried out their evil deeds is because they all either worship Satan or carry his or his fallen angel's DNA – or else have spent time thinking about committing some criminal or sinful act.

In these End Times it seems that huge numbers of people harbour latent evil that has been introduced into their minds by constant repetition of imagery in print, the media and films or by preachers in mosques, synagogues and secret societies linked to the New World Order and Old Paganism.

We never know where the next atrocity by these people will occur and so we Christians must go about our daily lives hoping we will remain alive until Jesus returns to rapture us off Earth while he eradicates all the evil ones.

Chapter Sixty Three

End Times Signs by Jesus and today. Gospel worldwide, Spread of Knowledge – this book! Ron Wyatt, Sodom and Ark, Jews refute Jesus.

The Bible lists these **Signs of The End Times:**

One: False Bible teachers would be money hungry. They would be smooth talkers, have many followers, and slur the Christian faith (2 Peter 2:1-3)

For several decades television has had programs with evangelists preaching what appears to be true gospel but their utter avariciousness is both breathtaking and disgust as they pursue their own idea of Prosperity Gospel. Their claim is the more money the congregation gives the more they will receive – which is basically true in the sense that the widow of Luke 21:4 gave her entire wealth though ordinary people are only asked to give a tithe – but when the evangelist is revealed as spending lots of the donated money on fancy houses cars, boats and jewellery it leaves a foul taste and give the atheists cause to sneer at GOD and Christianity. Paula White is a hard faced woman who urges her congregations to shove money into the collecting buckets. Crespo Dollar incites his congregations to throw money at his feet for him to trample upon and demanded and got his followers to donate $65 for the most luxury, fastest, business jet. Kenneth Hagin was just one of the many US preachers who started out as a genuine evangelist before being snared by Satan and becoming possessed hissed and flicked his tongue like a snake before moving into the audience and making them burst into demonic laughter with wild body motions and animalism. Joel Osteen preaches the prosperity gospel but refuses to condemn Mormons and other cults as being Satanic.

Two: Homosexuality would be increasingly evident at the end of the age (2 Timothy 3:3)

Christian church leaders now openly admit being homosexuals and eagerly carry out marriages between homosexuals and lesbians as well as divorced people instead of telling them all to read the Bible and clean their life up. Once again the pope and Catholics are downplaying the rampant pedophilia and homosexuality of the Catholic heirarchy. Christians who refuse to cater to homosexuals and lesbians are being taken to court and fined heavily to confirm the Bible warning that in the End Times bad will be seen as good and good as bad.

Three: Earthquakes would be in diverse places (Matthew 24:7)

Earthquakes and tremors are now so common that only the major quakes make the news. Even my flat here in UK was rattled by a slight tremor a few years ago.

Sinkholes are an increasing problem with reports of people and houses disappearing into holes that open beneath them. GOD used something like a sinkhole to kill Korah, Dathan and Abiram as they had been complaining about Moses being leader of the Israelites during their Exodus journey. But as that hole closed over the three and all their families including the young children we have to think it was not a regular sinkhole nor a a regular earthquake but a supernatural movement of the ground at GOD's command.

Most sinkholes fall into two categories: one, the subsidence of old mine workings or salt extraction wells: two, the subsurface loss of lime or chalk deposits by the action of acid waters soaking down over many years. The nineteenth and twentieth centuries industries and use of coal and oil for heating and general fuels created vast numbers of holes in the ground and practically none had much backfilling but were just left to crumble. A farmer friend of mine dug a sewage hole for a new cow barn he was building and down about eight feet his machine almost tumbled into ancient coal workings that no-one knew existed. Whole lakes and bayous have drained down into old caverns left after salt domes were extracted by dissolving the salt with plain water and sucking it out. Natural sinkholes are caused by acid rain from fossil fuels slowly soaking down to lime deposits and dissolving them so they run away leaving a small or large cavern – these are popular tourist attractions and I first went to one about 60 years ago. The cavern can link up and stretch many miles as each rainstorm flushes out more limestone. If the cavern is deep underground the land above may be safe but sinkholes can appear much closer to the surface where the limestone is exposed by an up- or down-lift from Earth resettling after The Flood so that the vast limestone sheets are exposed. Yesterday I watched a program about such a cave system and the explorer filmed the lines of baked jellyfish aka 'flints' in the cavern sides. I did think of contacting the explorer and correcting him about 'the prehistoric' flints being baked jellyfish but didn't as it would be a fruitless exercise.

Four: Stress would be part of living (2 Timothy 3:1)

Doctors in UK are now prescribing anti-depressants to huge numbers of people. Most of those people get depressed from boredom, lack of exercise, lack of initiative and spending hours watching ugly, stupid, ignorant, vicious, uneducated people in awful soap operas that portray nothing but depressing, boring life! Schools have to hire counsellors for children who claim to be depressed – as well they might be by daily watching many hours of the awful depressing garbage that British television has become in recent years as Satanic forces dictate and produce the programs.

Publication of this book was delayed for many months as I was so stressed from traffic noise, howling police vehicles sirens and noisy neighbours staggering home drunk in the early hours! Adam and Eve had nothing disturbing to listen to but the birds, bees and happy creatures!

Five: Many wars would erupt (Matthew 24:6)

Depending on how big a strife has to be before being labelled 'war' there hasn't been a war-free year for about 100 years. Manufacturing armaments is very big business with much illegal trade and criminals involved.

Six: People would forsake the Ten Commandments as a moral code, committing adultery, stealing, lying, and killing (Matthew 24:12)

The police and judiciary have decided that crime should go unpunished unless it is a Christian speaking Bible truths and then the law gets into top gear! Instances of this are the bakers pursued and pilloried for refusing to bake cakes for homosexuals and lesbians. Preachers, bishops and archbishops lying and condoning homosexual fornications and pedophilia. The public are sick and tired of seeing tax evasion and corporate greed and theft held up as good and honourable. The UK government in particular condones a vast tax evasion and avoidance industry and lack of tax means the UK is the slummiest and dirtiest country while being home to more billionaires than any other country!

Seven: There would be a cold religious system, in denying GOD's power (2 Timothy 3:5)

Most church services are a mockery of what a good service of worship and praise should be. The Bible is seen as old fairy tales of little use. Church services are boring and uninspiring with none of the joy seen when a passionate pastor leads.

Eight: Men would substitute fantasy in place of Christian truth(2 Timothy 4:4).

This is so evident at Christmas when the birth of the Saviour is lost behind the myth of Santa Claus. Easter, Whitsuntide and the Trinity Weeks all promote pagan myths or non-scriptural events.

Jesus is increasingly said to be just an ordinary man whose followers invented many stories of his healing power. He is seen as a charismatic hippy rather than the son of GOD.

Nine: Deadly diseases would be prevalent (Matthew 24:7).

The worldwide increase in AIDS deaths is almost inestimable. Over 160,000 Americans die of cancer each year. Constant reports of strange new diseases and mass deaths make the headlines. There is mounting evidence that many diseases are man-made with the intention of long term population control in order to achieve the 500 million population the Illuminati believe would allow them to once again become Nimrods and spend their days hunting.

Ten: The fact that GOD flooded the earth in Noah's day would be denied (2 Peter 3:5-6).

There is a mass of fossil evidence to prove this fact, yet it is flatly ignored by the scientific world because of its unpalatable implication. Huge numbers of teachers, researchers and museum staff depend on the EVONONSENSE fairy tale to keep their job and the research money flowing.

Eleven: The institution of marriage would be forsaken by many. (1 Timothy 4:3)

Apart from the cost of an actual wedding ceremony there is the problem of taxation and social security benefits reductions that make cohabitation better than marriage.

Twelve: There would be an increase in famines. (Matthew 24:7)

Famines and droughts have been a constant factor across the world for the last four millennium and as rains and snows fail to arrive more and more land dries up and cannot support a crop. Overgrazing especially by goats denude and depletes more land each year. In ironic contrast much land is lost each year due to excess watering and fertilising leading to buildup of salt and dust storms.

Thirteen: Increase in vegetarianism would increase. (1 Timothy 4:3-4)

While everyone is free to decide what they will eat many religious groups demand their adherents abstain from certain foods. Muslims and Jews refuse pork and seafood products despite farmed pork and seafood being cheap and nutritious. There is no reason not to eat pork or seafood and these two religions prohibit it because they are devilworshippers and prefer Satan's lie about the food being unacceptable instead of accepting that GOD had declared all creatures could be eaten. Obviously creatures that are wild scavengers ought to be avoided but no common meat or vegetable should be thought unfit food. Most fish and sea creatures can safely be eaten apart from the fact that much of these creatures are now threatened by overfishing. On the other hand GOD did design humans to be vegetarian and we can be sure that a wide variety of fruit, nuts and vegetables will make us healthy and attractive.

Fourteen: There would be a cry for peace. (1 Thessalonians 5:3)

Many national groups clamour for peace but on their own terms. Muslims especially claim to be peaceful but the atrocities they carry out against harmless civilians as they seek to expand Islam across the entire Earth proves Islam is Satan's own religion of hate. The pope speaks of the need for world peace but he is destined to introduce the dragon and beast who will murder any Christian who dares reveal the truth of Catholic worship of Ishtar and Tammuz.

Fifteen: The possession of Jerusalem would be international turmoil. (Zechariah 12:3)

2017 is the 100th anniversary of the British Balfour Declaration which the Jews took as being a binding agreement to allow them to regain possession of parts of Palestine as well as Jerusalem. In 2017 the Muslim president of Turkey urged Muslims to flood into Jerusalem specifically to cause trouble in order to foment unrest and civil strife. The Jews are planning to dispossess all Muslims and build a Third Temple in which they can turn the clock back and make blood sacrifices and burnt offerings despite GOD making it plain that such sacrifices were all made unnecessary after Jesus's own sacrifice.

Sixteen: Knowledge would increase (Daniel 12:4)

READ THIS BOOK AND STUDY THE BIBLE! HOWEVER DANIEL 12:10 READS: 10 MANY SHALL BE PURIFIED, AND MADE WHITE, AND TRIED; BUT THE WICKED SHALL DO WICKEDLY: AND NONE OF THE WICKED SHALL UNDERSTAND; BUT THE WISE WILL UNDERSTAND.

There is vast opportunity to study all aspects of life and religion but far too many people ignore the opportunity to learn or clarify what they know and belief as they prefer to watch or listen to rubbish soap operas and music. It is very hard to teach people once they pass beyond the first school years and reach the age of 12 or 13 as they have too many diversions competing for their time and interests. Children and teenagers have to be committed, dedicated and focussed on a career before they can be motivated to study hard and long.

Despite his harshest critics and detractors Ron Wyatt persevered in his quest to verify the Exodus from Egypt and The Life of Jesus and thanks to him we now have direct proof of the Exodus and the Israelite's Crossing of the Red Sea, The reality of Sodom and Gomorrah, the Ark of the Covenant being denied the Jews until Jesus returns to retrieve it and hold up the tablets of the Ten Commandment inscribed with the very fingertip of his father and ours – YHWH GOD!

Thanks to Ron and several others we know that the top of Mount Sinai is burned black from some intense fire – just as described in Exodus.

Thanks to myself and Charles Darwin we now know why there are so many barren hilltops, deserts and worked out fields around the world contrary to GOD's plan for Paradise of the Garden of Eden.

Thanks to me we can now know that many of the shallow craters on the moon and other planets were formed by big dollops of water or fluffy snow/ice and the smaller deep ones were formed by chunks of Earth rock. The evidence has been before the Evoquack's eyes ever since slow motion filming was invented – but the eyes of the Evoquacks are dimmed so they cannot see and therefore cannot understand or even think: "That crater looks like it as formed by water!"

Also thanks to me we can now know why vast shoals of fish died contorted in agony...

Also thanks to me we can know that the frozen mammoths of Siberia are like deep frozen meat and not fossilised.

Thanks to Walt Brown's intensive studies we now know that practically all astronomy and space sciences are total fairy tales. Just today I advised a local astronomy club that comets are not ancient solid bodies but are just like puffy old snowballs made from dirty water blasted out of Earth at the start of The Flood.

Ron, Walt and myself all have one thing in common: We believe the Bible to be a true account of GOD and Creation!

Seventeen: There would be hypocrites within the Church(Matthew. 13:25-30)

Archbishop Sentamu is such a typical case as he claims to be a Christian but then denies Creation! The pope now says Jesus is Lucifer and himself is higher than GOD and he can make rules above and beyond GOD's. Many pastors, priests and ministers now ignore GOD's rulings on homosexuality/lesbianism and pedophilia and openly advocate ecumenism to the point of joining other religions in worship of strange imaginary gods such as 'allah' or Ishtar and Tammuz.

The happy-clappy pentecotal movement is taking over many churches and offering new members an experience more like a rock concert than a worship. The Gospel preached at these events is a watered down travesty of what the disciples preached and died for. Many happy-clappy churches have come and gone but the present leader of the pop-concert religious movement is Hillsong Church.

Eighteen: There would be increase of religious cults/false teachers(Matthew 24:11 & 24)

It is practically impossible to keep track of all the cults in the world today. Cults are every religion except Fundamental Protestant Christianity.

Nineteen: The future would seem fearful to many. (Luke 21:26)

Many pressure groups petition or openly agitate for an end to warfare and strife in the hope of living endlessly in peace. It therefore seems strange that the constant violence of today has made many people immune to the troubles and prefer to blank off in front of the television and its endless garbage – 'opium for the masses' as someone once said!

Twenty: Humanity would become materialistic. (2 Timothy 3:4)

Until just a few decades ago only the western societies seemed to be excessively materialistic while much of the East and Africa remained at poverty and subsistence level but now practically all societies spend far too much on accumulating unnecessary products for home and on hobbies and vacations. The media has cultivated an intense demand for evermore complicated and expensive mobile phones and services that now eat up substantial amounts of money formerly spent on entertainment or more essential needs.

Twenty one: There would be many involved in travel. (Daniel 12:4)

The roads and airports are never quiet with thousands of aircraft in the air at any moment and millions of vehicles on the roads. Much of this travel is just an attempt to avoid sitting and thinking or socialising with neighbours; much travel is also a means of wasting excess income.

Twenty two: The Christian Gospel preaching as a warning to all nations (Matthew 24:14)

Even though it is a deadly vocation the preaching of the Gospel goes ahead in all countries despite Catholic, Islamic, communist and atheist leaders trying to prevent it.

Twenty three: Jesus said Christians would be hated "for His name's sake". (Matthew 24:9)

There are daily reports of Christians being hated for preaching the Gospel or just trying to live a Christian life. In the last four years ISIS has beheaded and crucified countless thousands of Christians in Syria. Pakistan regularly burns Christians to death. Saudi Arabia jails Christians or anyone owning a Bible – Satan hates the Bible and its promise of him being crushed by Jesus!

Twenty four: And there shall be signs in the sun, and in the moon, and in the stars; and upon the earth distress of nations, with perplexity; the sea and the waves roaring; Men's hearts failing them for fear, and for looking after those things which are coming on the earth: for the powers of heaven shall be shaken. (Luke 21:25-26).

Despite efforts to suppress all reports it is obvious there are many strange sights and sounds in the sky . Many can be traced to military experiments but others are perplexing and probably due to Satan and his demons.

Twenty five: Youth would become rebellious. For men shall be lovers of their own selves, covetous, boasters, proud, blasphemers, disobedient to parents, unthankful, unholy. (2 Timothy 3:2)

Youth has been rebellious for many decades but since the advent of television soaps including scenarios with youngsters being disrespectful and criminally inclined the situation has become much worse. Far too many do-gooders have ignored the need to discipline youngsters and have ignored bad behaviour until today no-one dare raise a voice or hand against a vicious youth for fear of being jailed for criminal assault. Television producers have undermined parental control and respect by featuring themes and storylines of youngsters disrespecting parents and legal authorities.

Twenty six: Men would mock the warning signs of the end of the age saying, "for since the fathers fell asleep, all things continue as they were from the beginning of the creation." (2 Peter 3:4).

The Bible even reveals the motivation of many people in this group. They fail to understand that a day to the Lord is as a thousand years to us. GOD is not subject to the time that He created. He can flick through time as we flick through the pages of a history book. The reason He seems to be silent, is because He is patiently waiting, not willing that any perish, but that all come to repentance. He constantly sends signs and highlights the hidden sins of the leaders but none take much notice imagining each event to be just one of a continuum from time immemorial.

Many TV news reports and unreported videos on the internet show many of these signs with the most dramatic being the flashfloods, earthquakes and tsunamis. The most deadly earthquake or recent years being the one centred off Japan that wrecked a nuclear powerplant as well as causing mass deaths as the shocks and tsunami raced across the oceans to be followed by a plume of radioactivity. The latest news on that powerplant is that the core is hating up and out of control. Russia's Chernobyl nuclear accident spread contamination over vast areas causing numerous deaths and a legacy of deformations in babies born to women living in the zone of contamination.

Here in Britain the power of nature to bring speedy devastation featured on the television news in 2004 when a flashflood wrecked Boscastle - one of the picture postcard perfect little English seaside towns hardly changed over hundreds of years with quaint old houses, little stream running down under an old bridge, and a steep beach in a tiny cove and all boxed in with high hills. On sunny days people would visit the place to look in the gift and craft shops or have a snack - in the very nice English manner. On 16th August 2004 heavy clouds blew in from the Atlantic and rising up the valley started to pile up, condense and start dropping their water in a very heavy rain. The valley is a catchment area and began to funnel the water down the small stream running through the village. The water quickly eroded soils and soon mud, stones and trees were thundering into the town. The floods washed many cars and caravans into the sea, destroyed several buildings and ruined many homes. The flood was written off as just a typical quirk of nature as happens regularly around the world.

Later, on seeking information about the occult Wiccan religion I found reference to Boscastle having a museum of Witchcraft in which is an idol of a horned beast bearing perfect likeness to the Baphomet goat the Freemasons worship. As the museum was a popular visitor attraction and its exhibits would no doubt influence many people to investigate the occult and thereby open themselves to demonic possession or Satanic suggestions and lies I do wonder if the flood may have been a warning from GOD.

Another instance of divine retribution seems to be the death of John David Roy Atchison in 2007. He was a US attorney working out of the Pensacola office and had gleefully pursued a false charge of racketeering against Creationist Kent Hovind who had unsuspectingly fallen foul of the law by taking cash out of his bank each month to pay staff wages and expenses. Atchinson had pursued the case and labelled Hovind an evil criminal and succeeded in having him sentenced to ten years jail. Two weeks later Atchinson was arrested by police after flying all the way to Detroit thinking he was going to have sex with a five year old girl after he had been looking at pedophilia websites.! He suicided to avoid the shame.

It seems that the mass murderers, the psychopathic killers, the sex maniacs, the raving homosexual, lesbian or and rapists, and especially the pedophile murderers, the robber barons who can bring poverty or prosperity to millions of people, the business and legal people who will do anything for money or favour, the vicious crook who hurts and steal without a moment of remorse over the victim's pain – all these may be carrying Nephilim DNA and could trace their ancestry back to that wife of Ham. Those three son's families now number 7.6 billion and it seems that at least one third of that number carry the deadly DNA and make up the criminal, unlawful and devious people who constantly make the headlines.

Most days the television will offer a selection of programs on UFOs, the occult, witchcraft and magic while the chat shows will feature people who claim to be reincarnated or be conduits for spirit communication with dead people from many previous civilisations.

Practically every town and city will have churches or events offering spiritualism phenomena or communicating with supposed dead. My town has a hotel that holds such events several times each year. I wrote to one inn that held these spiritualist functions and told them they were literally playing with death but I doubt they cared.

Youtube allows everyone to post all sorts of videos of real or fictitious events with true or manipulated images and film sequences showing all sorts of strange phenomena. Many of these videos are made by devotees of the various computer image manipulation programs and can look very convincing so viewers are made to believe what they see. Thus many people have been convinced that it is possible for humans to fly or materialise/dematerialise at will, that UFOs are crewed by real aliens from distant stars, that Big Foot is a real huge ape creature that lives in isolated places around the world, that signs of intelligent design have been seen as constructions on the moon and planets.

Gospel preached around the world.

Despite every effort of Satan and his puppets in Catholicism, Islam, Communism and atheism the Gospel is being preached around the word in these Last Days! Radio, television and the internet ensures everyone in the modern countries can have easy access to the written word or spoken sermons and Bible discussions. Poorer countries lacking infrastructure and disposable income hear the Gospel from itinerant preachers or by donations of free Bibles, pictorial tracts, or by inexpensive clockwork or solar powered digital players loaded with the New Testament. Actual printed Bibles are being secretly distributed to those countries where owning one is punishable by death.

Every day there is news of preachers and converts being harrassed or killed by the devilworshippers around them. This is most pronounced in Pakistan and North Korea – the former Islamic and the latter Communist.

Despite the freedom to worship in western countries the Christian church continues to lose worshippers as GOD, Jesus and religion are ridiculed and belief is seen as symptom of mental illness. Because I am a Fundamental Christian believing every word of the Bible to be true I was labelled possible mentally unstable by the probation officer who compiled a pre-sentence report on me after I tried evangelizing Muslims!

Satan is winning in developed countries but GOD is winning in the Third World just as prophesied!

Public and private groups and institutions are beginning to advise, coerce and demand all humans to carry a microchip supposedly to reduce fraud and aid police and health services but while those are reasonable and logical aims they need to be resisted because the real mark of the bast is Sun worship on Sun-day. Cats, dogs and other animals benefit from a chip as it ensures speedy recognition when the animals is lost, stolen or injured.

These End Times signs all lead to the irrefutable conclusion that the Tribulation will soon be upon us.

Chapter Sixty Four

Jesus Christ's Second Coming. Dead shall rise. RAPTURE, Jesus and Angels, Mass Slaughter.

When Jesus returns all eyes around the world will see him and know their time is up. It will not be like today when the news is restricted to trivialities of government and politicians or pop stars scandals – his returns will be on a cloud with an army of angels and will strike terror in everyone who has refused to accept he really is the son of GOD YHWH! His appearance will be such a shock that any evil people who have survived the pouring of the bowls of wrath will seek to hide under rocks!

Two days before being betrayed Jesus sat on the Mount of Olives with his disciples and told them what and when the End Times Tribulation would be:

Matthew 24: 14 "And this gospel of the kingdom shall be preached in all the world for a witness unto all nations; and then shall the end come.

15 When ye therefore shall see the abomination of desolation, spoken of by Daniel the prophet, stand in the holy place, (whoso readeth, let him understand:)

16 Then let them which be in Judaea flee into the mountains:

17 Let him which is on the housetop not come down to take any thing out of his house:

18 Neither let him which is in the field return back to take his clothes.

19 And woe unto them that are with child, and to them that give suck in those days!

20 But pray ye that your flight be not in the winter, neither on the sabbath day:

21 For then shall be great tribulation, such as was not since the beginning of the world to this time, no, nor ever shall be.

22 And except those days should be shortened, there should no flesh be saved: but for the elect's sake those days shall be shortened".

These 9 verses encapsulate the persecution of Christians in the end times, the introduction of worship of an abomination in the temple in Jerusalem, the need for Christians to flee and hide, the mass slaughter of all idolaters, liars, atheists, agnostics and the sinful – and then the halting of the tribulation by the return of Jesus to save the faithful Christians.

Naturally Satan's slaves in the Catholic, Anglican, Islamic, Buddhist and similar pagan churches and cults dismiss and downplay the tribulation as mere fable and pretend it really means there will be more of the continuous strife and trouble we have seen for hundreds of years – and as that trouble continues most people become immune to it and just try get on with their daily life whether of honest toil or preying upon other people.

Many people have tried to interpret the discourse but fail because they strangle themselves with inability to understand how long The Tribulation will really last and if there is to be a rapture of Christians and if so when it will be.

About 600 years before Jesus came into the world the prophet Daniel spoke with a glorious angelic being who told him what the End Times would be like and the dates of it. After revealing the information to him the angel told Daniel to write down the prophecy and seal the words up as no-one would understand them at that time:

Daniel 10:8 "And I heard, but I understood not: then said I, O my Lord, what shall be the end of these things?

9 And he said, Go thy way, Daniel: for the words are closed up and sealed till the time of the end.

10 Many shall be purified, and made white, and tried; but the wicked shall do wickedly: and none of the wicked shall understand; but the wise shall understand".

The angel was so right and today only a tiny percentage of humans can understand what the Tribulation really is going to be and how Christians will be Raptured off Earth for safety and to keep alive a nucleus of humans who will have to refill the Earth as GOD ordered way back in the Garden of Eden!

People sneer at me when I say that if Jesus were to return on his mission this moment he would slaughter perhaps 6 and possible 7 billion of the 7.6 billion people on Earth at this moment! As that angel told Daniel, many of those 5,6,7 billion do not understand they are wicked – but their wickednesses are listed at 1 Corinthians 6:9 and Revelation 22:15 for all to see!

I have a website on the computer in front of me and the author of that site took matters out of context and claims the Rapture will occur before the Tribulation! That author also writes the real destination of the living raptured people and the resurrected Christians is heaven itself when the Bible repeatedly says no human can see GOD's face and live! That author lacks wisdom in separating THE KINGDOM of HEAVEN from HEAVEN itself!

Few people can understand the fact those verses from Matthew, Daniel and Thessalonians confirm living Christians really will be lifted off Earth while deadly war consumes the entire world!

It is a sign of the End Time that pastors and preachers are afraid to educate their flocks about the reality of Tribulation and Rapture but instead speak lies about dead Christians flying off with angels to heaven and lead their congregation to the jaws of the beast by embracing Ecumenism and its bloodstained path to the popist worship of Baal, Ishtar and Tammuz.?

My Bible says at 2 Corinthians 6:14 'Be ye not unequally yoked together with unbelievers: for what fellowship hath righteousness with unrighteousness? and what communion hath light with darkness?' Despite that so many of today's pastors and priests are eagerly linking with pagan devilworshippers – perhaps just so they can seem trendy or file reports of large congregations after each joint service?

Here is the real meaning of these verses of Daniel, Thessalonians and Matthew:

Matthew 24:14 "And this gospel of the kingdom shall be preached in all the world for a witness unto all nations; and then shall the end come".

In these End Times missionaries will be using whatever means possible to preach the Gospel by word of mouth, distributing printed leaflets, magazines and Bibles, organising crusades or by giving away mini recorders loaded with the New Testament. True Christians will preach the Gospel far and wide and close to home while travelling, working or shopping.

Every person needs to hear the Gospel to decide for themselves if they want to accept Jesus is the Son of the True GOD YHWH who died as a perfect offering for sin or if they want to continue to worship some idol or false god or have no religion at all.

If they accept Jesus they are granted eternal life so long as they follow a few simple commandments but if they reject the offer and continue in their idolatries or atheism or agnosticism they will be killed during the tribulation.

Obviously Satan empowers his slaves in the pagan religions and governments to persecute and kill missionaries as we hear in the news every day but every effort is being made to get the Gospel to the most remote parts of the Earth.

24:15 "When ye therefore shall see the abomination of desolation, spoken of by Daniel the prophet, stand in the holy place", (whoso readeth, let him understand:)

In these End Times the Jews – many of whom are really pagans who worship Baal just as they have done ever since the Exodus – will have agreed with the New World Order or UN, USA, Catholics and Muslims to build a new tabernacle – probably not a complete temple - in Jerusalem.

This will not be the Third Temple all Jews speak of as that Third Temple will descend from Heaven.

For some years the Jews have been making the temple furniture and musical instruments and weaving the special garments for priests as described in Exodus 28; and training priests to carry out ritual slaughtering of animals just as was done in the First and Second temples. They are convinced they will be making blood sacrifices and burnt offerings to GOD despite the blood sacrifice of Jesus being the last and final sacrificial offering GOD demanded - and

He has said that He hates the stench of burning offerings.

In other words the Jews call GOD and Jesus liars and fully intend to carry out useless mock ceremonies just as Caiaphas did in the empty holy of holies! Don't forget that the Jews rejected Jesus, had him killed and agreed to take his blood on their hands! Instead of accepting him as a sacrifice for their sins they added unforgivable sins on themselves by killing him!

When the New World Order helps the Jews erect the new tabernacle built and have been performing these fruitless sacrifices and burnt offerings for 3.5 years Satan will step forward to halt the sacrifices and have Catholicism and worship of the sun replace the empty Jewish rituals.

Ezekiel writes of the Jews and their abominable worship of the sun-god Baal and use of hallucinatory burning twigs – like marijuana -in Ezekiel 8:15- 17.

Ezekiel 8:15 "Then said he unto me, Hast thou seen this, O son of man? Turn thee yet again, and thou shalt see greater abominations than these".

8:16 And he brought me into the inner court of the Lord's house, and, behold, at the door of the temple of the Lord, between the porch and the altar, were about five and twenty men, with their backs toward the temple of the Lord, and their faces toward the east; and they worshipped the sun toward the east.

8:17 Then he said unto me, "Hast thou seen this, O son of man? Is it a light thing to the house of Judah that they commit the abominations which they commit here? for they have filled the land with violence, and have returned to provoke me to anger: and, lo, they put the branch to their nose".

Today's Rastafarians, Sifis, Pakistanis, New Agers and pagans use burning weeds – the branch - in their rituals. This is a sure sign those religions are Satanic.

The prophet Zephaniah also spoke of the Jew's persistent worship of sun-god Baal and the sacrificial burning of babies to Moloch/Malcham/Molech – the bull headed hybrid god made of brass in which babies were thrown: the 'passing their children through fire' which astounded GOD who had never imagined such an evil thing!

Aaron made the Israelites a golden calf – a little Moloch - when they demanded an idol to worship while Moses was on Mount Sinai speaking with GOD.

The Bible many times records the pagans Canaanite and other tribes and the Israelites joining them in worshipping idols of Baal with accompanying Ashoreth stake or tree carved with images of demons, beasts and serpents – exactly as Jehovah's Witness put subliminal images in the trees in the pictures in their literature as an offering to their god Satan!

Zephaniah 1:4 "I will also stretch out mine hand upon Judah, and upon all the inhabitants of Jerusalem; and I will cut off the remnant of Baal from this place, and the name of the Chemarims with the priests;

5 And them that worship the host of heaven upon the housetops; and them that worship and that swear by the Lord, and that swear by Malcham".

The pope who already claims the title of god and believes himself higher than GOD will be aided by Satan and his ability to perform mighty miracles to kill all who refuse to worship the the pope and the sun. A still valid papal edict states anyone who refuses to believe that the bread and wine or the water and wine used in Catholic mass really is Jesus's flesh and blood has to be killed! This edict will be made law when the pope gets power as head of the New World Order.

Matthew 24:16, 17,18, 19.

16 "Then let them which be in Judaea flee into the mountains:"

17 "Let him which is on the housetop not come down to take any thing out of his house:"

18 "Neither let him which is in the field return back to take his clothes."

19 "And woe unto them that are with child, and to them that give suck in those days!"

These verses mean Christians around the world need to flee or keep a very low profile as Satan will have whipped up his Muslim, Catholic and other pagan slaves into a rampage of murder.

24:20 "But pray ye that your flight be not in the winter, neither on the sabbath day:"

This verse is a little strange. Hoping against a winter flight is logical in the time it was written though today winter need not be a great problem for many Christians in the developed world – unless by the time tribulation occurs the world economies have been crashed spectacularly as happened in 1929 as such a crash may make travel impossible. Also, if by the time that the Christian have to flee their lives had been made difficult by not being able to buy or sell then escaping could be arduous? The hope against a Sabbath emergency seems odd at first as Paul told Christians that we are not bound by the Sabbath prohibitions but then Jesus himself had shown the Sabbath was made for humans. Perhaps he gave the warning of the Sabbath as all the Jews kept the Sabbath to some degree in those days?

24:21 "For then shall be great tribulation, such as was not since the beginning of the world to this time, no, nor ever shall be."

The meaning of this verse would be understood by the listeners as in that time they were used to hearing the Books of Isaiah, Ezekiel and Daniel read in the synagogues and discussed daily. The people were all well aware those prophets had prophesied about the downfall of Jerusalem, and were familiar with the previous occasion when invaders had carried the Jews off to exile in Egypt. They were also familiar and believers in The Flood catastrophe and well aware of the meaning of the word 'tribulation'. They were also familiar with an End Times Resurrection to Judgement as Lazarus's sister revealed when she told Jesus her brother would surely rise in the Last Day – John 11:24 Martha saith unto him, "I know that he shall rise again in the resurrection at the last day".

The 24:21 verse warns that the tribulations of the seven bowls will be poured on Earth to afflict all those who do not have the seal of GOD on their forehead. These will be bypassed by the death angels just as were the Israelites who smeared lambs blood on their doorposts on the night of the Passover. Though the Christians will be spared the angel's bowls they will still be hunted by the armies of Satan and many will be beheaded.

24:22: "And except those days should be shortened, there should no flesh be saved: but for the elect's sake those days shall be shortened".

This is the important verse Christians can take comfort in! The elect are obviously the Christians and the remnant of the Jews who have remained faithful and are metaphorically marked on the forehead with GOD's seal. During the tribulation the elect will have been reduced by murder and especially beheading but there will still be a considerable number left alive. Being killed may seem terrible but the victim is sure of a resurrection to immortal life with Jesus.

The situation at the end of tribulation will seem hopeless and all life about to perish during what is called the Battle of Armageddonbut when all seems lost GOD will order Jesus to come and put an end to the terror and destruction by carrying out the promise of

Thessalonians 4:16.

Apostle Paul who had spoken to Jesus on the road to Damascus was inspired by holy spirit to write to the church in Thessalonica, Greece, praising them for their faith and efforts in preaching the Gospel but also to clarify the matter of the end times which had been troubling some of the congregation. He wrote:

Thessalonians 4:16 'For the Lord himself shall descend from heaven with a shout, with the voice of the archangel, and with the trump of God: and the dead in Christ shall rise first:'

This verse clearly explains that as the last evil people are gathering against Jerusalem the clouds will part and Jesus will be seen with his army of angels coming to slaughter all listed at Revelation 22:15. He will kill off every last sinner and call all the birds to feast on the corpses. And then will resurrect all those martyrs who were beheaded during the 6,000 years from Creation to Tribulation. All these faithful martyrs will be carried up to meet Jesus and receive their new white clothes as a symbol of their new immortality – they died once and cannot die again.

4:17 'Then we which are alive and remain shall be caught up together with them in the clouds, **to meet the Lord in the air:** and so shall we ever be with the Lord'.

THAT WILL BE THE RAPTURE! Immediately Jesus has raised all the martyrs he will seek out all those Christians and Jews who have successfully hidden from Satan's armies and lift them off the smoking Earth.

How rapturous a moment it will be to be lifted off the smoking blood-soaked Earth to meet Jesus in the air and be whisked away to some safe place until he and his angels have slaughtered all the evil people to make Earth safe to be set down upon again!

Better yet though the promise of being taken up to safety with Jesus for ever is wonderful – it will only be the start of an amazing 1,000 years – THE MILLENNIUM!

The resurrected martyrs and the preserved elect will not go to heaven – their destination is the renewed Earth!

A Christian's only fear of the tribulation is being caught by the murderous demon-possessed armies of the New World Order and dying an awful death. The one about to die will be consoled by knowing that 'a split-second later' they will wake to be with Jesus.

The soldiers and slaves of the NWO who impose the pope's laws on humans will have an awful time during tribulation and at the moment of death will realise that they face an eternal oblivion.

Armageddon perplexes most people with the account in Revelation and its links to Ezekiel's description of survivors spending seven months searching out and pinpointing human bones for burial parties to bury deeply. Most have the idea that a literal Gog and Magog and all the others mentioned in Revelation will battle over Israel. Jews themselves have the idea that GOD will send the Messiah to strike all the Gog and Magog armies and let the Jews rule over the world and live happily ever after making burnt offerings in a new and glorious Third Temple in Jerusalem.

Anyone who holds any of these ideas is wrong!

Jews are certainly wrong as they openly deny or refuse to accept Jesus as can be seen in a multitude of videos and printed statements! So how can they be saved by the one they despise and ridicule? And whose torture, crucifixion and blood they took upon themselves! The Jews howled for Barabbas to be released instead of Jesus and then openly declared: "His blood be upon us and our children!" -Matthew 27:25. They scorned Jesus and his message that he was the innocent lamb that had come to offer forgiveness for sin to all the world and not just Jews. Zephaniah 1:4-6 prophesies the hunting of Jews around Jerusalem by Jesus's angel army. Jews will be slaughtered en masse during Armageddon along with all manner of pagans and false Christians. However GOD has promised that a remnant of TRUE JEWS will be saved but as those true Jews have been dispersed around the world in the past 2,600 years they will come from all nations and many will not know they are Jews. Just how big that remnant will be is impossible to know but in the last days, weeks and months before Armageddon many will have been sought out by the holy spirit and helped clarify their beliefs.

The idea that 'Armageddon' actually means a single pitched battle in the valley known as Aramageddon is wrong as the whole world is to be cleansed and all the evil slaughtered. It is ludicrous to believe all the Muslims, atheists and agnostics around the world could be gathered into the small area of Armageddon. There are evil pagans, idolaters, atheists, agnostics and generally evil people all round the world with no way or means of travelling to the real Armageddon. A perfect example of this is the Muslims who have now grown to a population of 1.6 billion spread in all the countries of the world where they resist Christianity, commit atrocities, speak blasphemies and hate and instead preach their own devilworship of Baal. These will have to be slaughtered by the angels that accompany Jesus when he arrives in splendour – all the world will see him!

During Armageddon the angels will be busy hunting down sinners in every town or city around the world. The streets of many cities will run with blood during that time.

Slowly the sounds of battle and humans meeting agonising deaths will subside as the last guns fire, the last bombs fall and the last satellites plunge from the skies with their payloads of terrible weapons of mass destruction and powerful spying equipments.

The last sinful humans will succumb to their injuries helped along by fear of the shining angels coming towards them.

What horror all Evoquacks will have as they die during Armageddon.

How impossible it is to make today's 5,6,7 billion walking dead know that they must stop their sinning, devilworship and atheism and seek GOD YHWH and His son Jesus.

Paul who had been the murderous Jew delighting in hunting Christians for stoning before meeting the resurrected Jesus wrote at Acts 4:12: "Nor is there salvation in any other, for there is no other name under heaven given among men by which we must be saved."

Earth will be free of war for the first time since GOD willed it into existence.

Chapter Sixty Five
Jesus Rules Earth. Millennium. Satan's Little Season.

Earth smoked and steamed, red with blood, disfigured with vast quantities of rubble and bones - everywhere bones – human bones - held together with bits of flesh and sinew alive with maggots and flies. Dust and ash drifted out of the sky and was stirred as birds landed to peck at the bones. The scene was silent: no wind, no human voices, no calling animals; no squabbling birds – they had such a surfeit of food they had no reason to fight for morsels. It was the day after Armageddon.

After the angels have done their work on Armageddon Day the cities will resemble Berlin and Tokyo in 1945 – just masses of ruin and dereliction with no human life.

Jesus will bring the Raptured ones back to Earth and set up his rule in what remains of Jerusalem. The preserved elect will have the delightful task of helping Jesus cleanse, purify and replant Earth to make it the Paradise that The Garden of Eden was – and better yet they will be fully human.

His angels will drag the Ark of the Covenant out of the deep cave under Jeremiah's Grotto where it has been hidden for 2,600 years. He will pause to stare for a long moment at the stains of his own blood upon its Mercy Seat then open it and lift out the two tablets of stone on which his Father and ours wrote with His fingertip those 'Thou shalt nots...'

He will then read the ten commandments to all the people on Earth and emphasize how important it is that they all follow them laws. The resurrected martyrs millennial duty is to sit alongside Jesus and guide the elect - the Raptured - in the ways of righteousness.

The martyrs will be sexless as angels as they have lived and died their life once as Jesus explained at Matthew **30** "For in the resurrection they neither marry, nor are given in marriage, but are as the angels of God in heaven". These cannot die again as their first death was perfect atonement for their sins.

As the smoke and dust clears the climate will once again become warm and moist and softly lit with a pearly light filtered through a heavy layer of water droplets. The soil will be rich with the ashes and blood of all the dead to be fertile and with the aid of the worms will grow a layer of vegetation of all kinds – right to the tops of the newly lowered mountains. All those closed seashells on the peaks will disappear under grass the way a neglected grave does in an English churchyard. Each morning a mist will rise to settle as dew on the growing vegetation to wash it pure for any human or creature to graze. Springs of pure water will rise out of the land for all to savour.

Best of all, thanks to the newly fertile and productive Earth the humans will live on a rich vegetarian diet of fruit, seeds and vegetables just as GOD told Adam and Eve to eat 6,000 years previously.

That good food will make their bodies renewed so that each day they feel younger, fitter, healthier so they will walk and never get tired. The single ones will find partners to marry and with the married elect couples will be easily capable of having a baby each year for the 1000 year Millennium!

Even the elderly elect will find new vigour and produce babies – just as did the elderly Abraham and Sarah, John and Elizabeth, Elisha's Shunammite woman. And best of all the babies will be perfect with not a blemish to spoil either mind or body!

We know many babies will be born during the millennium as Isaiah prophesied:

Isaiah 11-8 'And the sucking child shall play on the hole of the asp, and the weaned child shall put his hand on the cockatrice' den'.

For the next 1,000 years the whole Earth will be a place of peace and harmony.

Sadly though, despite living through the tribulation and the millennium, some people will continue to harbour the foul DNA of the Nephilim and have a latent potential for sin and evil deeds. They are the carriers of Satan's seed! Just as Ham's wife carried the Nephilim DNA through The Flood so an unknown human will carry the toxic DNA through the Tribulation and Rapture. As the years pass by and that human produces generations of offspring the DNA will spread through the population just as plagues and pestilences spread across the Earth today. While Jesus rules these people will conceal and suppress their demonic nature just as so many similar ones do today.

As the millennium ticks its 1,000 years completion Satan will be released from his chains and will burst out of Earth with his gang of fallen angels – just as the pope's new throne show – and once again those evil angels will be allowed to fly about the Earth to attempt to make slaves among the humans just as they did formerly. They will succeed by finding those humans who still carry the seed – the DNA – of the Nephilim. These people will belong to the old tribes of Gog and Magog and be uncountable. Jesus will allow Satan and his new slaves to prosper for a very short season.

Despite all the evidence Satan will once again imagine he can be a powerful leader and lead his army of Gogites and Magogites and the fallen angels in besieging Jerusalem and the settlements of the Raptured. Hitler showed he was the seed of Satan as he mounted equally futile campaigns!

GOD will be watching Satan's developing army of evil until they threaten the peace and calm of Jerusalem before once again sending deadly fire from heaven to consign the evil humans, fallen angels to eternal destruction! Satan's seed will be wiped off the face of the Earth as GOD prophesied 7,000 years previously. Finally Satan will be cast into the lake of fire – and every spark of evil is destroyed for evermore!

As Satan disappears into the eternal fire Judgment Day commences.

GOD sat on a glorious throne will reveal Himself to the billions of humans that emerge from the soil they were buried in, the seas they drowned in; many will emerge from the very air that their essence disappeared into after they died in flame and explosion over the millennia. Not a single dead human will escape Judgement Day no matter how obscure, unnoticed or unrecorded their death was. They will resemble what they looked like before they died as GOD knows exactly how many hairs we all have on our heads and will easily materialise perfect bodies for all the resurrected.

When all the dead are stood before GOD He will open the books that record the lives of all the dead so that each person recorded as a sinner can be reminded of the sins they committed and especially how they delighted in sinning and exploiting other humans and creatures and refused to seek Jesus or live by GOD's standards. To their horror they will be told they cannot live on Earth with Jesus and will be cast into the lake of fire and total destruction – the second death!

Many of the resurrected will be ecstatic to find their names written in the Book of Life that lists those who lived by GOD's standards and commandments even though they lived in isolated lands or communities or bad situations and had no opportunity to hear of GOD, Jesus, Creation or the Gospel. As they were good people before they died GOD knows they will be good people for evermore and He will ensure they join Jesus to live for evermore in the new Jerusalem which is the perfect paradise that Earth will become under Jesus's rule.

We know who one of these happy people will be: that thief on the cross next to Jesus all those years ago! How delighted he will be to discover Jesus has kept his promise and raised him to everlasting life!

What joy that thief will have!

Chapter Sixty Six

ABHOT#2 -The Real History!

Hawking's original novel ABHOT#1 was put together by and for those people who cannot see truth from lies.

Practically everything you have ever read about EVONONSENSE is just Satan's nonsense.

Satan himself may have existed a long time before GOD and Jesus set themselves to creating the universe about 6,000 years but all the material universe we see has only existed the 6,000 years described in the Bible.

The simple proof of this is the fact that around the world there is a vast area of barren rock, barren sand and barren ice while in every area when there is rain and worms there is an ever deepening topsoil supporting lush vegetation for man and beast.

Add to this fertility the existence of great lakes of hot water under the crust and all theories of Earth forming from dust or meteorite impacts and the only conclusion is that Earth was created specially for the entire range of life we see on it today or in the fossil displays in museums.

Earth's time has stretched out to no more than 6,000 years – and it will stretch on for another 1,000 years until Satan's Little Season has come and gone.

Then by reversing the First Law of Thermodynamics of Day Four of Creation GOD will burn up this tied, polluted planet as described at 2 Peter 3:10 The Day of the Lord:

10 But the day of the Lord will come as a thief in the night, in which the heavens will pass away with a great noise, and the elements will melt with fervent heat; both the earth and the works that are in it will be burned up.

He will then make a new Earth with the glittering New Jerusalem promised at Revelation 21:
1 And I saw a new heaven and a new earth: for the first heaven and the first earth were passed away; and there was no more sea.
2 And I John saw the holy city, new Jerusalem, coming down from God out of heaven, prepared as a bride adorned for her husband.

3 And I heard a great voice out of heaven saying, Behold, the tabernacle of God is with men, and he will dwell with them, and they shall be his people, and God himself shall be with them, and be their God.

Chapter Sixty Seven
WORMS PROVE EARTH IS YOUNG! How worms and roots eat through concrete.

It was exploring old ruined farms on the local West Yorkshire hills that revealed how worms prove Earth is young! Some years previously I walked to one old ruin at which the builder had obviously needed a constant supply of water for washing fleeces after shearing the sheep on that high exposed land. He had constructed the building into the hillside just where a spring emerged so that the water passed through the wall and into a sink chiselled out a solid block of stone and then overflowed through a cutout. I was very impressed with the design and decided to take some photos to ask the local heritage agency if the sink could be put in a museum before someone 'salvaged' it for a garden feature. The agency never replied so I let the matter drop. Then about 5 years later I decided to go check if the sink was still there but was disappointed as I approached the site to not be able to see it. I then realised it was still in situ but now covered with a solid mat of grass and soil.

I was puzzled at how the grass and soil had developed so completely - before concluding that it was the grass sending out exploring shoots and roots to make shelter for worms and woodlice, and providing seeds for mice and birds to feed on. And all these creatures would simultaneously fertilise the roots to enable the grass to grow so fast on barren stone and over just a few years a thick mat of grass and soil built up. I also noticed how an unusable corner of the local children's school carpark had grown an impressive mat of grass, weed and trees in just a few years. Clearly my damp West Yorkshire is very conducive to worms and the formation of topsoil!

A few days later I passed by a paved footpath once used by workers taking a shortcut to some mills until the mills closed about thirty years previously after which no-one used the path with the result that grass and soil had built up and covered all but the centre of the path.

This grass and soil build up was one of the keys of how to prove Earth is young! EUREKA!

I had to wait three years before discovering how Hitler's Bunkers is the second key to unlock the mystery of how WORMS PROVE EARTH IS YOUNG!   EUREKA!

I was looking at a video of a team of urban explorers mapping the abandoned bunkers under Berlin when in one that had been stripped out then sealed off shortly after World War Two they reached a place where the roof was breaking into great chunks and collapsing. The leader explained the concrete was cracking purely by the action of tree roots. Hitler had a fetish for thick concrete and specified a roof thickness of 8 feet for bunkers. No World War 2 bomb could penetrate the soil over the bunkers let alone shatter the concrete but the tree roots were quietly eating the concrete until huge chunks fells off. EUREKA!

I had already read about roots converting rock into soil. EUREKA!

Many years earlier I had read about a man rejuvenating a Dust Bowl farm by using contour ploughing and on one occasion when excavated a drainage channel he had found a small wooden bridge buried under 40 feet of dust. Obviously vast quantities of soil had washed off the land during the worst years. He noticed that alfalfa could send roots down over twenty feet and each rootlet was surrounded with a dark brown matter. I was previously aware that some plants can send roots down great distances but not twenty feet. I have been a keen gardener most of my life and have seen how each tap root looks like fur from countless microroots that must be secreting rock dissolving chemicals and then sucking up the soup of dissolved minerals – collard greens are full of iron! - but some mushrooms can suck up dangerous quantities of cadmium!

I had to wait another two years before the final piece of the puzzle fell into place when I spotted the link between Genesis 2:6 and 9:12! EUREKA!

Genesis 2:6 say 'a mist went up from the earth and watered the whole face of the ground'. Genesis 9:13 says 'GOD said he has set His rainbow in the clouds'. The link between these two verses is the clue to the pre- and post- Flood climate. 2:6 means there was no direct sun, no rain or cloud but each morning had a mist of dew that wet the plants – and all the Earth had a mild moist climate perfect for fast luxuriant growth of trees and vegetation. During The Flood the Earth's crust shattered and released all the hot water while the waters in the sky condensed and fell as the great rains. After the waters had drained off the climate was radically altered: great masses of clouds now rolled across the sky with strong winds driving them. When the clouds became thick and dark with water and rain started to fall any sunbeams shining through gaps in the cloud would strike the raindrops and appear as a brilliant rainbow. **Adam and Eve never saw rainbows, clouds, winds or direct hot sunlight!** Noah's new world was hot, freezing, windy, bright – but he was the first human to see rainbows!

Now I had the **Twelve Parts Of The Key** to proving Earth is young!

One: Creation Day Four explains how GOD arranged piezo-electricity to warm the Earth.

He arranged the crust to be granite over quartz. As the sun and moon circles around they pull all the crust in turn and gently flex the quartz. It causes piezo-electricity that warms the granite and provide underground heating TO LIFT MIST that then condenses as dew! Earthquake electrical fireworks have been reported constantly but they will happen so quickly have rarely been captured on film.

Two: The warm soil encouraged seeds to germinate and vegetation to creep over the land with its roots penetrating warm soil and rock to dissolve out the minerals and simultaneously create a brown 'rhizosphere' of new soil. As the dews soaks down past the plant roots it dissolves nutrients and minerals for the hair roots to suck up.

Three: Worms and roots convert rock into soil and minerals to cover the ground for more vegetation to grow to nourish more creatures.

Four: Worms, woodlice and creatures eat the living and dead vegetation, mice and birds eat the seeds and all their droppings act as fertiliser to stimulate more growth.

Five: Worms follow the roots as the plant dies off and eat the root and the rich soil of the rhizosphere and eventually deposit it on the surface. Worms eat the dying plant roots and the rhizosphere and cast it on the surface where it fertilise more grass. The grit the worms bring up gradually raises the level of the land. Darwin's meticulous research revealed worms can bring up one inch of soil every five years. Earth should be 6,000 years bigger that at Creation!

Six: Root secrete acids that can dissolve the calcium carbonate bond between quartz and can penetrate solid concrete to weaken it and create paths for worms to follow and enlarge. Worm saliva can also dissolve the calcium that bonds quartz sand grains to allow the worms to eat through concrete and rock.

Seven: From Creation to Flood there was no rain, or sun or wind - only a mist which was sufficient to water a great covering of juicy vegetation.

Eight: GOD intended humans and all creatures to walk everywhere on a thick soft carpet of grass growing in a soft absorbent layer of fine soil - and that soil and grass would be getting ever deeper by at least the one inch in five years that Darwin calculated. There should be no stones or rocks visible anywhere on Earth!

Nine: The mild pre-Flood climate, plus worms and vegetation promoted great masses of vegetation and creatures that were all washed off Earth during The Flood and buried in layers we now see as fossil fuel and fossilised creatures.

Ten: After The Flood the Earth now had climatic zones as we have today. The top and bottom of Earth was cold and eventually developed great thick ice caps. The equatorial zone became too hot and dry for anything and again remains so today. Only the temperate zone had ideal conditions for vegetation and creatures and remains so today! However as the crustal flexing from the sun and moon has continued since The Flood the crust under north and south poles will be pushing a little heat into the land under the ice and be now causing the slow melting of ice caps and the tundras or Alaska and Siberia.

Eleven: Pre-Flood Earth did not have mountains. Low hills would exist but as the climate was warm and moist the worms, insects and birds would easily travel to the hilltops to burrow through the soil and fertilise the vegetation to make the entire hilltop green and fertile. This will occur again during The Millennium.

Twelve: Post-Flood the new winds, hot sun, snow and ice would make hilltops and mountains too cold and hostile for worms, insects, birds and small creatures. The high places were too cold and windy and lacking vegetation they eroded away and are still eroding today.

Darwin's book explained that worms can swallow quite large grit and when expelled some of it has been attacked by digestive juices. The worms then excrete their meals of vegetation and soils and grits as worm casts which is very fine particles and the gritty bits loosely bound with digestive juices but which generally quickly get soaked and dissolve in rain and spread out to release their nutrients of any local vegetation's roots. Darwin gives many examples of solid objects that have 'sunk' into solid topsoil by the action of worms eating and burrowing through the soil under the objects and then depositing their casts on the objects so that over time the object disappears under soil topsoil and lush vegetation – what I call the Time Team Phenomenon as that program constantly shows how abandoned ruins can be under several feet of topsoil in about 1600 years.

After several years worms will have removed all organic matter from the subsoil to make it dense and sterile as we find today when we dig about twelve inches/ 300 cms deep.

Darwin was well aware that worms seemed able to eat through concrete between brick wall and floor tiles but was not aware of how plant roots aided this.

To put Darwin into The Flood we have to visualize and understand that from the sixth day of Creation when GOD had put all the creeping things on the Earth and they had commenced breeding, feeding, burrowing and burying the topsoil in their rich compost the Earth was remarkably fertile and unbelievably productive. We know that a few worms can still reach giant proportions and as the pre-flood environment was very conducive to gigantism we have to accept that the worms would have helped the topsoil to grow very deep very quickly – with a corresponding bumper crop of all vegetation from mosses to giant tree ferns. This topsoil had been growing for about 1,600 years from Creation to the point GOD brought the flood: it could have been many feet thick and not just a few inches as we have now. Darwin cites various locations with very deep topsoil and pre-flood this would pertain over all the land.

Now it is perfectly obvious that much of the land is barren bedrock and sand – ignore the frozen areas – and the question must be asked: Why do these barren areas exist if worms and roots can produce topsoil at the rate of one inch in 5 years? In 1,000 years there should be 200 inches of topsoil but obviously there isn't! In fact hilltops are becoming more barren and unproductive – just as Darwin found. I recently watched a video about how feral goats in New Zealand are ruining the hills as they graze the grass away despite the climate providing plenty of rain. Despite my **Twelve Parts** Key it is obvious that worms are not succeeding in building topsoil! So are my **Twelve Parts** Key wrong?

No, all **Twelve Parts** are easily and scientifically verifiable and Darwin himself had considered all twelve without spotting the links. There must be something preventing soil building up - and Darwin himself investigated the problem but his denial of GOD and Genesis blinded him to the truth of why topsoil was being lost and not increased!

Darwin's childhood and teenage years are the key to his inability to recognise why the very productive worms cannot build topsoil to great depths. His mother died when he was eight and must have been a traumatic experience. His father enrolled him in an Anglican school where each day would start and end with prayers. He worked as assistant to his doctor father treating the poor people who lived in such poor housing with bad water supplies and hygiene that caused a 50% death rate for children under 5 years of age. He worked as a taxidermist and so was familiar with all the common creatures and their bodies. He joined a group debating materialism versus religion. His father enrolled him at college to study to become an Anglican minister. He was impressed with the evidences for intelligent design but took the faulty view that the superb design of all life was evolution from more primitive designs. During his teenage years the first full Ichthyosaur, Plesiosaur and Pterosaur fossils had been found and widely publicised. These discoveries and the great ages given to the fossils caused him to have doubts of the truth of Genesis. Rather than continue studying to become the minister he accepted an opportunity to sail on HMS Beagle as the ship had been ordered to make accurate charts of South America but it then travelled on to Australia and returned home via Africa. In Australia he saw the kangaroo and duck-billed platypus and saw giant fossil bones. On the Galapagos he saw different birds on each island and imagined they had evolved their difference over millions of years. On the ship he had read Charles Lyell's 'Principles of Geology,' which set out uniformitarian concepts of land slowly rising or falling over immense periods – or the textonicz plates woozle today's Evoquacks prattle on about. He should have known that his Bible told of the mountains rising but by that stage his eyes had become blinded to the truth as Jesus said at Matthew 13:14 "And in them is fulfilled the prophecy of Esaias, which saith, By hearing ye shall hear, and shall not understand; and seeing ye shall see, and shall not perceive".

All this life experience and anatomical studies added to the nonsense of Lyell's book led Darwin away from believing the Bible and fantasising Earth was billions of years old. He was thus blinded to the real reason why the worms he so meticulously studied failed to build up great depths of topsoil despite his own direct studies proving that deep topsoil should cover the entire Earth!

If he had imported an inch of topsoil onto his own gardens and lawns every five years he lived there eventually his paths would be trenches with about eight inches of soil either side of them! His house is still standing 135 years after his death and he moved in in 1842. In theory the lawns and flower beds should now be 175 divided by 5 = 35 inches higher than when he first saw them. Obviously they are not much higher than in 1842 – so something is wrong! What that is is simply the gardeners obsession with removing weeds and grass in the interests of 'prettiness' and so both starving away the soil creatures and depleting the soil by removing the skeletal remains of creatures and plants that bulk up soil.

Despite Darwin's intense study of his gardens he could not see was the link between Genesis 2:6 and 9:12 – MIST VERSUS HEAVY RAIN !

If Darwin's eyes had not been blinded by Satan he would have realised that in the perfection of the Garden of Eden the worms would be happily aerating the soil and bringing up nutrients to fertilises the grass and the dew would be unable to wash the wormcasts away! The land would grow more fertile everyday as the worms and plant roots extracted minerals from the soil and rocks and deposited on the ground to fertilise the grass. EUREKA!

He knew rain washed worm droppings down hill and after a heavy rain there would be dirty water running down the fields – it was dirtied by the dissolved wormcasts! But having lost his faith in GOD and the Bible Darwin could not accept that Adam and Eve lived in a very mild climate with only mist/dew to water all the vegetation! WARM GENTLE DEW CANNOT WASH WORMCASTS AND SOIL AWAY BUT COLD HEAVY RAIN CERTAINLY CAN!

If The Flood had never occurred Earth today would be entirely covered in deep rich topsoil and vegetation.

If Adam and Eve had faithfully followed GOD's orders and extended the Garden all over the Earth it would be totally green all over right to the mountaintops with vast numbers of tame friendly animals to bring delight to perhaps 50 billion humans! The mountaintops will drip with sweet wine and the plough will follow the harvester.

It will be so fertile at the end of The Millennium of Christ's reign!

THE EVIDENCE THAT WORMS PROVE EARTH IS YOUNG IS THE FACT THAT GIVEN A MILD CLIMATE WITH ONLY MIST AND DEW FOR MOISTURE THE WORMS AND PLANTS ROOTS WILL STRIVE TOGETHER TO TURN ROCK, SAND, SUNLIGHT, CO2 AND WATER INTO RICH TOPSOIL COVERED IN AN EVER THICKENING MAT OF LUSH VEGETATION AT THE RATE OF ONE INCH EVERY FIVE YEARS. THE LACK OF SUCH A DEPTH OF TOPSOIL AND THE BARREN HILLTOPS ALL OVER THE WORLD IS CLEAR PROOF EARTH IS YOUNG AND THE FLOOD WAS A REAL CATASTROPHE VERY RECENTLY! EUREKA!

www.ingramcontent.com/pod-product-compliance
Lightning Source LLC
Chambersburg PA
CBHW071254220526
45468CB00001B/125

* 9 7 8 1 9 8 3 7 4 8 9 4 3 *